Programming Guide for ROBOTIS DREAM with R+Block, R+Task & Python

By Chi N. Thai

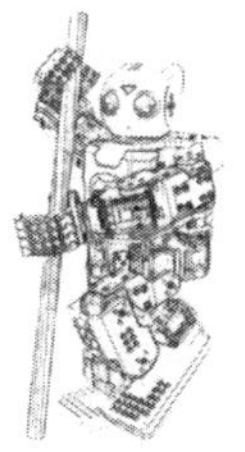

CNT Robotics LLC, Buford
2022

Author: Chi N. Thai
roboteer@comcast.net
www.cntrobotics.com

Programming Guide for ROBOTIS DREAM with Chromebook (2022)

ISBN 978-0-9993918-8-4 (Paperback).

Library of Congress Control Number: 2022905155.

Published by CNT Robotics LLC, Buford, GA, U.S.A.

Printed by Amazon KDP.

To my parents for giving me life.
To Christine and Emily for their enduring love and support.
To Jenson and all my grand nieces and nephews.

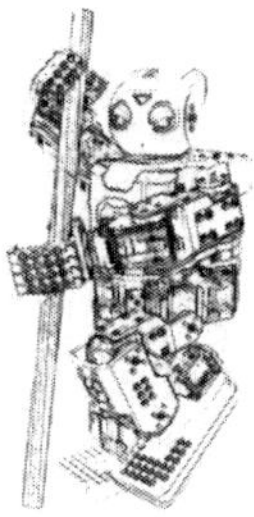

Preface

The DREAM II™ (School Set) programmable robotic kit was released by ROBOTIS® in Spring 2018 and it is recommended for users at age 8 or older. It comes with instructions to build 23 programmable robot examples and it can be interfaced with several programming tools on a Chromebook:

1. The first interface is R+BLOCK which uses a Graphical Block Programming IDE suitable for young beginning programmers, and it accesses the robot's actuators and sensors directly from the Chromebook at runtime.
2. The second interface uses R+TASK which uses a Menu-driven IDE which lets the user choose various commands to interact with the robot's actuators and sensors, via a compiled program which operates on the robot's Controller CM-150/151 at runtime.
3. The third interface uses the Linux Virtual Environment of the Chromebook to develop and run Standard Python programs via a "helper" Python module named "CNT_Block.py". This module contains Utility Functions that have a one-to-one match with the R+BLOCK commands that access the robot hardware. This feature is designed to help R+BLOCK programmers transition to Python usage. This Python module is designed to work with a companion TASK code (CNT_Block.tskx/tsk3) that works as an extension of the CM-150/151 Firmware.

This book uses a unique approach in developing the same 8 DREAM II robot projects in three ways (R+BLOCK, R+TASK and Python+TASK) to expose the reader to the principle/idea that "Robotics Programming" needs to go beyond the physical robot and seeks to integrate/adapt/optimize the "Supervisory Platform" through its Hardware and Software capabilities and limitations.

Who is this book for?

This book is for you if you are a young robotics enthusiast looking at achieving on your own a firm foundation in Robotics Programming, or if you are an adult investigating the possible use of the DREAM II School Set to help children learn about Robotics Programming.

What do you need?

At the minimum, you need to purchase the ROBOTIS DREAM II School Set which is widely available via several web stores.

For the Chromebook, the author is recommending one with 32 MB RAM (if the reader plans to program in Python), otherwise 16 MB suffice for programming with R+BLOCK and R+TASK.

The user needs to additionally purchase the BT-410 receiver to put on the robot when using R+BLOCK and R+TASK, and also the BT-410 USB Dongle to be used on the Chromebook side when programming in Python via Linux.

Two IRSS-10 sensors are also needed for some projects described in this book (https://www.robotis.us/ir-sensor-irss-10/).

All software tools needed can be downloaded for free from the web (more details in Chapter 1).

What is in this book?

This book consists of 4 chapters:

- Chapter 1 presents an overview of the DREAM II hardware and software capabilities, and its relationship with the Chromebook environment. The Sense-Think-Act paradigm used in developing the contents of this book is also described in this chapter.
- Chapter 2 is written for beginners who want to learn Block Programming. It will guide the reader from R+BLOCK Programming Basics (Sequence, Function, Condition, Loop, Remote-Control and Autonomous-Behavior) to a Structured Approach for Project Analysis and Solution based on the Sense-Think-Act Paradigm, using 8 projects with increasing complexity.
- Chapter 3 is written as a parallel and independent path for beginners who want to learn TASK Programming instead. It will also guide the reader from R+TASK Programming Basics (Sequence, Loop, Function, Condition, Remote-Control and Autonomous-Behavior) to a Structured Approach for Project Analysis and Solution based on the Sense-Think-Act Paradigm, using the same 8 projects as Chapter 2 but yielding better runtime performances.
- Chapter 4 expects the reader to know some basic Python programming. It starts as a guide for using the Module "CNT_Block.py" as a black-box to redevelop in Python the solutions to the same 8 projects demonstrated in Chapters 2 and 3. Chapter 4 ends with a "Look Under The Hood" of the inner processes between Module "CNT_Block.py" and its companion TASK code "CNT_Block.tskx/tsk3" to provide the reader with a more expansive "Robotics Programming" outlook beyond the Chromebook and DREAM systems.

This book also provides appropriate source codes and tutorial videos (via YouTube®) to illustrate the presented concepts. Please visit www.cntrobotics.com/dreamcb for access options to the source codes and tutorial videos.

How should you use this book?

All readers should first read Chapter 1.

If the reader is a younger person and new to Computer and Robotics Programming, the author recommends continuing with Chapter 2 and then on to Chapter 3. Next, this reader would likely need to pick up some basics in Python Programming and Linux usage elsewhere, before getting back into Chapter 4.

If the reader already has some experiences with a Structured Computer Language but is new to Robotics Programming, the author recommends skipping Chapter 2 and dive into Chapter 3. For Chapter 4, this reader may have to learn "on the job" some Python Programming and basic Linux usage.

And please, do let the author know about your experiences in using the materials in this book, this will help the author improve the next edition and the companion resources. Please visit the author web site www.cntrobotics.com for updates and other robotics materials. Please send your comments and suggestions to roboteer@comcast.net.

Trademark Attributions

Content in this book made references to the following trademarks:

ROBOTIS product names are trademarks or registered trademarks of ROBOTIS, Inc. Used with permission by ROBOTIS Co. Ltd.

Android®, YouTube®, Chrome®, Chromebook® are registered trademarks of Google Inc.

SCRATCH® is a registered trademark of the Scratch Team at the Massachusetts Institute of Technology.

Windows® is a registered trademark of Microsoft Inc.

Bluetooth® is a registered trademark of Bluetooth SIG, Inc.

Linux® is a registered trademark of the Linux Foundation.

Buford, GA, USA   Chi N. Thai

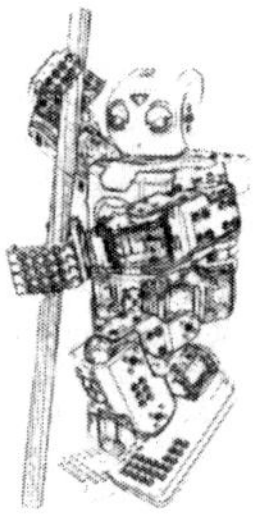

Contents

Chapter 1: Foundational Concepts

1.1 ROBOTIS DREAM II System Overview

ROBOTIS offers quite a large range of educational robotics kits to fit different user's ages and skill sets, please review this YouTube® video (https://youtu.be/Rxmgfose8wE) for an older (2017) survey of the ROBOTIS "family" of robots. It did not include the ENGINEER Kits 1 and 2 (https://www.youtube.com/watch?v=dJ0kP0mVPwE) which were released later in 2019-2020.

The DREAM II system (http://www.robotis.us/dream/) offers an extensive range of sensors such as Touch, Color, Magnetic, Passive IR and Temperature sensors. It can operate 2 different types of actuators (Continuous-Turn or Position-Control). Its Controller can be the CM-150 (https://emanual.robotis.com/docs/en/parts/controller/cm-150/) or CM-151 (https://emanual.robotis.com/docs/en/parts/controller/cm-151/) (see Fig. 1.1). Either Controller can be programmed with several software packages such as ROBOTIS' tools (R.TASK/R+BLOCK/R+PLAY700), SCRATCH (from MIT), or EDBOT (from Robots in Schools) for use with Python (see Thai, 2018 and Thai, 2019).

The present book was written to leverage special features of the Chromebook so that 3 Robotics Programming approaches could be presented to readers/users with different skill levels:

1) "Block Coding" with R+BLOCK for the youngest beginners.
2) "Menu-Guided Programming" with R+TASK for users with intermediate skills.
3) Standard Python Programming with THONNY IDE and Chromebook's LINUX Virtual Environment for more advanced users.

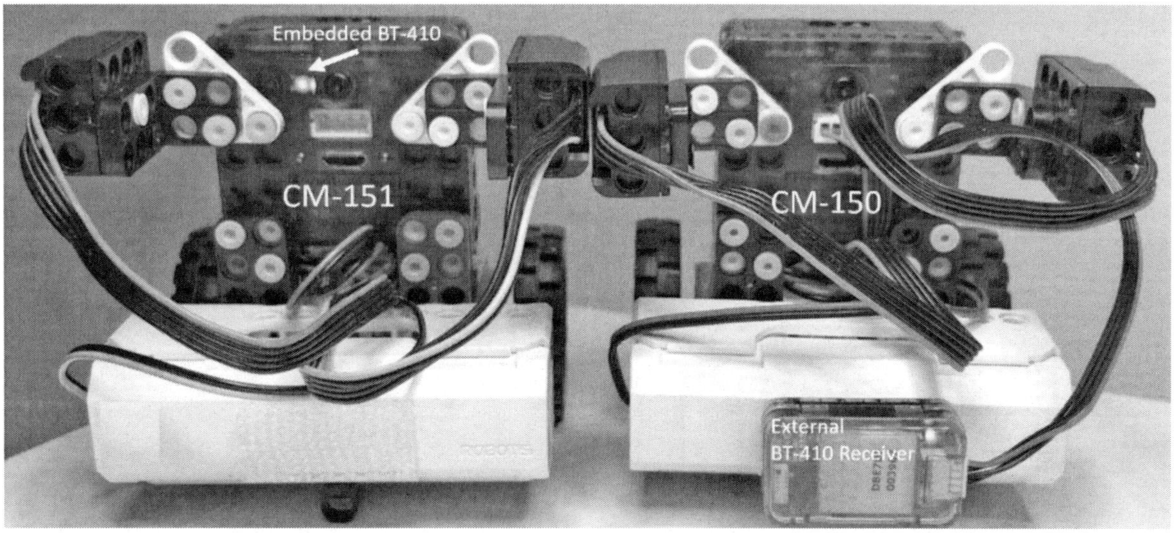

Fig. 1.1 Hardware Controllers for DREAM II kits: CM-151 (left) and CM-150 (right).

The DREAM II system shares into a common ROBOTIS System Design Paradigm which considers a typical ROBOTIS robot to be a Computer Network with four major components:

1) A Hardware Controller, CM-150/151 (see Fig. 1.1), is the "Main Brain" for the robot. It contains the user's instructions for the robot to execute varied tasks as programmed by the user. Outwardly, there is no physical difference between CM-150 and CM-151, but the CM-151 has about 40% improvements in processing speed and sensor throughput. It also has an embedded BT-410 receiver with much improved wireless communications reliability (no packet losses for indoors usage per author's experiences). The CM-150 uses an external BT-410 receiver via its UART port.

2) The Sensors component (such as NIR or Touch sensors – see Fig. 1.2), which helps the Hardware Controller get information about its surroundings. Sensor data are usually obtained as analog signals that need to be digitized by the Hardware Controller before the programmer can make use of the sensor inputs.

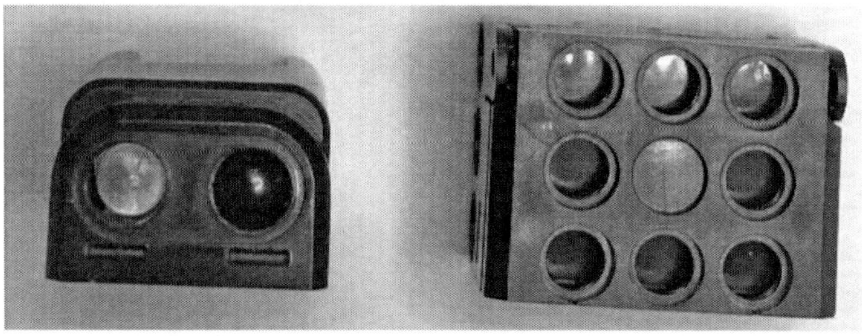

Fig. 1.2 Selected Sensors: NIR Sensor (left) and Touch Sensor (right).

3) The Actuators component (e.g., Servo Motors or LEDs – see Fig. 1.3), which allows the Hardware Controller to perform the appropriate robot actions upon the real world as programmed by the user. Advanced actuators such as servo motors have their own micro-controllers to work in concert with the Hardware Controller.

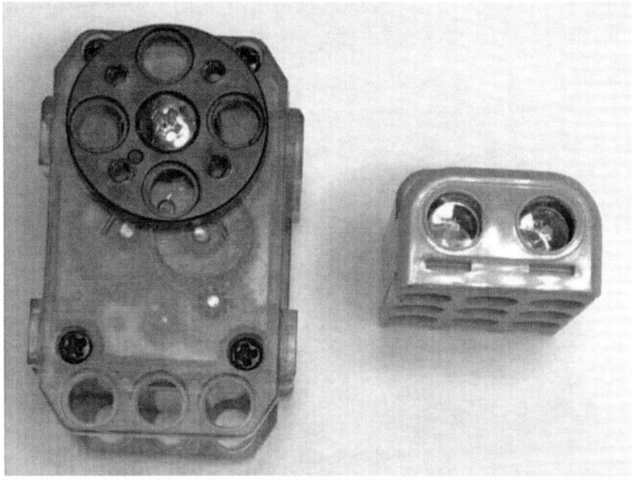

Fig. 1.3 Selected Actuators: Servo Motor (left) and LED Display (right)

4) A Remote Device, which can be an RC-100B module or a smartphone (see Fig. 1.4), or even a Windows PC or Chromebook via TASK's Virtual Controller. The Remote Device has a more flexible role depending on the specific device used and on its current role in this network of robotic components. For example, if the basic RC-100B is used with the DREAM system, it essentially functions as a NIR-based Sensor (attached to the user) informing the Hardware Controller about which "Up/Down/Left/Right" buttons had been pressed by the user. The basic RC-100B would require line-of-sight access between the physical RC-100B and the NIR receiver (IR-10) attached to the robot. When a smartphone (running the R+m.PLAY700 App) is used with the CM-150/151, it can act as a Bluetooth Sensor telling the CM-150/151 about which Touch Area on the phone screen had been pressed by the user, or it can act as an Actuator displaying appropriate graphics or videos as commanded by the CM-150/151 according to its programmed logic (see Thai (2019)).

When a Chromebook (CB) is used as a Remote Device, the issues become more complex. For example, if the TASK's Virtual Controller (see Fig. 1.5) is used, then the CB acts as a glorified Touch Screen informing the CM-150/151 about which virtual RC button had been pushed by the user. However, if the "LINUX+THONNY" tool chain is operating on the CB (see Chapter 4), then the CB takes on the role of a "Supervisory Controller" which is responsible for the main robot control logic, implemented via a Python program executed on the CB. While the CM-150/151 now acts only as a "simple" Hardware Controller handling wireless commands sent from/to the CB via the ROBOTIS communications protocol named "Remocon" (https://emanual.robotis.com/docs/en/parts/communication/rc-100/#communication-packet), and to pass them through to the robot's Actuators and Sensors, via a TASK program running on the CM-150/151. With this approach, the CB can even be configured to control multiple DREAM II robots.

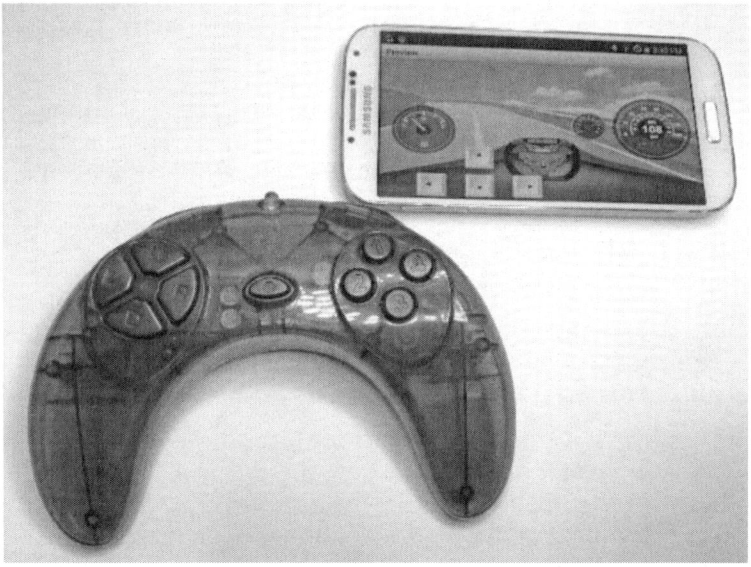

Fig. 1.4 Selected Remote Devices: RC-100B (Left) and Smartphone (right).

Fig. 1.5 TASK's Virtual Controller Interface.

1.1.1 DREAM II - Levels 1-5 Edition

Since Spring 2018, the complete DREAM II System is available for the U.S.A. market and it includes 5 levels (http://www.robotis.us/dream/) which work in a cumulative scheme: i.e., users would have to purchase/start with Level 1, then go to Levels 2 through 5 consecutively:

1) Besides the various frame parts, the major components of Level 1 include 1 Low-Speed Geared Motor, 1 Micro-USB cable and 1 LiPo battery unit. Level 1 showcases 11 motorized but non-programmable robots.

2) Level 2 adds 1 CM-150/151 Controller, 2 Geared Motors, 2 Rubber Tires and more frame parts. Level 2 showcases 11 programmable robots; some are wheel-based, and some are linkage-based (i.e., "walking" robots) and all as autonomous behavior applications.

3) Level 3 adds 1 Servo Motor, 1 Remote Controller RC-100B and its NIR Receiver IR-10, and more frame parts. Level 3 showcases another 11 programmable robots using the Servo Motor and illustrating Remote Control techniques.

4) Level 4 adds 2 more Servo Motors, 2 External NIR Sensors, 2 Touch Sensors, 1 LED Module, and of course more frame parts. Level 4 showcases 12 additional robots with more sophisticated mechanical designs and behaviors, both autonomous and remote controlled.

5) Level 5 adds a set of Master and Slave BlueTooth modules to be used with the RC-100B, 1 more LiPo battery unit, and even more frame parts, including a Track system. Level 5 showcases 12 final wheel-based and track-based robots with autonomous and remote-controlled designs. Some applications use the more sophisticated Callback function which is a type of hardware-timer interrupt function.

1.1.2 DREAM II - School Set

The School Set is customized for the USA market and roughly corresponds to a bundle of the major components of Levels 1, 2 and 3 combined. Thus, both Autonomous-Behavior and Remote-Control applications can be achieved with the School Set (23 robot designs). This book is specifically written for the School Set (Fig. 1.6), but users need to also realize that extra components are available via the ROBOTIS e-Shop Network (https://www.robotis.us/ and https://www.robotis.com/).

Fig. 1.6 DREAM II School Set (Courtesy of ROBOTIS).

1.2 Sense-Think-Act Paradigm

Usually, the younger robotics students were surprised when I started my robotics short courses by mentioning that they had been using the Sense-Think-Act paradigm in everyday activities already. I used Fig. 1.7 to explain how humans use our Senses to let us know about the external World (i.e., Perception) and apply those Sensations into our Cognitive process (i.e., Thinking) to devise proper Actions/Reactions to the situation at hand. These Actions/Reactions in turn would change some aspects of the external World resulting in new Sensations triggering the next World-Human-Interactions cycle and so forth.

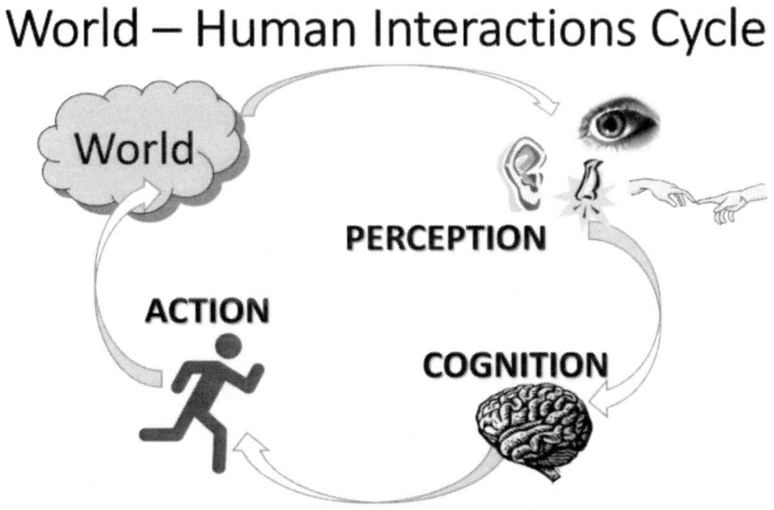

Fig. 1.7 World-Human Interactions Cycle.

Thus, in a way, doing Robotics is just trying to reproduce this interactions cycle onto a robot as shown in Fig. 1.8. Mechanical or electronic Sensors would convert their interactions with the World into Input Data to be sent to the Robot/Computer which was previously Programmed by the user to deal with Events deemed important to the operation of the Robot. Using this programmed Logic, the Robot Outputs appropriate commands to its Actuators which then could change the Sensors' perspective onto the World and therefore generate new Input Data for the Robot/Computer to process anew, through another interactions cycle. The robot can also be set into a Remote-Control mode, or an Autonomous-Behavior mode, or a mix between these two modes.

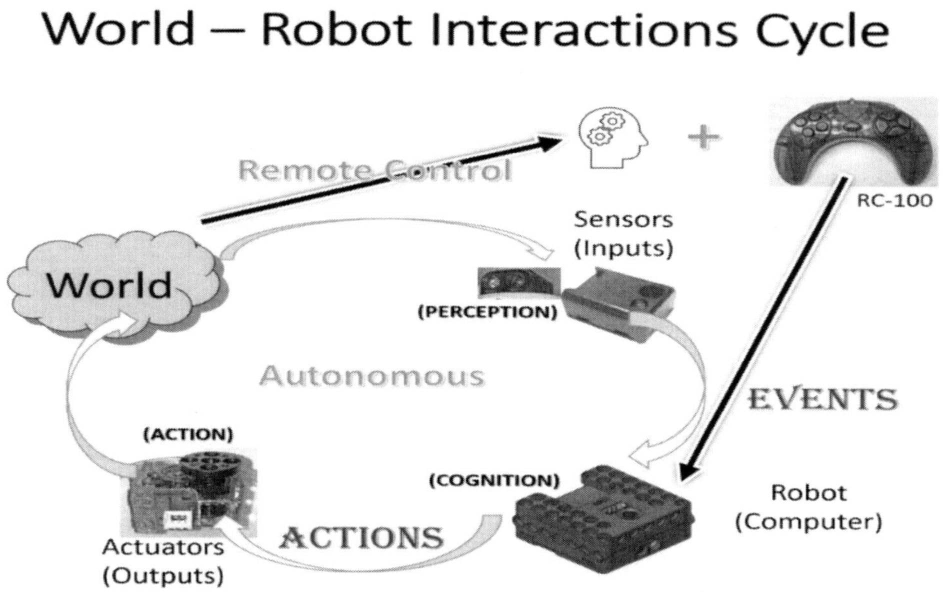

Fig. 1.8 World-Robot Interactions Cycle.

The Sense-Think-Act Paradigm will be revisited with more details in Chapters 2, 3 and 4 to illustrate its adaptations to specific hardware/software environments used for different robot systems.

1.3 Using DREAM II with Chromebooks

If the reader is using Chromebooks for the first time, the author is recommending this book "The Ultimate Chrome OS Guide" by Keith Meyers (2021-2022). It is available in Kindle format at this link https://www.amazon.com/s?k=ultimate+chrome+os+guide&crid=17COI4WVHVCE, and it has different versions depending on the actual Chromebook model that the reader already possessed or plans to purchase. The author used three different Chromebook brands in developing DREAM II projects for this book (see Appendix A for discussions about a "median" Chromebook to use for materials presented in this book).

Currently (April 2022), all ROBOTIS software tools such as BLOCK, TASK V2/3 and other mobile Apps were developed for devices running on Android OS, so their uses on Chromebooks are still "quirky". Appendix A lists possible use issues of the DREAM II Hardware and Software suite and the author's "work-arounds" for those issues. **The author recommends readers to read Appendix A before continuing with Chapters 2, 3 and 4.**

Out of its shipping box, the DREAM II School Set is fully functional on a Windows PC platform equipped with a USB port using the provided micro-USB cable, but this option is not workable at present on Chromebooks, as USB devices are not yet fully supported in Chrome OS. This means that Chromebook users will need to buy extra BT communications devices from ROBOTIS to use programming tools such as R+BLOCK, R+TASK or Python:

- R+BLOCK and R+TASK work with either BT-210 (https://www.robotis.us/bt-210/) or BT-410 (https://www.robotis.us/bt-410/) receivers connected to the CM-150/151 UART Port.
- On the Chromebook's side, procedures for BT connections are "quirky":
  - The BT-210 can be paired via "Settings" in Chrome OS, but **not** for the BT-410.
  - But inside R+BLOCK and R+TASK V2 and V3, both types of BT receivers can be scanned for, paired, and be used in programming the CM-150/151.
  - Additionally, when programming in Python via the Linux Virtual Environment, both BT-210 and BT-410 receivers are not recognized by the Linux OS. But if a BT-410 USB dongle is used on the Chromebook's side in conjunction with a BT-410 receiver on the CM-150/151 side, then THONNY IDE would work fine with the DREAM II system (https://www.youtube.com/watch?v=FdRiB44DvDk).

In summary, for programming DREAM II robots with Chromebooks, the author recommends readers/users to purchase a set of BT-410 USB dongle and BT-410 UART receiver in case of a CM-150, and only a USB BT-410 dongle for the CM-151 as it already has an embedded BT-410 receiver (see Fig. 1.1).

It is rumored that a newer USB BT Dongle will be available to work with the CM-151 to take advantage of its higher baud rate (1 Mbps) by the end of 2022.

Additional ROBOTIS modules are needed for projects described in this book and they are listed below:

- Two IRSS-10 sensors (https://www.robotis.us/ir-sensor-irss-10/).
- One LED Module LM-10 (https://www.robotis.us/led-module-lm-10/).

This web site (https://emanual.robotis.com/docs/en/software/#r-20) contains download links for all needed DREAM II software tools. In Chapter 2, we will need to use the R+m.Design App which can be installed from this web link https://play.google.com/store/apps/details?id=com.robotis.mdesign.

1.4 References

Myers KI (2021) The Ultimate Chrome OS Guide: Acer Chromebook 315.

Thai CN (2017) Exploring Robotics with ROBOTIS Systems (2nd Edition). Springer Publishers International, New York.

Thai CN (2018) Learning Robotics with ROBOTIS DREAM System. CNT Robotics LLC Publisher. Printed by Amazon KDP.

Thai CN (2019) Advanced Application Programming with ROBOTIS DREAM System. CNT Robotics LLC Publisher. Printed by Amazon KDP.

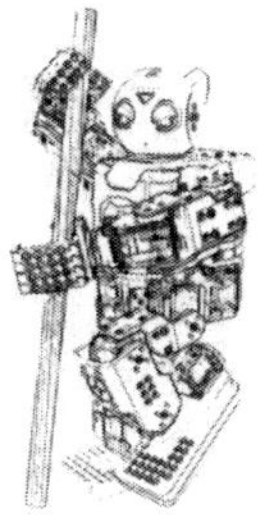

Chapter 2: Using R+ Block

Block Programming/Coding tools are well suited for the youngest beginners (5+) in computer programming, in general, and in robotics programming, in particular (https://childhood101.com/block-coding-websites-kids/, https://www.shaperobotics.com/block-based-programming/).

2.1 R+SCRATCH and R+BLOCK

For the CM-150/151, ROBOTIS offers two tools to work with Block Programming/Coding:

- R+SCRATCH is a helper application that runs on Windows PCs only and is designed to work with the MIT SCRATCH 2 Offline Editor (https://emanual.robotis.com/docs/en/software/rplus2/scratch/, https://scratch.mit.edu/download/scratch2). If the reader is interested in using this tool, please refer to Thai (2018). Currently, R+SCRATCH works with CM-150 only.
- R+BLOCK is based on Google Blockly and is originally designed for Android Mobile Devices (https://developers.google.com/blockly, https://emanual.robotis.com/docs/en/software/rplus2/rplus2_block/). For a comparison between Blockly and Scratch, the interested reader can visit this link https://www.robotlab.com/blog/blockly-vs-scratch-whats-best-for-me).

The R+BLOCK App was originally designed for Android Mobile Devices, but it is also usable on Chromebooks (https://play.google.com/store/apps/details?id=com.robotis.rblock). However, some issues remain and are discussed/resolved in Appendix A and later in this Chapter. True to its namesake, R+BLOCK uses a BLOCK-based interface to ease interaction and learning for the youngest beginners in robotics programming/coding (see Fig. 2.1). **Please install this App on your Chromebook if you have not yet done so**.

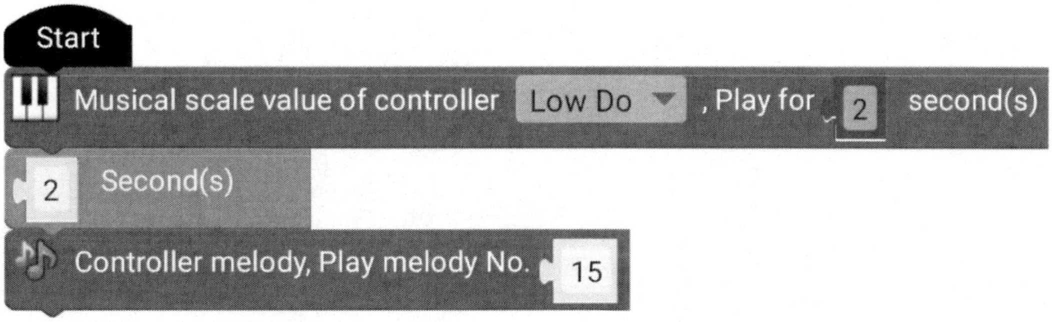

Fig. 2.1 A Basic R+BLOCK program/code.

This ROBOTIS web site contains good information for using the R+BLOCK App from software installation to how to navigate and how to use its Menu System (https://emanual.robotis.com/docs/en/software/rplus2/rplus2_block/). The current version of R+BLOCK is V.1.1.3 (c. 2020). **The author recommends that readers go over these materials and practice on their own Chromebooks to get some familiarity with the Menu system of R+BLOCK.** Readers can also watch this Video Tutorial for a "minimal" How-To Primer about R+BLOCK on Chromebooks (https://youtu.be/Fn-j2-X602Q).

Currently (Spring 2022), R+BLOCK works well with the CM-150 on Chromebooks, and only partially with the CM-151 as this is a newer controller, however this issue will likely be taken care of by ROBOTIS in the near future.

Chapter 2 is written for a complete beginner in robotics and computer programming, and we will be using the "Basic Avoider" platform and its variance (Fig. 2.2) to learn how to use the R+BLOCK App. A "Spiral" learning approach (Bergin, 2012) will be used for instructional purposes where the author would show how to use the R+BLOCK tool with projects having increasing scope and difficulty.

2.2 Construction of Avoider/Follower Robot

First, let's build a Basic Avoider/Follower robot as the demonstration platform (see Fig. 2.2) for various software projects presented in upcoming sections of this Chapter 2. Please watch Video at https://www.youtube.com/watch?v=TORJnsplPYI about how to use the R+DESIGN App to construct this robot, but only up to **Step 201 of the DREAM L2 Avoider design**. The final steps are to add two IRSS-10 sensors to the Back Side of this robot and to Port 3 and Port 4 of the CM-150/151 (also see Fig. 1.1 for a view of the finished robots).

These IRSS-10 sensors (https://www.robotis.us/ir-sensor-irss-10/) will have to be purchased separately as they do not come with the DREAM II School Set. Technical details for the IRSS-10 Sensor are available at https://emanual.robotis.com/docs/en/parts/sensor/irss-10/.

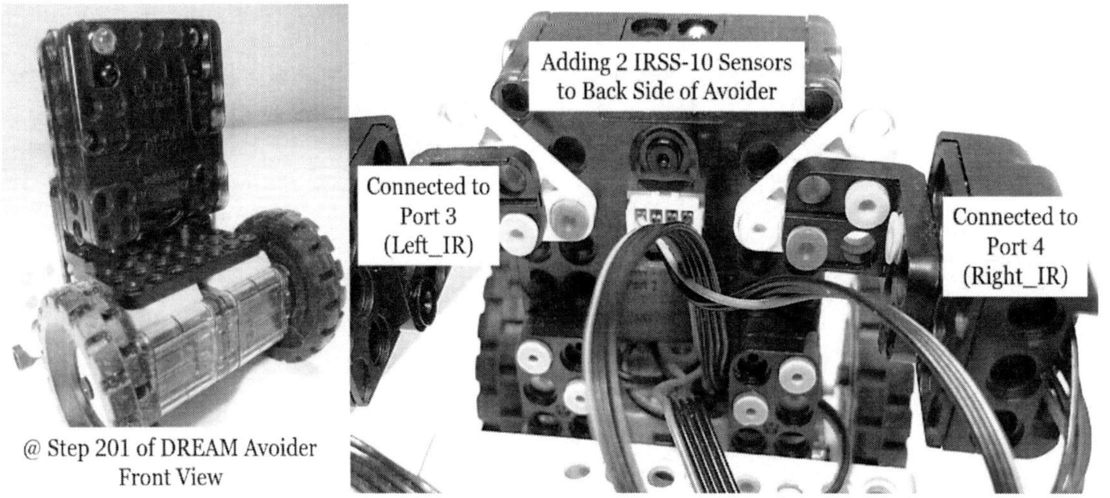

Fig. 2.2 Basic Avoider demonstration platform.

We are now ready to learn how to use the R+BLOCK App and most importantly how to learn basic Robotics Programming Concepts and Techniques using this Avoider platform. Please keep this web link (https://emanual.robotis.com/docs/en/software/rplus2/rplus2_block/) open for quick references when needed.

2.3 R+BLOCK Menu System

Fig. 2.3 depicts the Start-Up Window for R+BLOCK in Chrome OS:

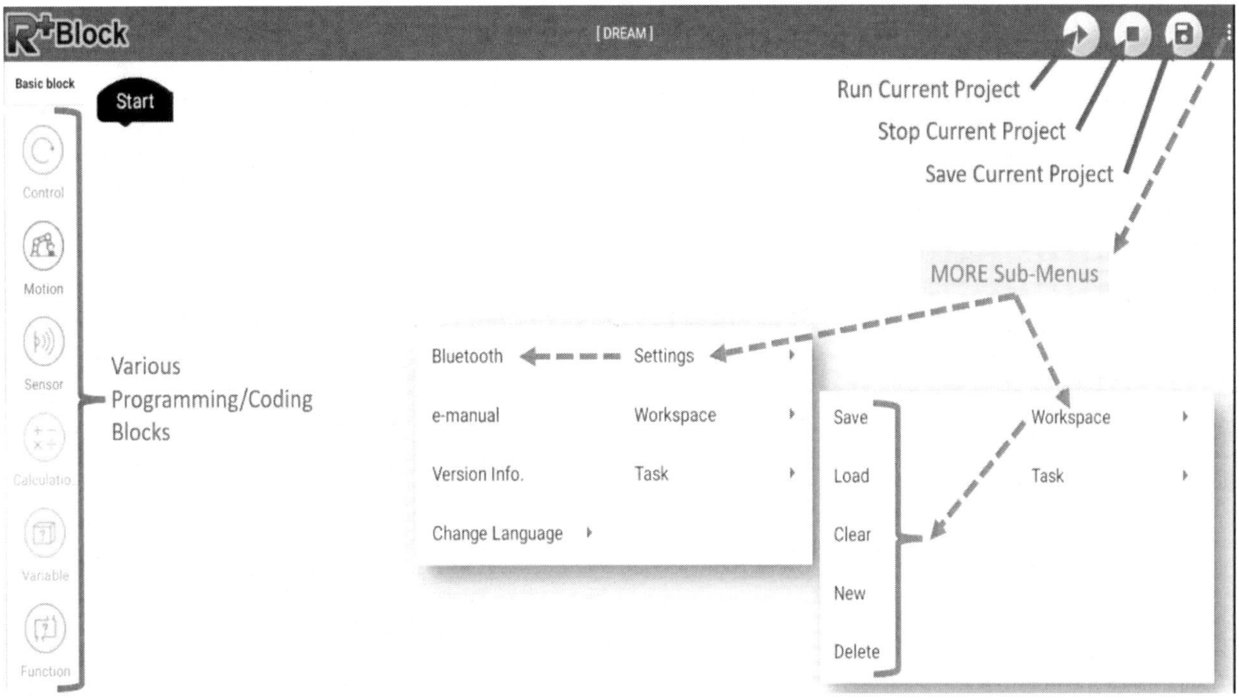

Fig. 2.3 R+BLOCK's Start-Up Window.

- On the Left Edge, ones can see the various types of Programming/Coding Blocks that can be used in a R+BLOCK Project: **Control, Motion, Sensor, Calculation, Variable** and **Function**.
- On the Top Right corner, ones can see 3 Buttons involving the Current BLOCK Project: **Play, Stop** and **Save**, and the **More** Button (3 vertical dots).
- The "**More**" Button will allow the user to access 3 more sub-menus: **Settings, Workspace** and **Task**. With **Settings** chosen and drilling down to **Bluetooth**, the user can scan for the BT-210/410 receivers connected to the CM-150/151. With **Workspace** chosen, the user can select various actions related to Projects (*.rpb files): **Save, Load, Clear, New** and **Delete** (see Appendix A for the exact location of the saved project files, i.e., the **DREAM** Folder).
- The **Task** sub-menu allows the user to download the Current BLOCK Project directly into the CM-150/151 as a TASK binary code, or to convert it into a TASK V.2 format/file (*.tskx) saved to the same **DREAM** Folder as for BLOCK Projects. This option will be discussed in more details within a later sub-section.

If the user selects **More** >> **Settings** >> **Bluetooth**, then the Bluetooth Scan Panel would pop up (see Fig. 2.4):

- Currently (4/2022), the information displayed vis-à-vis BT-410 and pre-pairing of BT-210 at the Chrome OS Settings level is **incorrect**, because the R+BLOCK's internal BT scanning tool works fine for BT-210/410 receivers used on a CM-150 and even for the embedded BT-410 found on a CM-151.

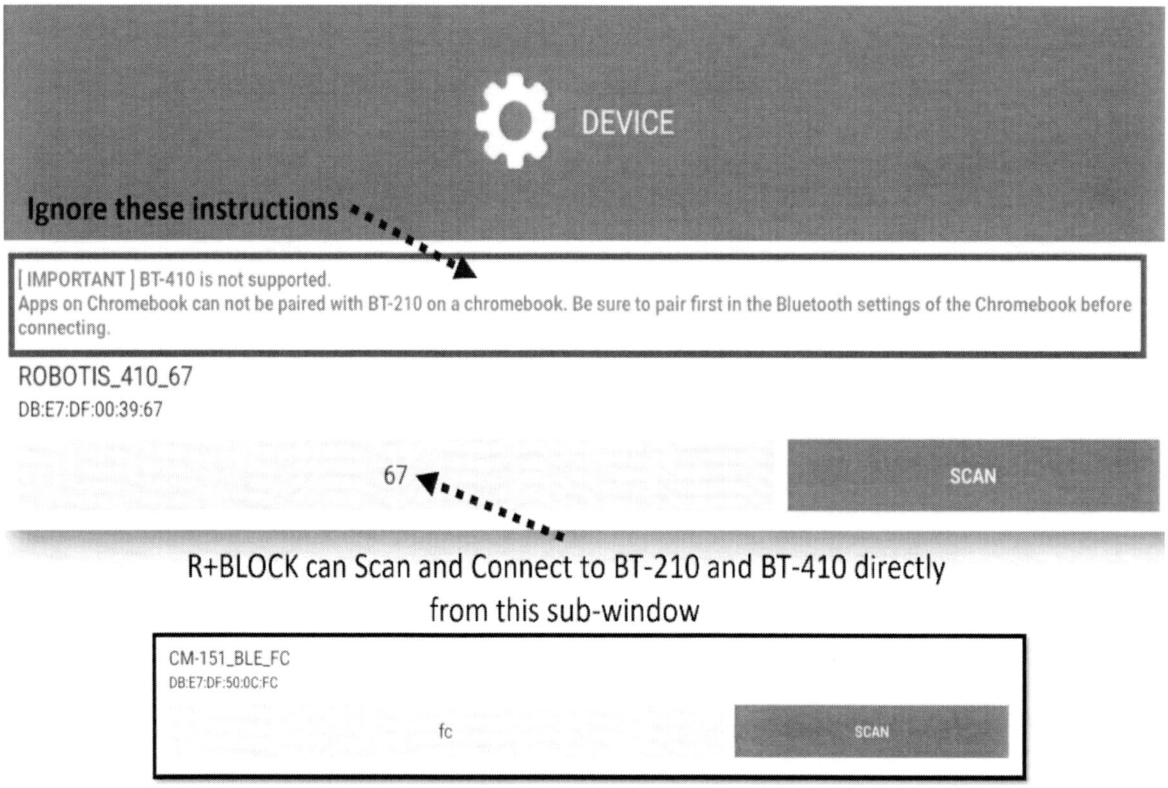

Fig. 2.4 R+BLOCK's Pop-Up BT Scan Panel.

- However, if the reader uses a BT-410 USB Dongle in combination with a BT-410 receiver on the robot, this internal BT scanning tool will fail (because R+BLOCK and Chrome OS do not support this type of USB device yet). The combo BT-410 Dongle+Receiver only works properly with the Linux Virtual Environment (i.e., when we do Python Programming in Chapter 4).

Fig. 2.5 shows the actual Programming/Coding Blocks that are of the **Control** type (Left Panel), and from the top they are:

- TIME DELAY Block for user-set number of seconds.
- ENDLESS LOOP as a Forever Repeat Do Loop.
- COUNTER-BASED Do Loop.
- CONDITION-BASED Do Loop.
- Simple IF structure.

- IF-ELSE structure.
- CONDITION-BASED Wait Loop.

Fig. 2.5 also shows the actual Programming/Coding Blocks that are of the **Motion** type (Right Panel), and from the top they are:

- ACTUATING Geared Motor connected to Port 1 or Port 2.
- ACTUATING Servo Motor (in Continuous Turn mode) connected to Port 3 or Port 4.
- ACTUATING Servo Motor (in Position Control mode) connected to Port 3 or Port 4.
- ACTUATING LED Module connected to Port 3 or Port 4.
- PLAY Buzzer's user-set Musical Scale and for user-set Time Interval.
- PLAY Buzzer's Melody with user-set Number.

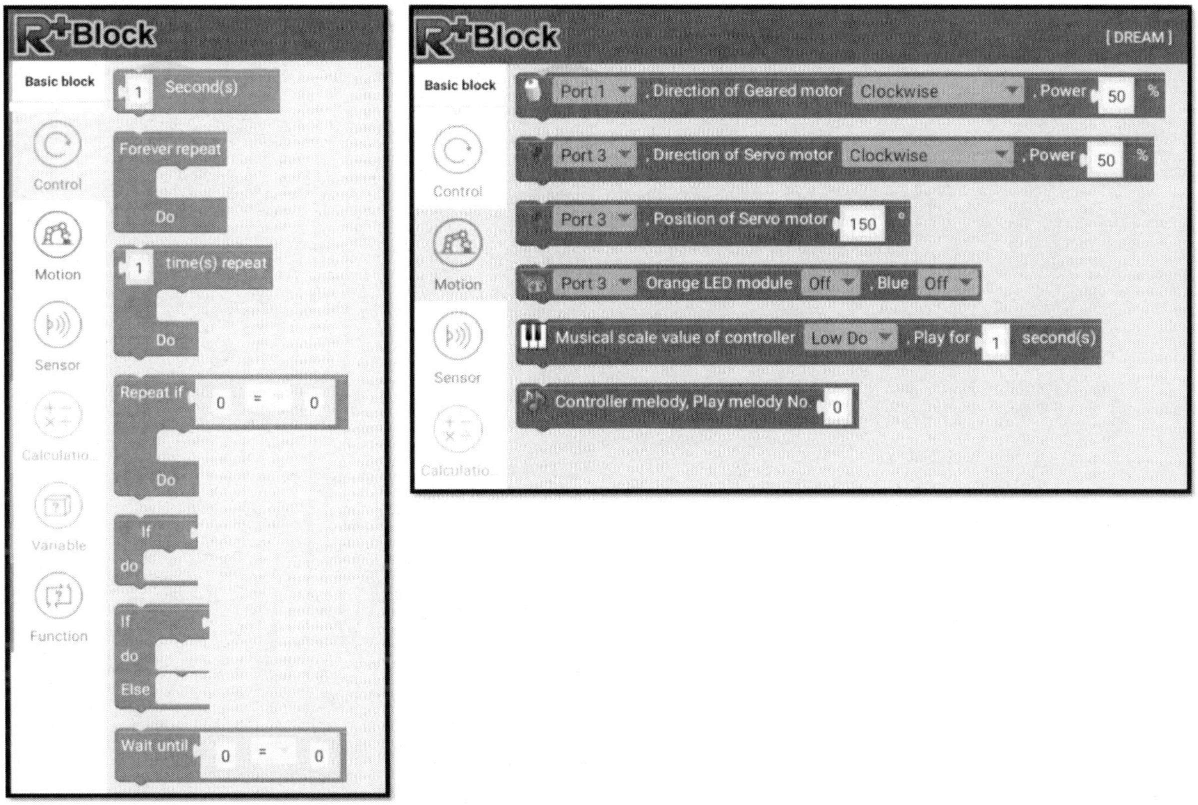

Fig. 2.5 Control and **Motion** types of Programming/Coding Blocks.

Fig. 2.6 shows the actual Programming/Coding Blocks that are of the **Sensor** type (Left Panel), and from the top they are:

- READ Left/Middle/Right Internal IR Sensor and COMPARE it to a user-set THRESHOLD.
- READ Real-Time Sound Counter Value and COMPARE it to a user-set THRESHOLD.
- READ Final/Resulting Sound Counter Value and COMPARE it to a user-set THRESHOLD.

- READ External IR Sensor connected to Port 3 (or 4) and COMPARE it to a user-set THRESHOLD.
- READ External Touch Sensor connected to Port 3 (or 4) and COMPARE it to a user-set THRESHOLD.
- PRINT a STRING (or Value of a Variable).
- READ Left/Middle/Right Internal IR Sensor.
- READ Real-Time Sound Counter Value.
- READ Final/Resulting Sound Counter Value.
- RESET Resulting Sound Counter Value to zero.
- READ External IR Sensor (or Touch Sensor) connected to Port 3 (or 4).

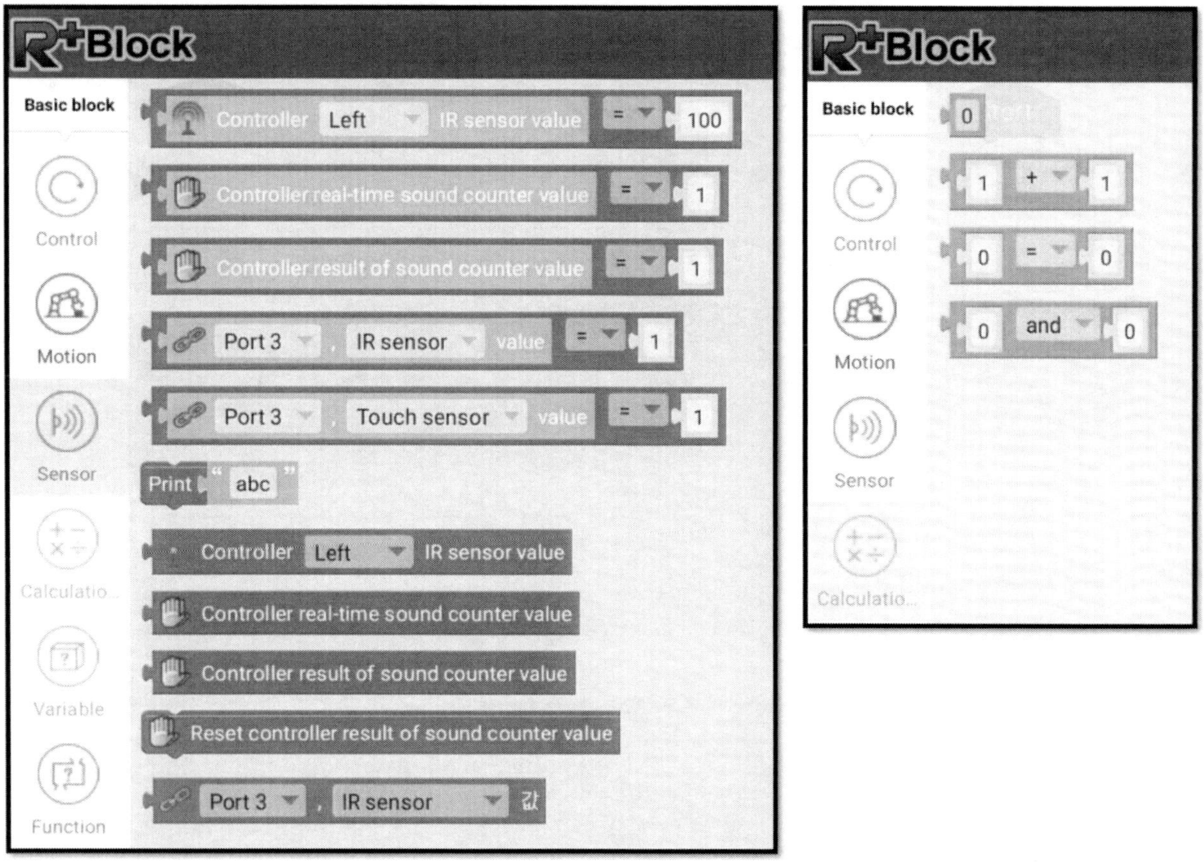

Fig. 2.6 Sensor and **Calculation** types of Programming/Coding Blocks.

Fig. 2.6 also shows the actual Programming/Coding Blocks that are of the **Calculation** type (Right Panel), and from the top they are:

- NUMERICAL CONSTANT.
- NUMERICAL ALGEBRAIC Expression.
- NUMERICAL COMPARISON Expression.
- LOGICAL AND/OR Expression (0 means NO, non-zero means YES).

Fig. 2.7 shows how to create and use the Programming/Coding Blocks that are of the **Variable** type:

- At start, when no VARIABLE had been defined, only the "NEW VARIABLE" Block is shown (Left Panel). The user needs to left click or touches this Block to create a NEW Variable (each time as needed).
- Once a VARIABLE has been defined, the SET/CHANGE/GET Block types will become available for each defined VARIABLE. In Fig. 2.7, "Claps" and "Speed" were used by the author.
- Currently, there are some "bugs" in renaming VARIABLES, as the New-Named Variable is created, but the Old-Named Variable is kept in the VARIABLE listing. Also SET and GET cannot be "reused" with another Variable Name, causing R+BLOCK to immediately crash.

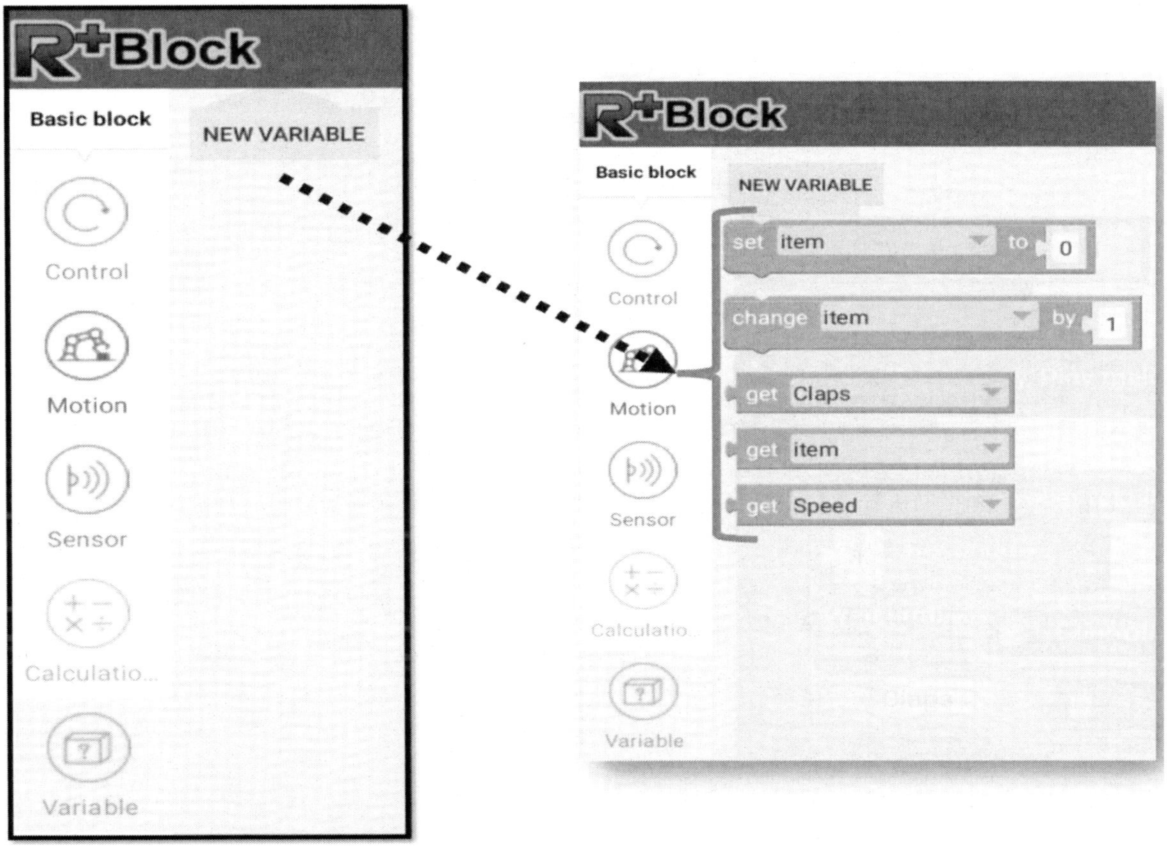

Fig. 2.7 Variable type of Programming/Coding Blocks.

Lastly, Fig. 2.8 shows the usual Blocks used to CREATE and CALL FUNCTIONS, along with the indispensable RETURN Block (that most programmers forget or don't want to use).

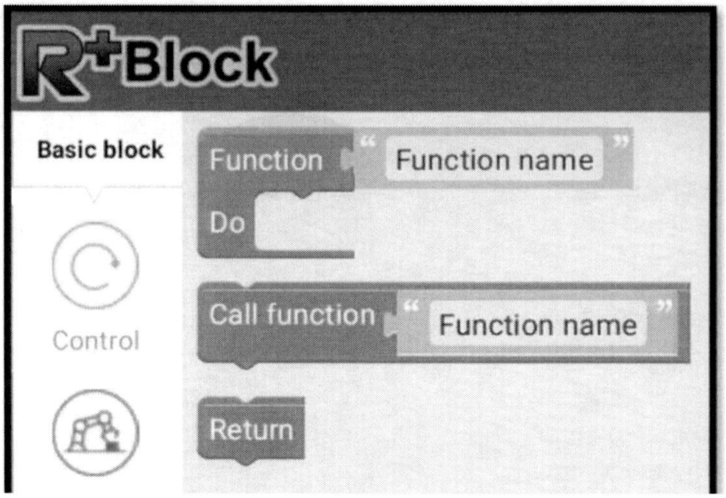

Fig. 2.8 Function type of Programming/Coding Blocks.

The reader needs to note that R+BLOCK is designed for the youngest beginners so it does not support features such as Remote Control via Virtual or Physical RC-100B, or other Sensors such as Temperature, Passive IR, Color, etc. that TASK V.2 and V.3 would readily support (see Chapter 3).

2.4 Learning R+BLOCK Programming Basics

At present, R+BLOCK (V.1.1.3) works properly with the CM-150 and only partially correct with the CM-151: for example, the Buzzer functions work the same way with both controllers, but internal NIR sensors are reporting their correct values with the CM-150, **but not with the CM-151**.

The following sub-sections, 2.4.1 through 2.4.5, were written for a "complete beginner" reader in mind, but they could be used as "review materials" for readers with more experiences in Block Programming from elsewhere.

The reader is recommended to first view https://youtu.be/CvgenQ5BXXs for a quick working tour of the R+BLOCK application and to learn about the editing basics for R+BLOCK by following the author's steps in creating the example R+BLOCK project named "SequenceControl_0.rpb" (see Fig. 2.9).

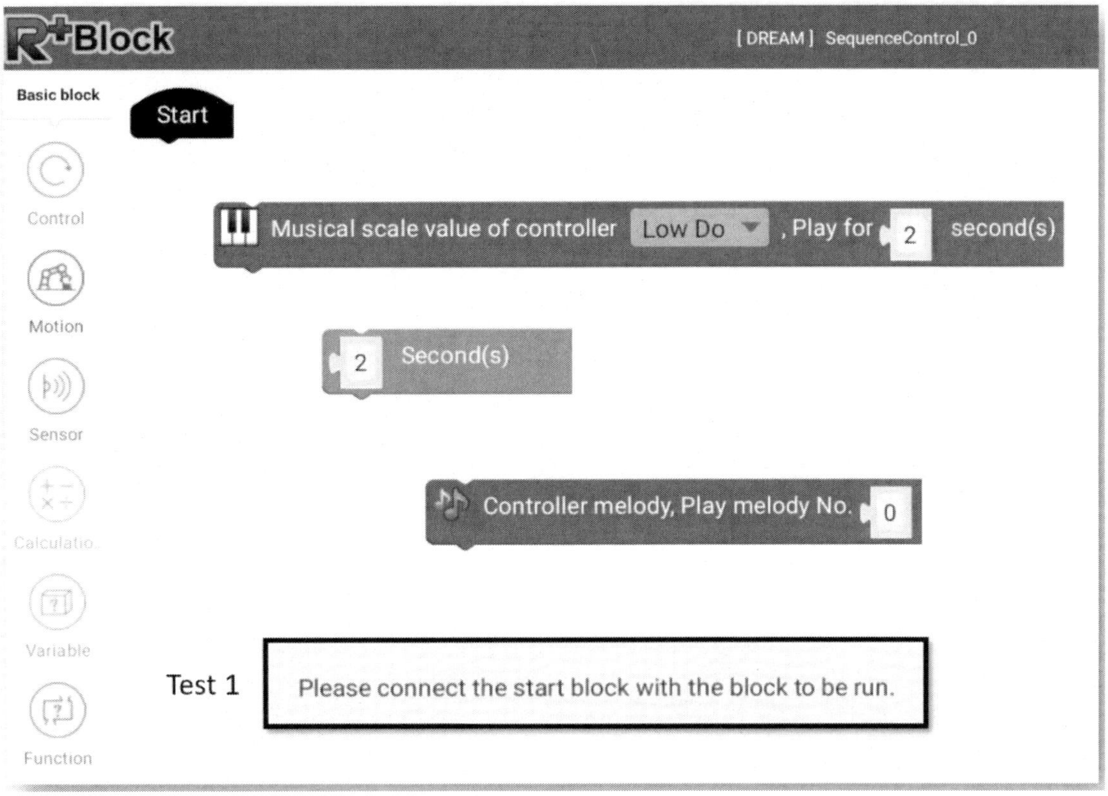

Fig. 2.9 Project "SequenceControl_0.rpb" and Test 1.

At this point, if the reader had been watching the previous video and had been practicing along with this video, then the reader is ready to continue to Sub-Section 2.4.1. If not, please start R+BLOCK on your Chromebook and load up the project named "SequenceControl_0.rpb" before going on to Sub-Section 2.4.1.

2.4.1 Sequence Control

In this unit, we are going to learn that **by default** the CM-150/151 controller behaves **sequentially**, i.e., it can do only one thing at a time. First, we are going to perform a few "tests" with "SequenceControl_0.rpb":

- **Test 1** (Fig. 2.9) – Just click on the **Play Project** button found on the Top-Right area of the current R+BLOCK Window (also see Fig. 2.3). Then the user should see a notification message appearing at the bottom center of the window saying that the user should connect at least one Coding Block to the Start Block to run this (and any other) **rpb** projects. Here, if readers have used MIT SCRATCH before, the author would understand disappointments from these users who are accustomed to be able to "run" each Coding Block separately and instantaneously, but this feature is not available with R+BLOCK.

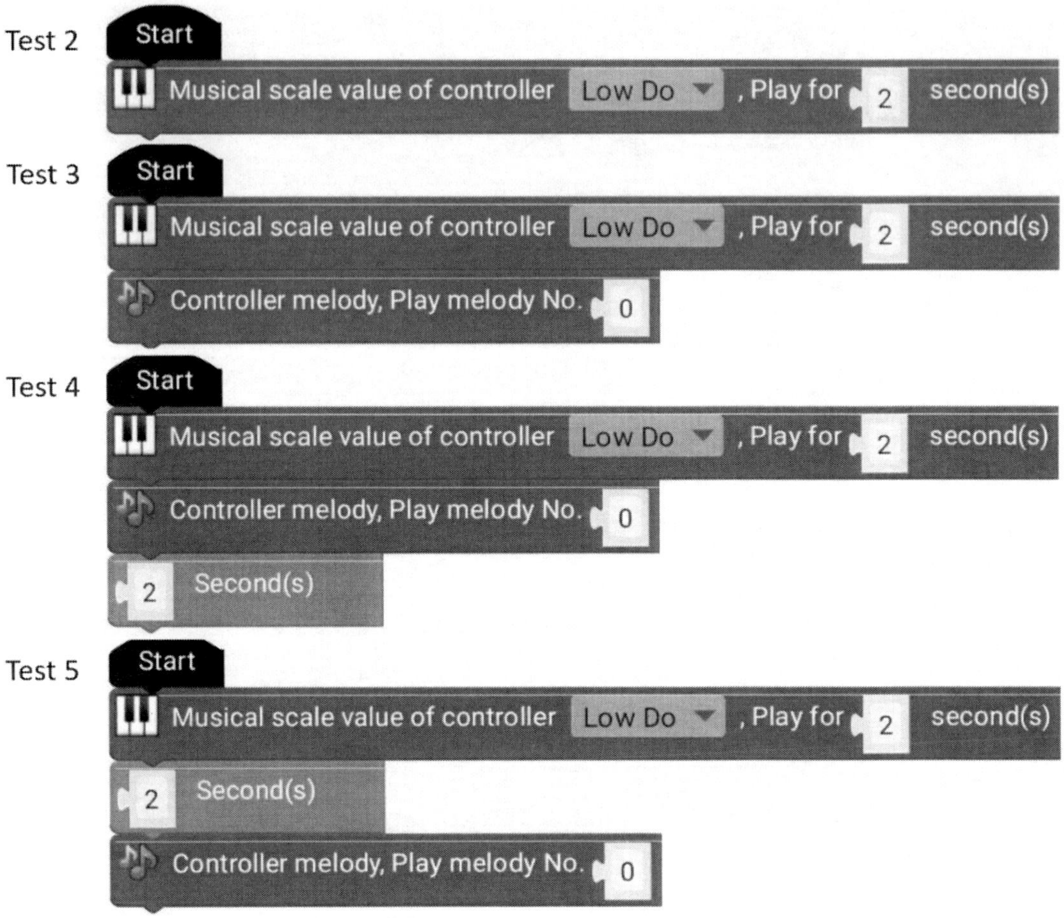

Fig. 2.10 Test 1 of Project "SequenceControl_0.rpb".

- **Test 2** (Fig. 2.10) – So let's just connect the "Music Scale" Block to **Start** and click on **Play Project** again. This time, we can hear the Buzzer sound out a "Low Do" note for 2 seconds, as expected. If the reader had kept his/her eyes on the current Coding Blocks while pressing on the **Play Project** button, the reader would have noticed that the "Music Scale" Block was colored "Orange" very briefly, this feature is called "Algorithm Animation" and it is designed to let the programmer know which Block is being activated during Play/Run Time.
- **Test 3** (Fig. 2.10) – Let's say that after playing the "Low Do" note, we want the Buzzer to play "Melody 0". Thus, we just connect the "Play Melody 0" Block below the previous "Music Scale" Block, as this would seem the logical step to do. Then, let's click on **Play** and we can hear a "Low Do" note for 2 seconds, just like before for Test 2, then **nothing else**! If the reader replays this project again but this time pays attention to the Algorithm Animation feature, one will notice that the "Play Melody 0" Block did get activated, but the CM-150/151 seemed to have ignored that second command to the Buzzer! So, what is going on? Did we have "Robot Disobedience" already? In a way, we did:
    - When the Buzzer receives the command for playing "Low Do" from R+BLOCK via Bluetooth, it started to perform this assigned task right away and for the 2 seconds as ordered. Now, back to R+BLOCK on the Chromebook: after sending

out the "Music Scale" command, R+BLOCK moves on next to the "Play Melody 0" command and sends it to the Buzzer. As the Chromebook's CPU is very fast, this is done within a few milliseconds from the previous command, and the CM-150/151 did receive this new Buzzer command, but it was still performing its first task because the 2-second buzzing interval had not been reached yet. So this second command was simply "lost/ignored" as the Buzzer was not designed to retain several consecutive commands at any one time. This is an important side-effect that an R+BLOCK programmer needs to account for, when there is a need to send different **consecutive commands to the same device** located on the CM-150/151. One simple solution is to use a Time Delay in the R+BLOCK code as shown in **Test 4** and **Test 5** in Fig. 2.10.

- **Test 4** of course is not the correct solution in this case, as the Time Delay Block is placed at an inappropriate location. So, the user would only hear the "Low Do" scale for 2 seconds with **Test 4** (like with **Test 3**).
- **Test 5** is the appropriate solution to provide the sound effects originally wanted by the user. The reader is encouraged to try other values for the Time Delay such as 1 or 3 seconds to hear the resulting effects. **The "Algorithm Animation" feature is also better illustrated in this example as it has "more" blocks to animate.**

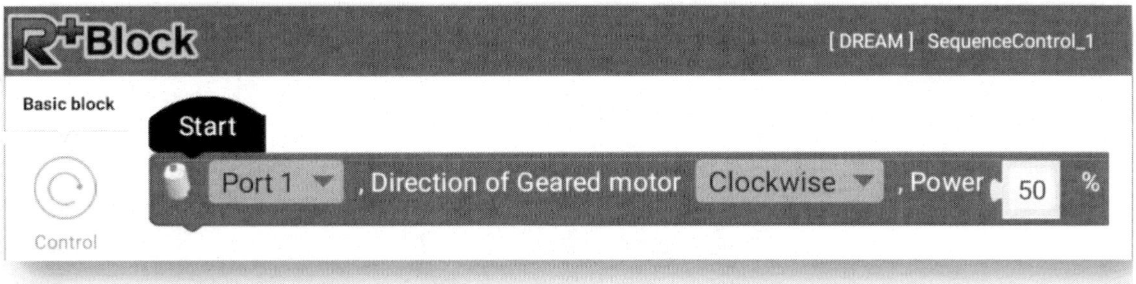

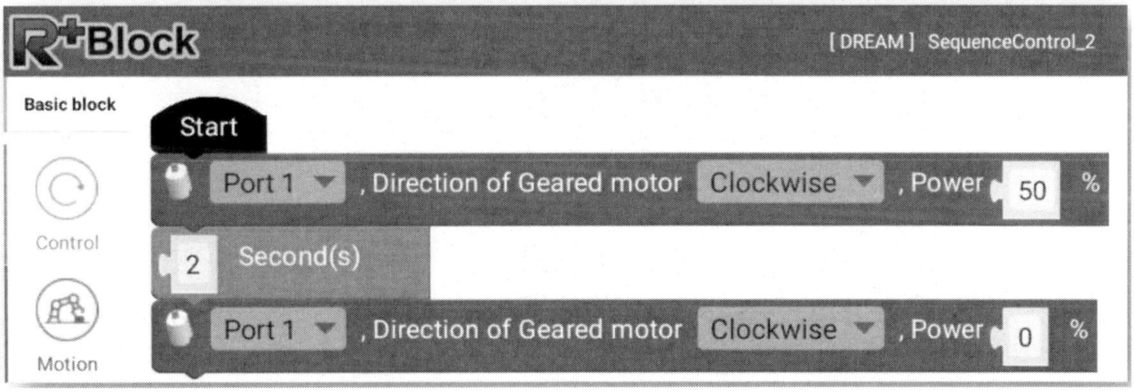

Fig. 2.11 Projects "SequenceControl_1.rpb" and "SequenceControl_2.rpb".

Next, let's see how the motors on Ports 1 and 2 behave under Sequential Control, so let's next load up "SequenceControl_1.rpb" and run it (see Fig. 2.11):

- The reader would readily notice that Port 1 Motor would turn CW at 50% power/speed as commanded. But it did not stop although the end of the R+BLOCK program had been reached. The user would have to click/touch on the **Stop Project** Button to shut down this motor for sure.

- In practice, the type of situation can happen to all users (beginners or advanced) when the robot happens to be in motion, and something may have happened to the Bluetooth communications facility, or that the robot's program has a bug, etc., then the robot would be running amok. The only thing that anyone can do at that time is to catch the robot safely, for the robot and for the operator, and shut its power down!

Thus, in "SequenceControl_2.rpb" (see Fig. 2.11), the last Block is used to shut down the motor(s) when the end of the program is reached under "normal" circumstances.

Fig. 2.12 displays the code for "SequenceControl_3.rpb" where both Port 1 and Port 2 motors are used to make the robot **Spin Left for 2 seconds then Stop**. The reader is encouraged to try different combinations of CW and CCW for each Port 1 and 2 to see the resulting effects on the movement of the robot. This is also a good time for the reader/user to do a final check the construction of the Avoider robot, whereas the Left Motor should have been connected to Port 1 and the Right Motor should have been connected to Port 2. **All subsequent example codes from this book assumed this configuration.**

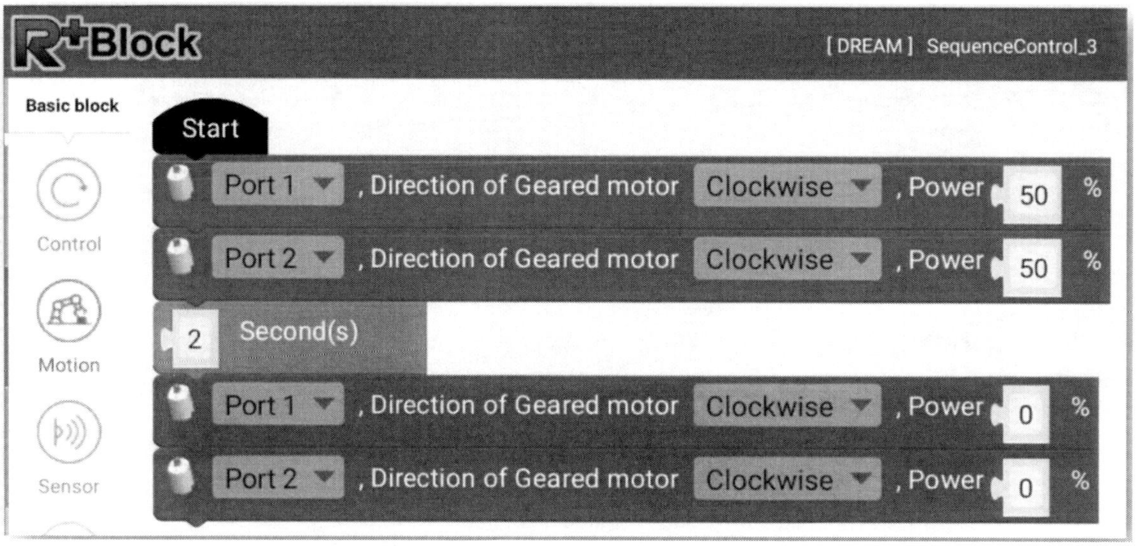

Fig. 2.12 Project "SequenceControl_3.rpb".

2.4.2 Program Modularization with Functions

Project "SequenceControl_3_F.rpb" (Fig. 2.13) makes the robot perform the same exact maneuver as Project "SequenceControl_3.rpb", except now it uses the R+BLOCK **Function** facility (please review Fig. 2.8):

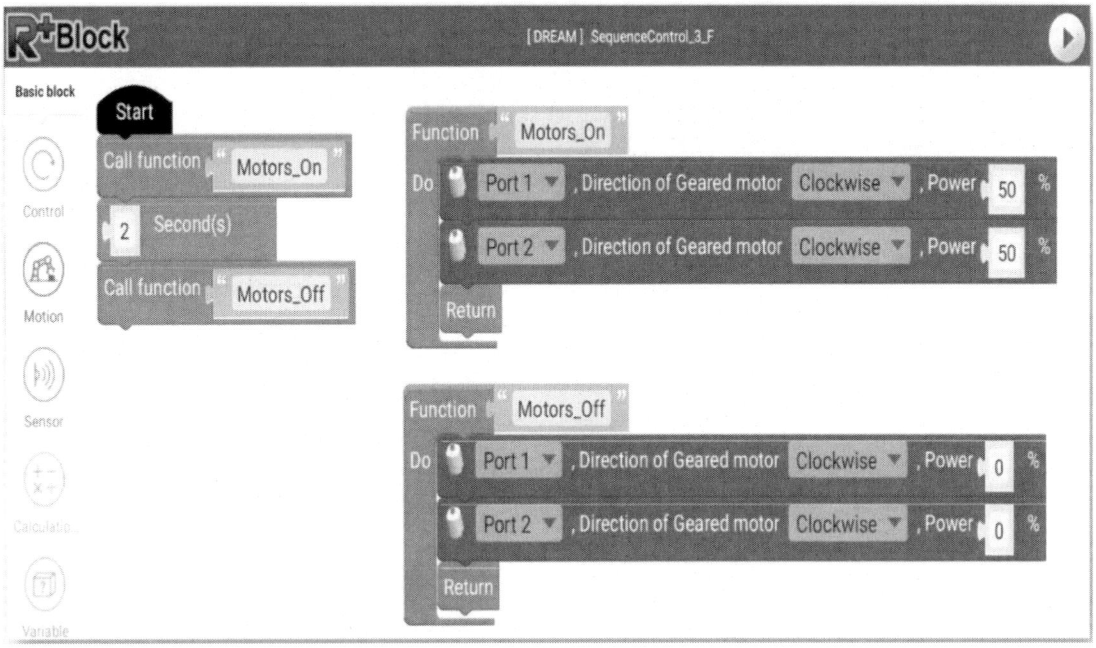

Fig. 2.13 Project "SequenceControl_3_F.rpb".

- Each **Function** essentially groups a finite number of Blocks that would perform a certain job when it is **Called** from the **Main Function** which is the **ONE** and **ONLY ONE** with the **START** Block.
- Function **Motors_On()** makes the robot Spin Left at 50% power, while Function **Motors_Off()** sets them to 0% power to stop the robot.
- Please note the **Return** Block used at the end of each Function, and this is the customary location for them. However, on more complex Functions, there may be several **Return** Blocks needed, and they can be anywhere within the Function as program logic requires (more in Chapter 4 when programming with Python).
- Blocks found within **Functions** are the only Blocks that can exist and be activated at run-time without having to be in **direct connections** with the **Start** Block. But of course, they are in **indirect connections** with the **Start** Block of the **Main Function** via the **CALL** Blocks. The R+BLOCK "Algorithm Animation" feature would offer a great way to visualize this process, but this feature only applies to blocks within the **Main Function** and does not extend into each Function's interior blocks. So instead, let's perform a "thought experiment" or a "mental stepping-through" for this program logic/algorithm:
    - When the programmer clicks on the **Start Project** Button, the **Main Function** executes its first Block which is "Call Function **Motors_On()**" thus Program Execution transfers to the first Block inside Function **Motors_On()** which is to turn Port 1 Motor CW at 50% power, then the next Block is executed to turn Port 2 Motor CW at 50% power. So, the robot is now Spinning Left. Then the Return Block is executed, and this "returns" Program Execution back to the **Main Function**.
    - The Time Delay Block is then executed, so it turns "Orange" for 2 seconds.
    - Next, the **Main Function** executes its third Block which is "Call **Function Motors_Off()**" thus Program Execution transfers to the first Block inside Function

Motors_Off() which is to set Port 1 Motor to 0% power, then the next Block is executed to turn Port 2 to 0% power. So, the robot is now Stopped. Then the Return Block is executed, and this "returns" Program Execution back to the Main Function which reaches the end of its coding.

- At present with R+BLOCK, if the programmer needs to track the progress inside Functions, the simple solution is to use Print Blocks placed at strategic locations within the Functions' bodies - essentially back to the early days of computing!

Project "SequencedManeuvers.rpb" (Fig. 2.14) is a more substantial application of Functions usage to help develop a longer sequence of robot actions that can be read and understood easily in the **Main Function**, while relegating the nitty-gritty specific actions required for the motors, to be specified inside their respective Functions. This permits the R+BLOCK programmer to develop algorithms at different levels of granularity or focus. For example, the **Main Function** in Fig. 2.14 is essentially a list of maneuvers (L/R/F/B/S) for the robot to perform sequentially, it also defines how long each maneuver should last, i.e., algorithmic thinking at a high level. While the Function associated with each robot maneuver implement lower-level algorithmic thinking at the individual "actuator" settings (i.e., Port No., Turn Direction and Power Level), while accounting for the final robot's behavior that the programmer had aimed for.

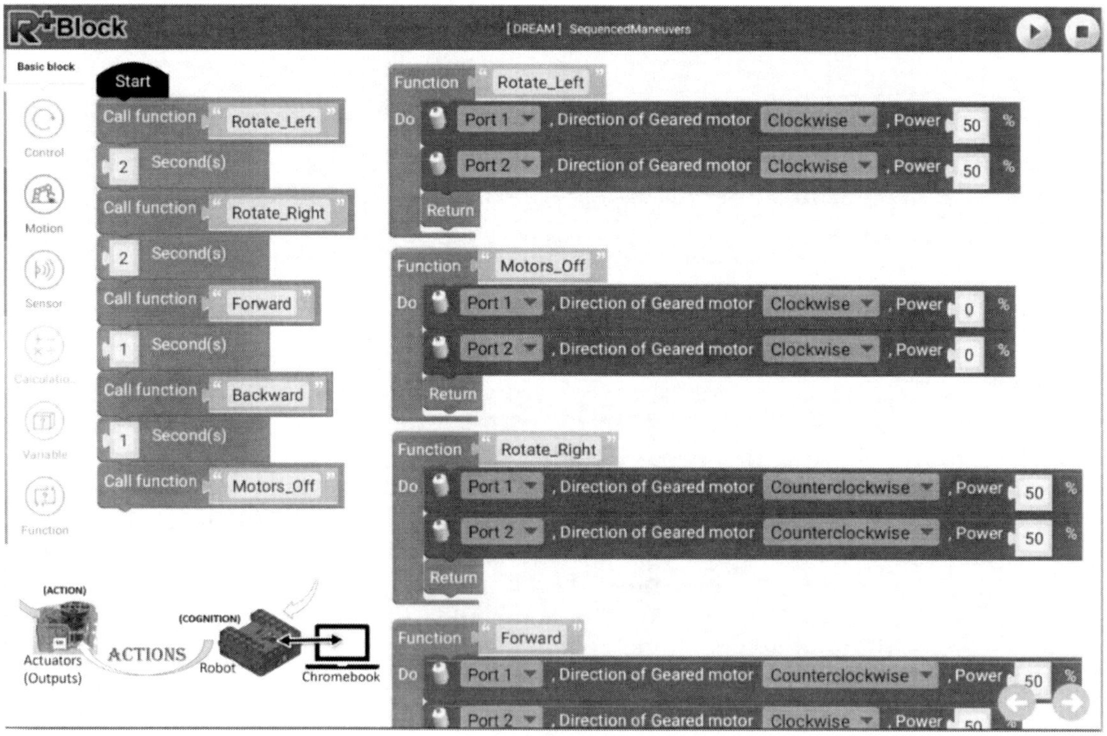

Fig. 2.14 Project "SequencedManeuvers.rpb".

Later projects in this Section will also demonstrate the re-usability of Functions.

Let's now run/play this R+BLOCK Project, and the author needs the reader to look carefully at the behaviors of the robot when it starts to go Forward (or Backward) and when it is about to end its Forward (or Backward) motion:

- In the case of Forward Motion, the reader should have noted that the robot "jerked" to the Right before rolling forward, and when it was about to stop, it would jerk to the Left.
- And in the case of Backward Motion, the robot "jerked" to the Left before rolling backward, and when it was about to stop, it would jerk to the Right.
- This was due to a "tiny but unavoidable" Time Delay (in ms) that happened between the first command to Port 1 and the next command to Port 2 inside Function **Forward** (see bottom right of Fig. 2.14). This Time Delay resulted in the activation of Port 1 Motor (Left) slightly ahead of Port 2 Motor (Right). This "Sequential Control + BT Communication" effect yielded the Right jerk at the beginning of a Forward motion, as Port 1 Left Motor turned on first, but Port 2 Right Motor was still off, thus making the robot veer Right (just for a few ms). Then, Port 2 Right Motor caught up with Port 1 Left Motor, so the robot is going straight afterwards and for about 1 second. Then, Port 1 Left Motor was commanded to stop a few ms before Port 2 Right Motor was commanded to stop, this resulted in the robot's Left jerk found at the end of the Forward maneuver. The reader is encouraged to apply the same analysis approach to understand the robot's jerking behaviors when it is commanded to go Backward.
- This Time Delay has several "contributors": refresh rate of R+BLOCK GUI, time taken for R+BLOCK/Chromebook to prepare and send its commands to the CM-150/151 via Bluetooth at 57.6 Kbps, time for the CM-150/151 to receive, decode and act upon the received commands (see bottom left of Fig. 2.14). Right now, there is no viable option for the programmer to tweak this chain of communications between the CB and CM-150/151 to help improve this situation. Later in Chapter 4, when we can program "under the hood" using Python on the CB and TASK on the CM-150/151, we'll look at ways to distribute the computing and communication loads between the CB and CM-150/151 in a more balanced way, to improve the run-time performances of both systems.
- Next, let's click on the **More** Button (at Right of Main Menu) and drill down to **Task >> Download** and click on the **Download** item. This procedure converts the current R+BLOCK code into an equivalent TASK code (in the background) and starts to download this converted code into the CM-150/151 working memory. When this job is finished, the CM-151 would play a short melody and turn itself off (the reader will have to turn off the CM-150 manually). Now, because the CM-150/151 already contains a "working" TASK program in its memory, the next time that the reader turns on this robot, it would immediately execute this TASK program and thus perform its Sequenced Maneuvers (L/R/F/B/S) on its own (without involving the **Play Project** Button in R+BLOCK). The reader should then notice that there are no more jerking moves at the beginning and at the end of the Forward or Backward motions!

Interesting result, isn't it? Here, we have the same physical robot, with the same behaviors coded in, but the "location" of the "run-time" codes was different between the two cases:

- The R+BLOCK code runs on the Chromebook (CB) which sends the actual Motion commands via Bluetooth at 57.6 Kbps to the CM-150/151 which then acts upon its Port 1 and 2 Motors as commanded by the CB.
- The converted TASK code runs on the CM-150/151's local memory which is integrated to Ports 1 and 2 via specific hardware addresses on theirs Control Tables (please view these web links for more information - https://emanual.robotis.com/docs/en/parts/controller/cm-151/#control-table or https://emanual.robotis.com/docs/en/parts/controller/cm-150/#control-table). In this situation, the activation command is still sent to Port 1 before Port 2, so there is still a "theoretical" Time Delay between these two commands, but it is much smaller now as motors and MCU are

"hard-wired" to each other and therefore this delay is "not detectable" by the human operator anymore.

In summary, all robotic systems have an Output Systemic Time Delay (OSTD) between the time when a command is issued by the controller's software and the time when the robot's actuators perform this command. The Chromebook/R+BLOCK/CM150-151/Robot system had been shown to have a larger OSTD than the one for the TASK/CM-150-151/Robot system. If the reader encounters similar runtime performance issues with his/her future R+BLOCK projects, it may be a signal to move on to TASK and/or Python programming.

With reference to Fig. 1.8, so far with Sub-Sections 2.4.1 and 2.4.2, we have been exploring the (Robot/Computer) and (Actuators/Outputs) components and the ACTIONS link between them.

Another practical issue that the reader needs to be aware of is about the Geared Motors GM-10 used on Ports 1 and 2. It is common to find that one GM-10 is weaker/stronger than the other one, thus for example, with one being powered at 50%, but the other one may need to be set at 55% in order for the robot to go Forward in a straight line. Sometimes, this correction does not apply to going Backward, although it is needed for going Forward!

2.4.3 Using Variables & Conditional Structure IF

In Sub-Section 2.4.3, we shift our focus to the (Sensors/Inputs) component and its EVENTS link to the (Robot/Computer) component (see bottom right of Fig. 2.15), using the "SetVariables.rpb" Project.

Fig. 2.15 shows some typical Input/Output manipulations often used with Sensor Data:

1. The first Block after **Start** shows how to access an **Internal Sensor** named "Middle IR Sensor", read its current Digitized Value, and save it into a Variable named "**Speed**".
2. The second Block gets/retrieves the **Numerical Value** that is stored in Variable **Speed** and **Print** it out on R+BLOCK's Runtime Output Panel.
3. The third Block gets Variable **Speed**'s **Numerical Value** and multiplies it by 2. The result of this multiplication is then saved into Variable **Power**. The multiplication factor of 2 was used because it suited the particular Internal IR sensors on the author's robot, readers can/should use other values to suit their situations.
4. As we plan to use Variable **Power** as a setting to Motors connected to Ports 1 and 2 later, we should aim for a number between 0 and 100, to match with the expected data type (a percentage) for these Motors. For this job, a Conditional Structure (simple **IF**) is used (4th and 5th Blocks):
   a. This **IF** structure uses a Conditional Expression **(Power's Value $\geq$ 100)** to decide what to do next (4th Block).
   b. At code run time, if Power's Value turns out to be less than 100, this Conditional Expression evaluates to FALSE, then Program Execution **jumps to the Block following the DO "bracket"** (the 6th Block in this case) and thus Prints out the current Value of Variable Power (which logically should be less than 100).

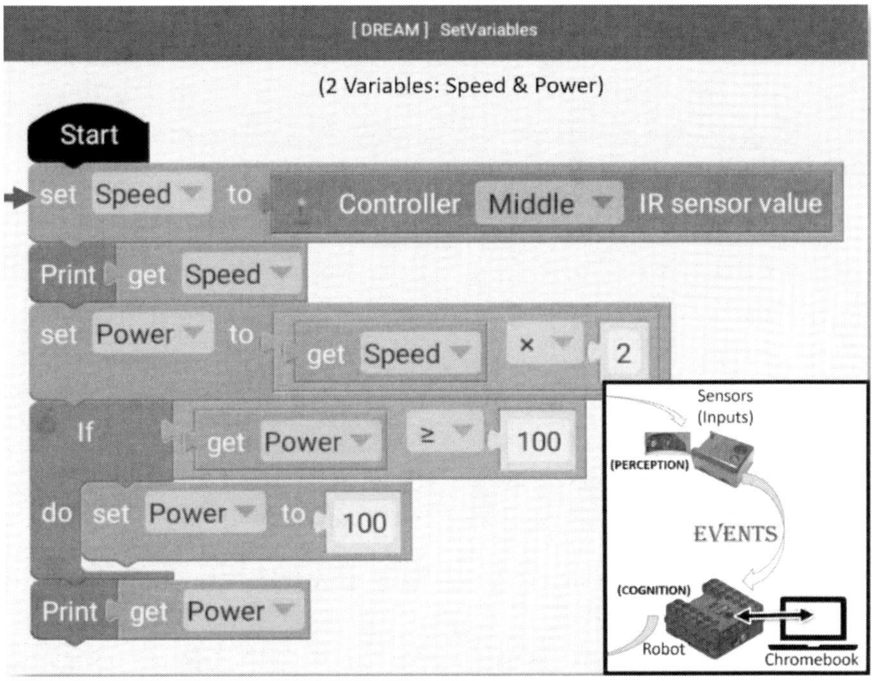

Fig. 2.15 Project "SetVariables.rpb".

c. However, if at run time Power's Value turns out to be greater than or equal to 100, this Conditional Expression evaluates to TRUE, then Program Execution "enters" the **DO "bracket"** and starts executing sequentially all the Blocks that are inside this **"bracket"**. In the current case, there is only 1 Block there and it sets/resets the Numerical Value stored in Variable Power to 100 (no matter what it was before this IF Structure is executed). Next, Program Execution jumps to the 6th Block and Prints out the current Value of Variable Power (which logically should never be greater than 100).

d. R+BLOCK also supports the **IF-ELSE-IF** Conditional Structure which will be showcased later in this book.

Let's run this Project "SetVariables.rpb" to see what kind of outputs are obtained for different situations at run time:

1. If the reader clicks on the **Play Project** Button, and nothing is put in the Line-Of-Sight of the Middle IR Sensor, then the reader would see two lines of output, both displaying **"0"**. This is because if the Middle IR Sensor did not detect any object in its Line-Of-Sight, it would report **0** to the Controller and therefore **Speed** is set to **0** (i.e., the first 0 printed). Subsequently, **Power** is also set to **0** (i.e., the second 0 printed).
2. If the reader puts an object in the Line-Of-Sight of the Middle IR Sensor and then clicks on the **Play Project** Button, then the reader would see two "non-zero" lines of output - depending on how close the object to the Middle IR Sensor is, for example:
    a. 30 and 60 (when Power turns out to be less than 100).
    b. 60 and 100 (when Power turns out to be greater than 100 – Power was first set to 120, then the IF structure clamps it down to 100).

Thus, at run time, the range of numerical values for **Power** is limited to **[0-100]** for the remaining R+BLOCK code, even though the range of numerical values for Speed can exceed **50**, depending on how physically close the object was to the Middle IR Sensor Port.

In other words, Project "SetVariables.rpb" illustrated a simple example of "Input Data Conditioning", other Input Data Conditioning techniques will be demonstrated later.

2.4.4 Using Main Endless Loop

R+BLOCK supports the usual types of Repetition Structures: Endless/Forever Loop, Conditional Loop, Counting Loop, and Wait Loop.

The Sense-Think-Act Loop (see Fig. 1.8) is an example of an Endless Loop. We will be using several Projects to demonstrate this most important concept:

- "IR_Throughput.rpb" to illustrate the "Input Systemic Time Delay" (ISTD) issue.
- "SetVariables_F.rpb" to troubleshoot the use of Variable Power for activating Ports 1 and 2 Motors.
- "Spinning_Top_1.rpb" and "Spinning_Top_2.rpb" to show an example of Simple Autonomous Behavior controlled by the Middle IR Sensor.

Previously, the Project "SequencedManeuvers.rpb" was used to identify the issue of Output Systemic Time Delay. This time, we are going to investigate a similar issue but for Sensor Input Data with the Project "IR_Troughput.rpb" using the 3 Internal IR Sensors.

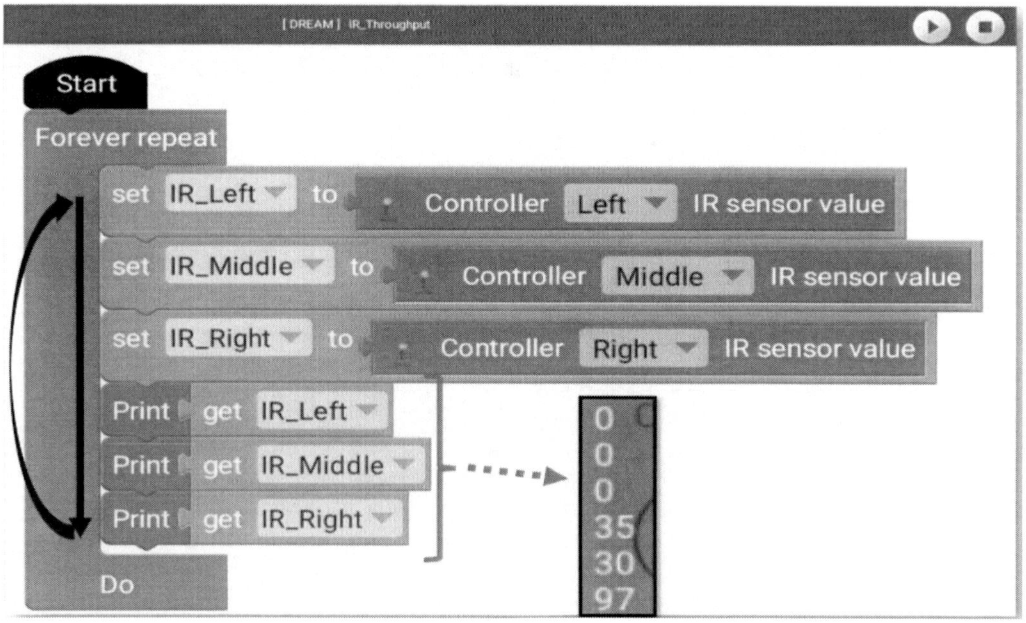

Fig. 2.16 Project "IR_Throughput.rpb".

Fig. 2.16 shows a typical implementation of an Endless Loop, whereas all the BLOCKS included in its Body would be **executed sequentially and in the order that they were created in**:

- The First Block reads in the Left IR Sensor's current **Value** and saves it into Variable **IR_Left**.
- The Second Block reads in the Middle IR Sensor's current **Value** and saves it into Variable **IR_Middle**.
- The Third Block reads in the Right IR Sensor's current **Value** and saves it into Variable **IR_Right**.
- The 4th through 6th Blocks prints out the respective and current Values of **IR_Left, IR_Middle** and **IR_Right**.
- Then the whole process would restart with the First Block and forever, or until the operator clicks on the **Stop Project** Button to stop the execution of this program.
- There is no time delay specified in this Endless Loop, so we are expecting that this loop will execute as fast as it can at run time, with Chrome OS support provided to R+BLOCK.

When the reader practices on this Project at home/school, the author would like to suggest the following procedure to investigate the issue of Input Systemic Time Delay (ISTD):

- "Play" this Project as normal, and the reader would see "0" scrolling down the computer screen on its left edge, if the reader did not block any of the 3 Internal IR Sensors.
- Next, use one's hand to block some or all IR sensors at once, and the reader would see "non-zero" values scrolling down the computer screen on its left edge. These values would correspond to the "latest" responses of the 3 IR Sensors.
- Then, remove one's hand as fast as possible from blocking the 3 IR Sensors while keeping a close watch on the "numbers" scrolling down the computer screen. The reader would again see the "0s" scrolling down, but only after a short delay (of around 1 cycle of the Endless Loop). This reveals that there is also an unavoidable **Input Systemic Time Delay** (ISTD) for the Chromebook/R+BLOCK/CM150-151/Robot system.
- Once again, the reader can try to Convert and Download this R+BLOCK Project into an equivalent TASK code which can be run via the Output Monitoring Tool of TASK V.2/V.3. The reader would witness a large increase in run time performance.

In short, all robots in this book would exhibit ISTD and OSTD symptoms, and these values are larger on the Chromebook/R+BLOCK/CM150-151/Robot system than on the Chromebook/TASK/CM150-151/Robot system. Thus, for **"Fast-Changing-Sensor-Data"** situations combined with **"Required-Actuator-Quick-Response"** demands, an R+BLOCK solution may not be able to reach the run-time performance needed, and TASK/Python solutions must be sought for.

In the example programs package prepared for this book, the author included the "Port_IR_Throughput.rpb" file for the reader's practice: it uses the IRSS-10 sensors connected to Ports 3 and 4 on the CM-150/151.

In the next Project "SetVariables_F.rpb", the author's goal is to notify the reader of an undocumented issue regarding the use of Variables as Power Setting for the Motors on Port 1 and Port 2. Previously, the author used 3 Variables in "IR_Throughput.rpb" without any problem, but when he decided to use two separate Variables as Power Settings for Maneuver-type Functions in "SetVariables_F.rpb", that was when he discovered an "**undocumented restriction**"!

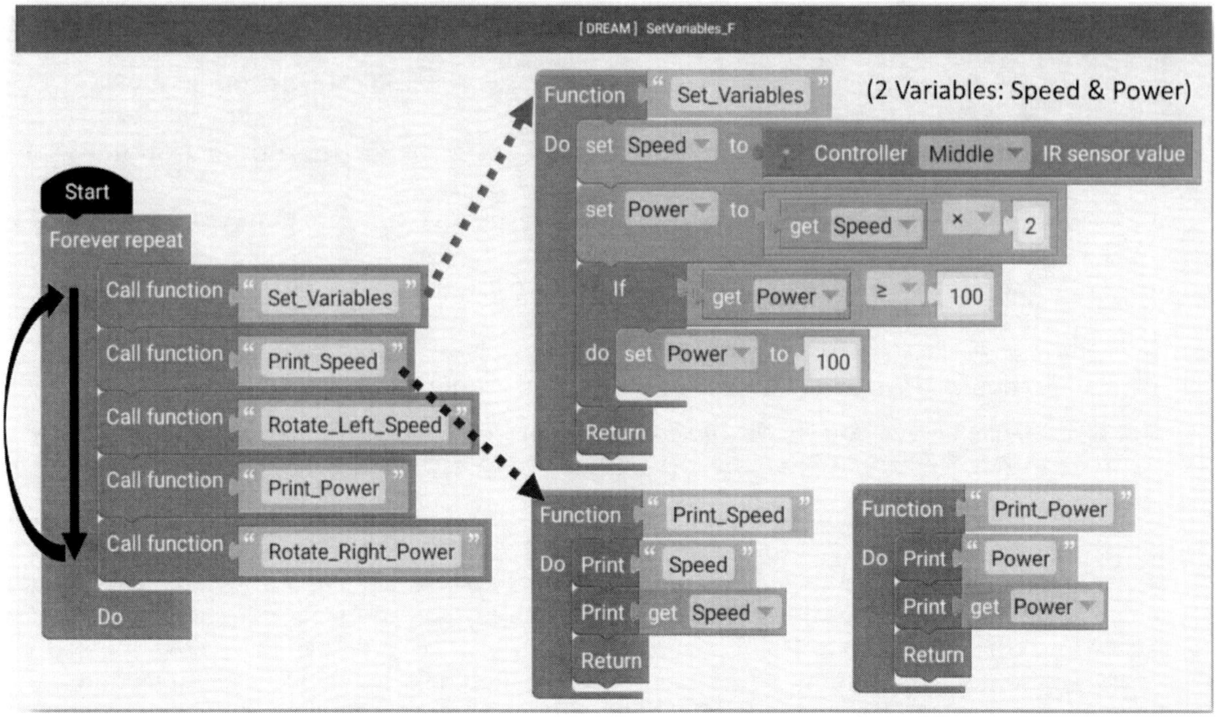

Fig. 2.17 Part1 of Project "SetVariables_F.rpb".

The Project "SetVariables_F.rpb" is described in Figs. 2.17 and 2.18. Fig. 2.17 lists the Main Function and 3 "regular" Functions named **Set_Variables(), Print_Speed()** and **Print_Power()**:

- Function **Set_Variables()** essentially contains the Blocks used in the Project "SetVariables.rpb" already described in Sub-Section 2.4.3, so that information is not repeated here.
- In Functions **Print_Speed()** and **Print_Power()**, the main point for the reader is to distinguish between two types of PRINTING to the Computer Screen at run time:
  - **Printing Text/String** is done via using such Blocks as [Print "Speed"] and [Print"Power"], which respectively prints out the strings "Speed" and "Power" on R+BLOCK Output Screen at run time.
  - **Printing Numerical Values of Variables** is done via Blocks such as [Print(get(Speed)] and [Print(get(Power)], which respectively prints out the current **numbers** stored in Variables **Speed** and **Power**.

Fig. 2.18 displays the Functions named **Rotate_Left_Speed()** and **Rotate_Right_Power()**:

- Function **Rotate_Left_Speed()** uses Variable **Speed** as Power Setting for Motors on Port 1 and Port 2.
- Function **Rotate_Right_Power()** uses Variable **Power** as Power Settings for Motors on Port 1 and Port 2.

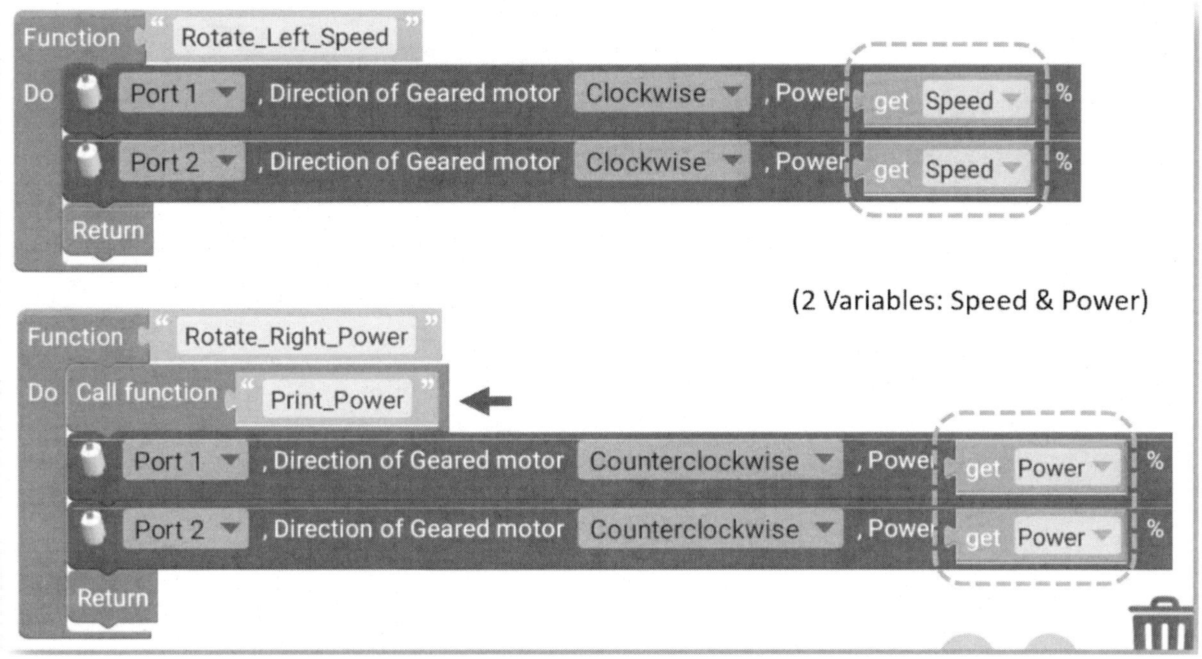

Fig. 2.18 Part 2 of Project "SetVariables_F.rpb".

Although, the Syntax used in Fig. 2.18 seemed appropriate enough and this R+BLOCK program ran without any "complaints" from R+BLOCK or Chrome OS, but the author noticed that, at run time:

- Function **Rotate_Left_Speed()** executed properly with varying speeds depending on the distance between an object and the Middle IR Sensor, and the correct numerical values for Variable **Speed** were appropriately printed on the computer screen.
- Function **Rotate_Right_Power()** executed properly, and the correct numerical values for Variable **Power** were appropriately printed on the computer screen. But the robot never Turned Right when it was time for it to do so, it was just staying still during this time period!

Emails sent to ROBOTIS Technical Support confirmed the author's findings that:

- **It is possible to set a Variable Power Setting to Motors connected to Ports 1 and 2, in Maneuver Functions for example, but it must be the FIRST VARIABLE defined by the programmer when creating his/her R+BLOCK program.**
- Please note that this issue is not present in TASK programs at all!

To verify these findings the author created the Project "Spinning_Top_1.rpb" which used only 1 Variable (**Speed**) in all its Maneuver Functions, while the Main Function is only a one-time-only Sequential Control algorithm (see Figs. 2.19 and 2.20).

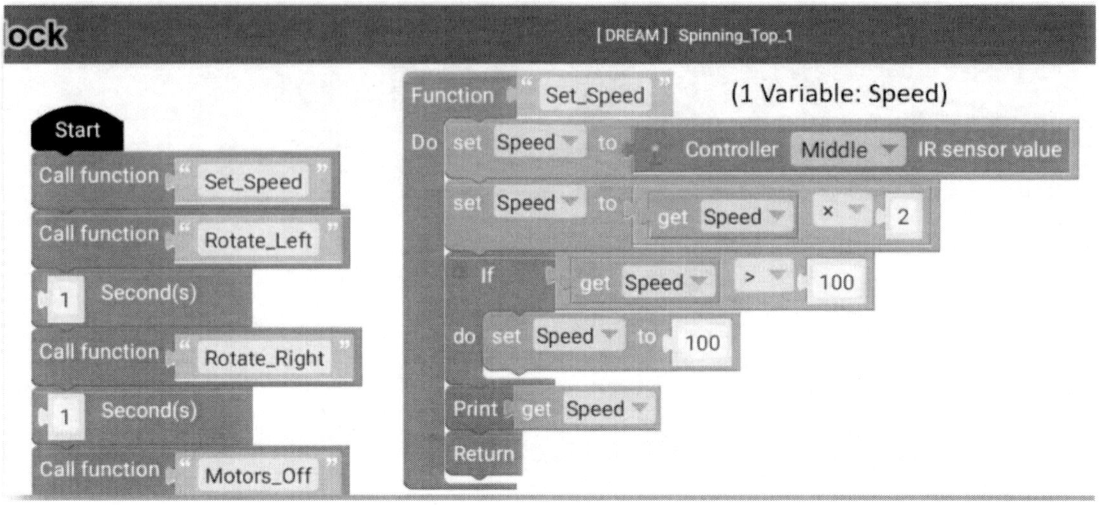

Fig. 2.19 Part 1 of Project "Spinning_Top_1.rpb".

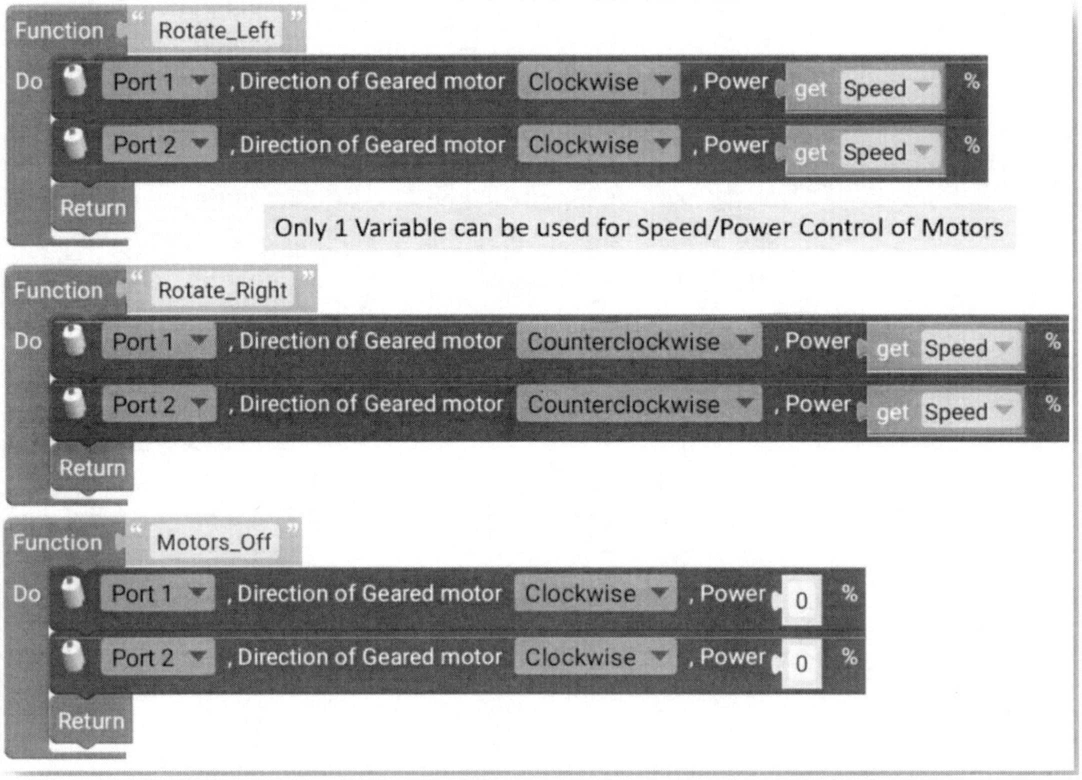

Fig. 2.20 Part 2 of Project "Spinning_Top_1.rpb".

In Fig. 2.19, Function **Set_Speed()** is defined practically the same as in Project "SetVariables_F.rpb", except for the 2nd Block from its beginning, whereas the Initial Value of **Speed**, as set from the 1st Block, is multiplied by 2 and then reassigned back to **Speed**.

The Main Function in Fig. 2.19 shows that after setting the appropriate value for **Speed**, it goes on to make the robot Rotate Left for 1 second, then make it Rotate Right for 1 second, then stops the robot as the final step.

Fig. 2.20 shows that both Left and Right Maneuvers used Variable **Speed** as a variable power setting.

At run time, the robot performed as programmed in the "Spinning_Top_1.rpb" code.

Next, Project "Spinning_Top_2.rpb" incorporates the code from Project "Spinning_Top_1.rpb" to create a **Simple but Complete** implementation of the "Sense-Think-Act" paradigm as an Autonomous Control application (see Fig. 2.21).

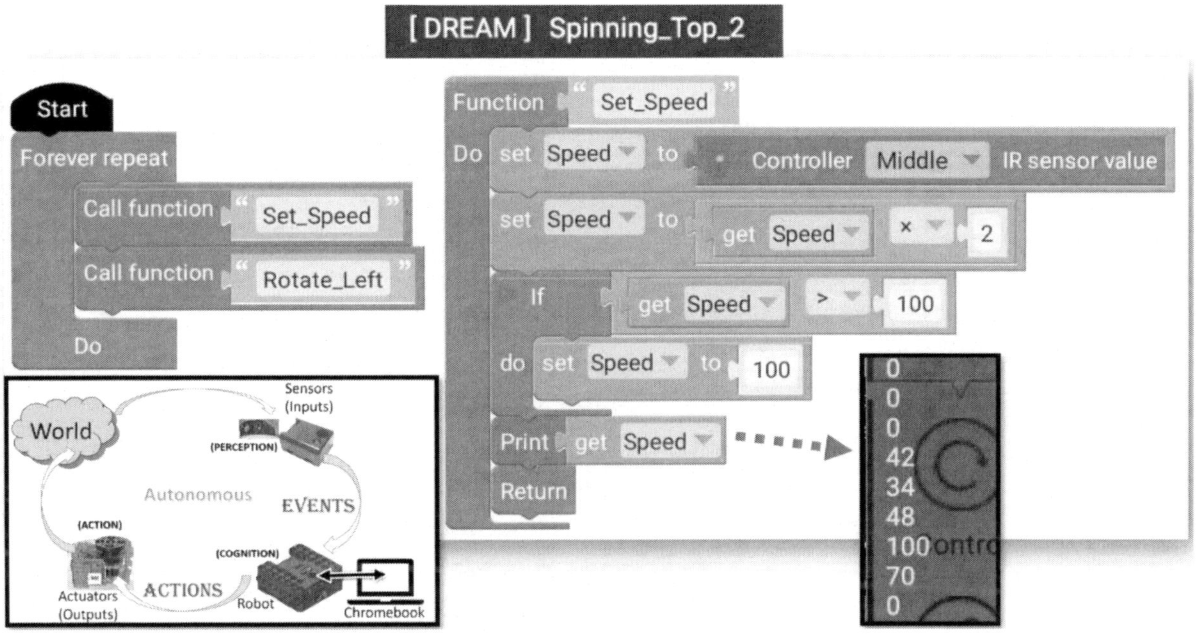

Fig. 2.21 Project "Spinning_Top_2.rpb".

At run time, Project "Spinning_Top_2.rpb" makes the robot Rotate Left at a speed inversely proportional to the distance between a blocking object and the Middle IR Sensor: the closer is the object to the sensor, the faster the robot would spin, but of course with a Time Delay caused by ISTD and OSTD as discussed earlier. To "see" this "Time Delay" effect, the reader can start bringing the blocking object closer and closer to the Middle IR Sensor, causing the robot to rotate faster and faster. Then, at a chosen time, the reader should suddenly remove the object from this sensor's Line-Of-Sight (LOS), the reader would see that the robot kept rotating for a second or two before stopping. As an example, this situation corresponded to the display of the "70" value between the "100" value and the "0" value (see darker panel at bottom right of Fig. 2.21).

2.4.5 TASK Conversion Tool

We had been using the R+BLOCK's TASK Conversion Tool since Sub-Section 2.4.2, but we did not examine the converted TASK code in detail yet. Thus, R+BLOCK was used to convert "Spinning_Top_2.rpb" to "Spinning_Top_2.tskx" which is next compiled and downloaded to the CM_150/151. This converted TASK program had great run-time performances as the issues of ISTD and OSTD were minimized, because the robot controlling code was then running at the CM-150/151 level (and not at the Chromebook level as in Sub-Section 2.4.4).

However, the author noticed a "bug": once the two motors reached the Power Setting of 100, the robot got "stuck" at this setting, even though the author had removed the blocking object from the IR Sensor's LOS at that point in time. Interestingly, if the author had removed the blocking object before the Power Setting reached 100, the robot would stop spinning as programmed.

Furthermore, and from the author's point of view, the main issue for this conversion tool is the approach used for converting R+BLOCK's Block Statements into TASK statements as shown in Fig.2.22:

Fig. 2.22 Converting R+BLOCK's Block Statement into TASK's statements.

- The reader can readily see that a single BLOCK, used for replacing **Speed** from its original value to a doubled value, was converted to a set of 4 TASK statements using 3 temporary (but hard coded) variables inside the equivalent TASK code.
- The overall "cost" of this approach resulted in the use of 20 variables for this converted TASK program (see Error List at bottom panel of Fig. 2.22), out of the maximum allowed of 50. If the author had created this TASK code from scratch using proper TASK syntax, it would have used only 1 Variable, namely **Speed**.
- Furthermore, when used on more complex R+BLOCK projects, this TASK Conversion Tool would produce some syntax errors and/or would exceed the limit of 50 variables used rather quickly, and consequently TASK would refuse to compile the converted TASK code.

These issues provided the incentive for the author to create Chapter 3 for readers interested in programming in TASK V.2 or V.3.

2.5 Beginner's Approach for R+BLOCK Projects

In this section, the author strives to provide beginning "roboteers" with a general approach to thinking about robotics projects, as well as concrete sequential steps that ultimately result in an efficient R+BLOCK program that can perform the tasks/behaviors that are required of the DREAM II robot for a given project.

2.5.1 Sense-Think-Act Paradigm revisited

Fig. 2.23 is just a duplicate of Fig. 1.7 and represents the Sense-Think-Act Paradigm (previously described in Sub-Section 1.2) which will be used a General Thinking Tool for solving various robotics projects from this point forward.

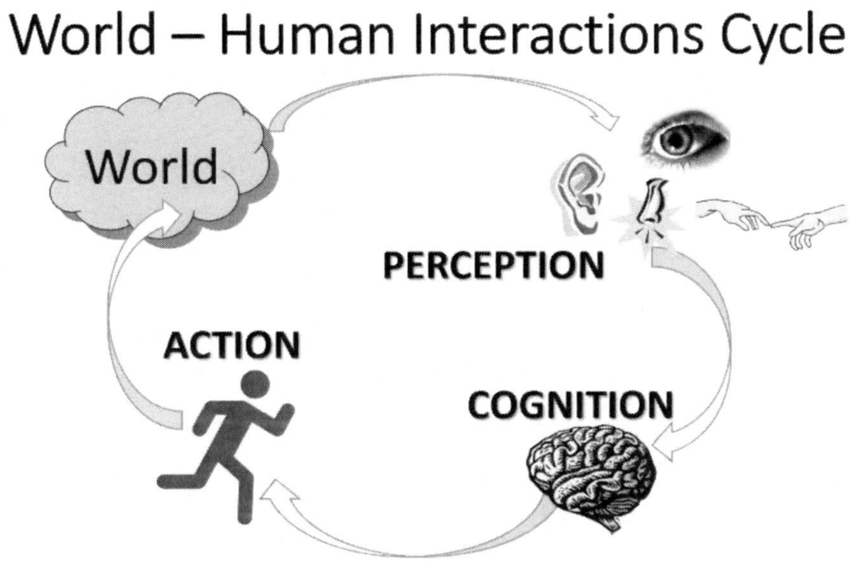

Fig. 2.23 "Sense-Think-Act" (STA) Paradigm.

As applied to our current use environment of "R+BLOCK + Chromebook + DREAM II Robot", the Human STA Cycle can be "redrawn" as shown in Fig. 2.24:

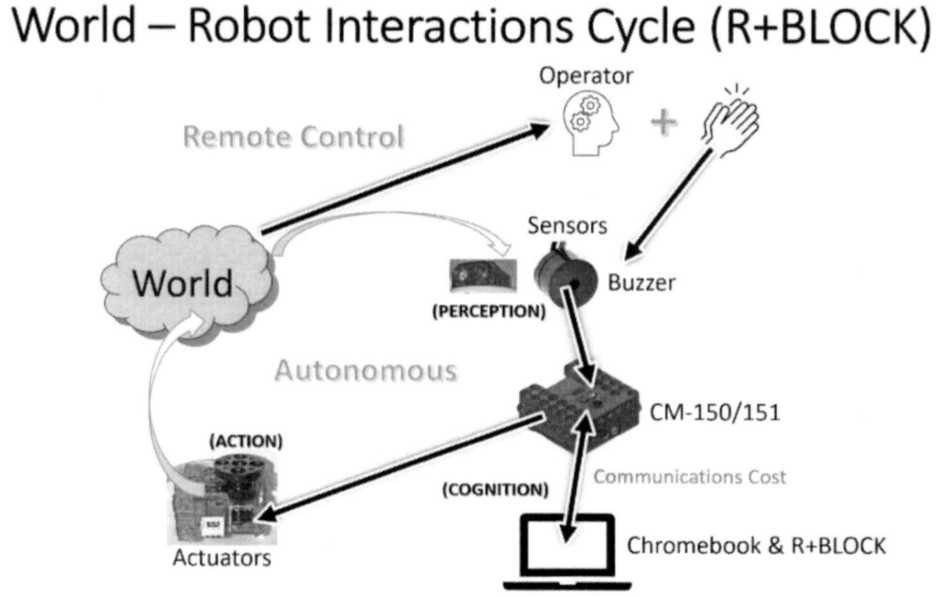

Fig. 2.24 World-Chromebook-DREAM II Robot Interactions Cycle.

- Fig. 2.24 displays the Autonomous Control Cycle, via the use of IR Sensors, that had been used previously in Section 2.4.
- As R+BLOCK does not currently support the use of a Remote Controller module, the author opted to use the Buzzer in detecting hand claps from the human operator to implement a Remote-Control mode with the Avoider/Follower robot.
- Fig. 2.24 also identifies the "Communications Cost" issue, previously described as ISTD and OSTD in Section 2.4.

We will be using the "Sense-Think-Act" Paradigm to helps us keep the "big picture" of any robotics project in mind, as we start our "drilling down" journey into this paradigm to get to the actual robot-controlling code that works on the Chromebook+DREAM system.

We will be developing two projects using this approach: one in Autonomous-Behavior mode ("Spinning_Top_3_XXX.rpb"), and the other in Remote-Control mode ("Clapping_RC_X.rpb").

Next, let's spell out the goals/rules for a new project "Escaping Spinning Top" that can lead to several example codes, labeled as "Spinning_Top_3_XXX.rpb":

- The Middle/Center IR Sensor will be used to detect how "far" a blocking object is from it.
- If no obstacle is detected, the robot should stay still.
- If an obstacle is detected "far", the robot should spin right.
- If an obstacle is detected "mid-range", the robot should spin left.
- If an obstacle is detected "too close", the robot should run away, i.e., go forward at high speed.

2.5.2 Reactive-Control Approach

Let's see how we can use the "Sense-Think-Act" (STA) Paradigm to help us program the DREAM II robot to perform the previous set of tasks. Fig. 2.25 is just another illustration of STA, but we'll be concentrating on the dashed red boundary in its lower right corner (i.e., zooming in on the Computing Hardware components).

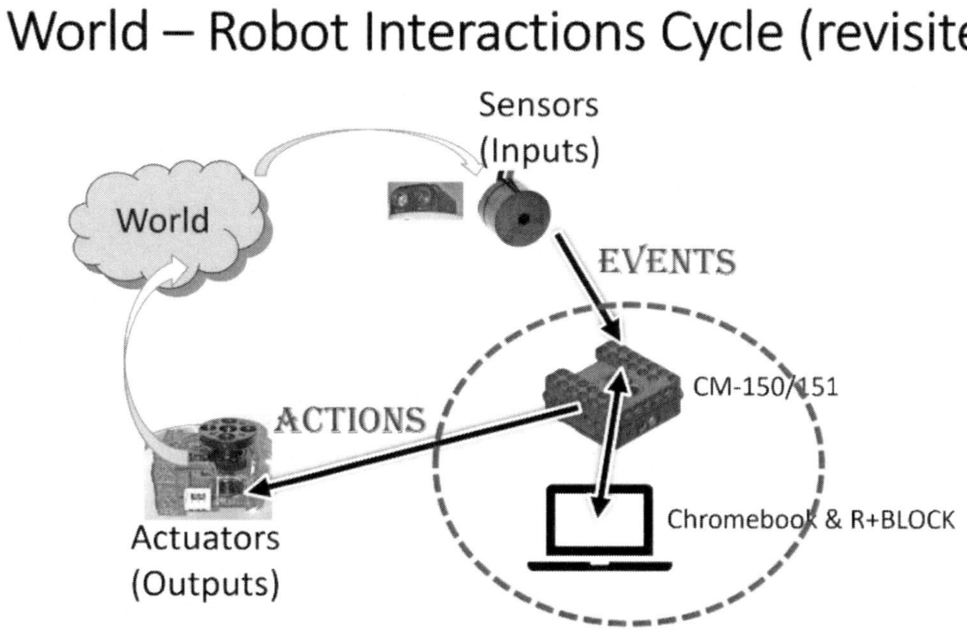

Fig. 2.25 World-Chromebook-DREAM II Robot Interactions Cycle (revisited).

STA still applies at this level, but different robotics practitioners would start using different labels such as "Events", "Actions" and "Reactive Control" as they are more useful in the next steps that we need to take to create the actual computer code to control the DREAM II robot. Interested readers should read Matarić (2007) for a more complete description of other robotics control approaches, or Ben-Ari and Mondada (2018) for more details about "Reactive Control" and "Finite State Machines".

2.5.3 Event-Action Pairs

With the "Reactive Control" (RC) approach, the **first step RC1** is to translate each project goal from the above list into a specific Condition/Event matched with an appropriate Action for the robot to do (see Fig. 2.26).

Reactive Control Approach

Given Condition (Event) >> Appropriate Robot Action

Condition/Event	Action
No Obstacle	Robot Stays Still
Obstacle Detected "Far"	Robot Spins Right
Obstacle Detected "Mid-Range"	Robot Spins Left
Obstacle Detected "Close"	Robot Runs Away

Fig. 2.26 Event-Action Pairs.

At this stage of the process, the reader needs not consider any hardware or software solutions in mind at all and just performs a "thought experiment" or "visualization exercise" whereas the reader would **imagine to be "one" with the robot and wait for a specific event to happen, and then would perform the appropriate maneuver**.

Before we go on to the next step, a closer look at these four Event-Action pairs listed in Fig. 2.26 should show that they are **mutually exclusive**, meaning that at run time, one and only one of these Event-Action Pair can be active at any one time.

2.5.4 Sensor-Actuator Pairs

The **second step RC2** is to build upon the results of the previous step **RC1** by mirroring its Event-Action pair into a matching Sensor-Actuator pair as shown in Fig. 2.27. **The goal is to specify the kind of Sensors/Actuators (in a general sense) to be used and to prescribe how they are to be used.**

Input-Output Table

Given Input Sensor(s) >> Activate Appro. Actuator(s)

Input Sensors	Output Actuators
Center IR value <=20	Both Motors Off
Center IR value between 20 & 50	Motor 1 CCW & Motor 2 CCW
Center IR value between 50 & 80	Motor 1 CW & Motor 2 CW
Center IR value > 80	Motor 1 CCW & Motor 2 CW

Fig. 2.27 Sensor-Actuator Pairs corresponding to Event-Action Pairs listed in Fig. 2.26.

The reader can see that some hands-on experimentations must be done to "translate" the "general" definitions of "No Obstacle", "Object Detected Far", "Object Detected Mid-Range" and "Object Detected Close" into "concrete" threshold values such as 20, 50 and 80 that are used in Fig. 2.27. One approach is to use the previous Project "IR_Througput.rpb" and a white sheet of paper as the "blocking object" to obtain those "numbers". The reader most likely would obtain different threshold values for his/her own setup.

However, the author used a different approach by using Function **Set_Speed()** (Fig. 2.21) which essentially transforms the Middle/Center IR Sensor's reading into a "number" between 0 and 100. The reader can now "understand" the author's choice of the 3 threshold values of 20, 50 and 80!

Please note that the previous procedures do not require the knowledge of any computer programming language and the author highly recommends that beginner roboteers to always start with these procedures, instead of jumping directly into coding or try to adapt someone else codes, as these procedures would help the roboteer refine in his/her mind the possible situations that the robot may encounter and the appropriate tasks to be performed by the robot, before committing them into hard codes. This "exercise" will save the programmer lots of time in the long run.

In the "Computational Thinking" practice, the previous steps would correspond to the "Decomposition" phase (Beecher, 2017; Denning and Tedre, 2019).

In real practice, the user should not expect to come up with a complete list of Event-Action and Sensor-Actuator Pairs in the first trial - the author never did in his many years of practice! Thus, refinement steps are "normal"! Furthermore, in more complex situations, the Event-Action pairs may have to follow a hierarchy of "cascading" events, as shown in more advanced projects later in this book.

2.5.5 Event-Programming Approach

Another benefit for using the Event-Action and Sensor-Actuator Pairs is that they will help with the actual coding (as shown later in Section 2.5.6), as the R+BLOCK syntax would support practically a one-to-one correspondence of all Event-Action Pairs A to Z shown in Fig. 2.28 into similar "physical" locations within the final R+BLOCK code, except for two aspects that had not been discussed yet:

1. A Main Loop structure allowing the Computing Component to go through the list of possible Events A to Z that may occur during run time, as it "listens" to what the Sensors tell it about the current state of the World. This overall loop represents the Interactions Cycle as shown in Figs. 2.23, 2.24 and 2.25. We have "casually" used this Forever/Endless Loop structure before, but now the reader can understand why it is absolutely "required" in all robotics programs.

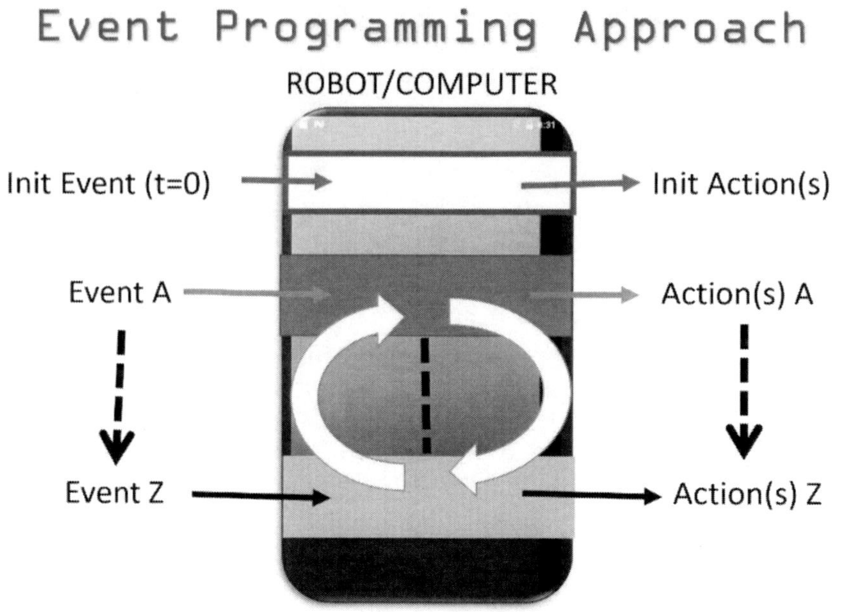

Fig. 2.28 Event-Action Pairs matching locations in R+BLOCK code.

2. The Init Event-Action which occurs once at time=0, i.e., when the **Play Project** Button is pushed inside R+BLOCK, or when power is applied to the robot in case of a TASK application. The Init Event-Action would usually result in a known state for the robot. For this project, we can just have the robot play a musical melody as its Init Action to let the operator knows that it is "ready to play".

In "Computational Thinking" practice, these steps would correspond to the "Pattern Recognition" phase.

2.5.6 Application to Escaping Spinning Top

The next step is to transform the list of Sensor-Actuator Pairs into logical structures that can be implemented on the Computing Component, usually in pseudo code first, then in the programming language chosen by the user.

From Fig. 2.27, two possible R+BLOCK solutions can be developed: one is using a PARALLEL-IFs construct, the other is using an IF-ELSE-IF construct. Both can be laid out in pseudo code as shown below:

- PARALLEL-IFs (can be called CONSECUTIVE-IFs also)
  - *If (Event 1 is TRUE) then (Activate Robot Maneuver 1)*
  - *If (Event 2 is TRUE) then (Activate Robot Maneuver 2)*
  - ...
- IF-ELSE-IF (designed for mutually exclusive Event-Action Pairs)
  - *If (Event 1 is TRUE) then (Activate Robot Maneuver 1)*
  - *Else If (Event 1 is TRUE) then (Activate Robot Maneuver 2)*
  - ...

In "Computational Thinking" practice, these steps would correspond to the "Abstraction" phase.

Some readers would already notice that each "PARALLEL-IF" seems to match each row shown in the Table illustrated in Fig. 2.27 and they are right! Fig. 2.29 displays the Init Action and the Main Endless Loop used in Project "Spinning_Top_3_Par_IFs.rpb":

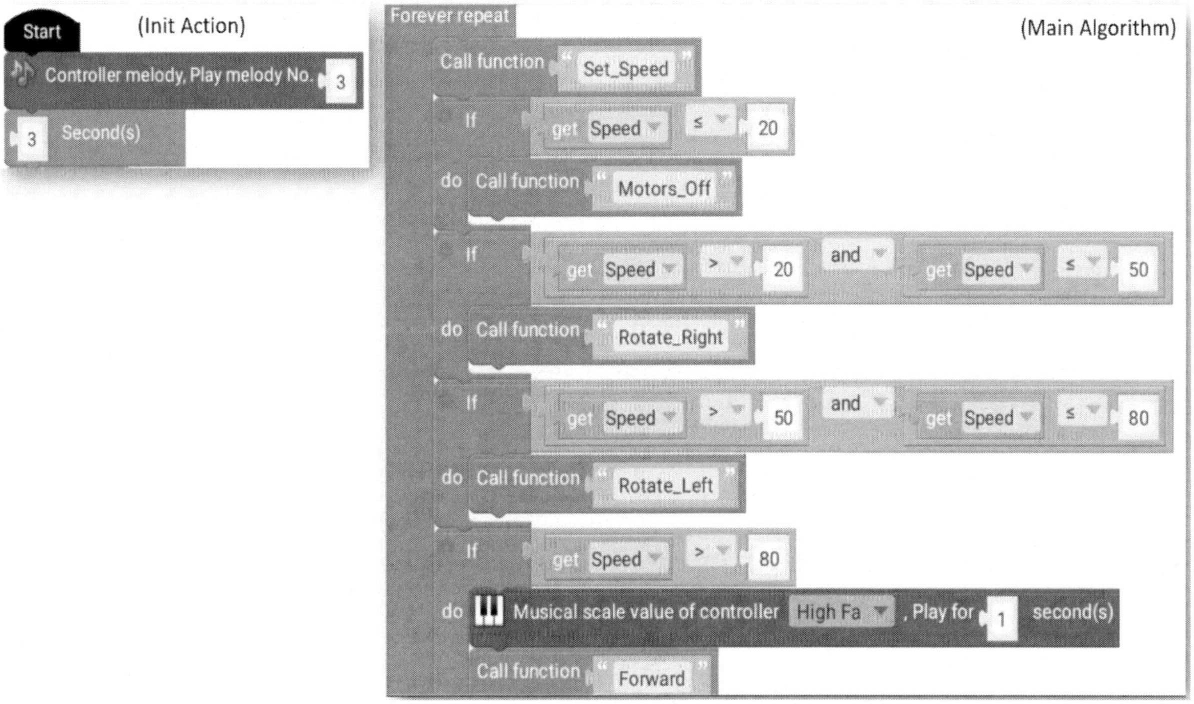

Fig. 2.29 Init Action and Main Endless Loop used in "Spinning_Top_3_Par_IFs.rpb".

- The Init Action is just to Play Melody 3.
- The Main Endless Loop starts by Calling Function **Set_Speed** which is the same used in Fig. 2.21. After this step, Variable **Speed** contains a Modified Value of the current Value reported by the Middle IR Sensor. This Modified Value is in the [0-100] range.
- The First IF-DO construct matches with Sensor-Actuator Pair #1, i.e., when the robot should stay still.
- The Second IF-DO construct matches with Sensor-Actuator Pair #2, i.e., when the robot should spin right.
- The Third IF-DO construct matches with Sensor-Actuator Pair #3, i.e., when the robot should spin left.
- The Fourth IF-DO construct matches with Sensor-Actuator Pair #4, i.e., when the robot should run away by going forward. The author added the playing of a High Fa for dramatic flair when the robot is "essentially" running away from the blocking object "coming down too close" to it.

A sharp reader may remember that all Maneuver Functions use Variable **Speed** as a Power Setting to the Motors on Ports 1 and 2. This means that at run time, let's say that the operator starts with the blocking object out of the detection range of the Middle IR Sensor, so the robot would stay still at first. Next, let's say that the operator keeps on lowering the blocking object upon the robot's Middle IR Sensor, he/she would see the robot spin right at a slow rate, then spin left at a higher rate, and lastly shoot out from under the blocking object at its highest speed. But, at that time, the robot would detect "no obstacle", so it would stop its forward maneuver and it is then ready for another "play" session!

In "Computational Thinking" practice, these steps would correspond to the "Algorithm" phase.

Other computing "side effects" that the reader should be aware of when the PARALLEL-IFs construct is used:

- At run time, all four IF-DO structures are "tested" for each cycle of the Main Endless Loop, although only one of them would be fully activated at any one time. So, there would be some "waste of computing cycles" in this situation.
- The programmer does not have to follow the order of the Sensor-Actuator Pairs (i.e., the IF-DOs) as listed in Fig. 2.27, and the run time behavior of the robot remains the same! This is left to the interested reader to try out on his/her own.

In "Computational Thinking" practice, these steps would also correspond to the "Algorithm" (Refining) phase.

Next, the reader may recall the previous discussion at the end of Sub-Section 2.5.3 which found that this project deals with **mutually exclusive Event-Action Pairs**, thus the IF-ELSE-IF construct should have been used as it can be more efficient "computation-wise", but it does require more "analysis work" from the programmer.

Referring to the pseudo-code listed for IF-ELSE-IF at the beginning of this Sub-Section 2.5.6, readers can consider the IF-ELSE-IF construct as a listing of all possible Event-Action Pairs that the robot may encounter at any point in time when this program is executed. There is a specific way that the Computing Component (CC) will execute this IF-ELSE-IF structure at run time:

- **Event 1** is always checked first, and if it happens to be **TRUE**, then Maneuver 1 is performed, and CC jumps out of this IF-ELSE-IF structure from this spot and goes on executing the rest of the R+BLOCK code (i.e., starts a new cycle of the Main Endless Loop in the current project). If Event 1 happens to be **FALSE**, then CC goes on to **check Event 2 next**.
- If **Event 2** happens to be **TRUE**, then Maneuver 2 is executed, and CC jumps out of this IF-ELSE-IF structure from this spot and goes on executing the rest of the R+BLOCK code. If Event 2 happens to be **FALSE**, then CC goes on to **check Event 3 next**.
- The same steps as described above apply to **Event 3**, and **Event 4**.

In "Computational Thinking" practice, these steps would correspond to the "Pattern Recognition" phase.

When R+BLOCK is used, the IF-ELSE-IF construct consists of **nesting** the equivalent IF-DO-ELSE structures into each other, yielding the Project "Spinning_Top_3_IF_ELSE_IF_1.rpb". Fig. 2.30 lists only the Main Endless Loop (all other parts remain the same as in Fig. 2.29).

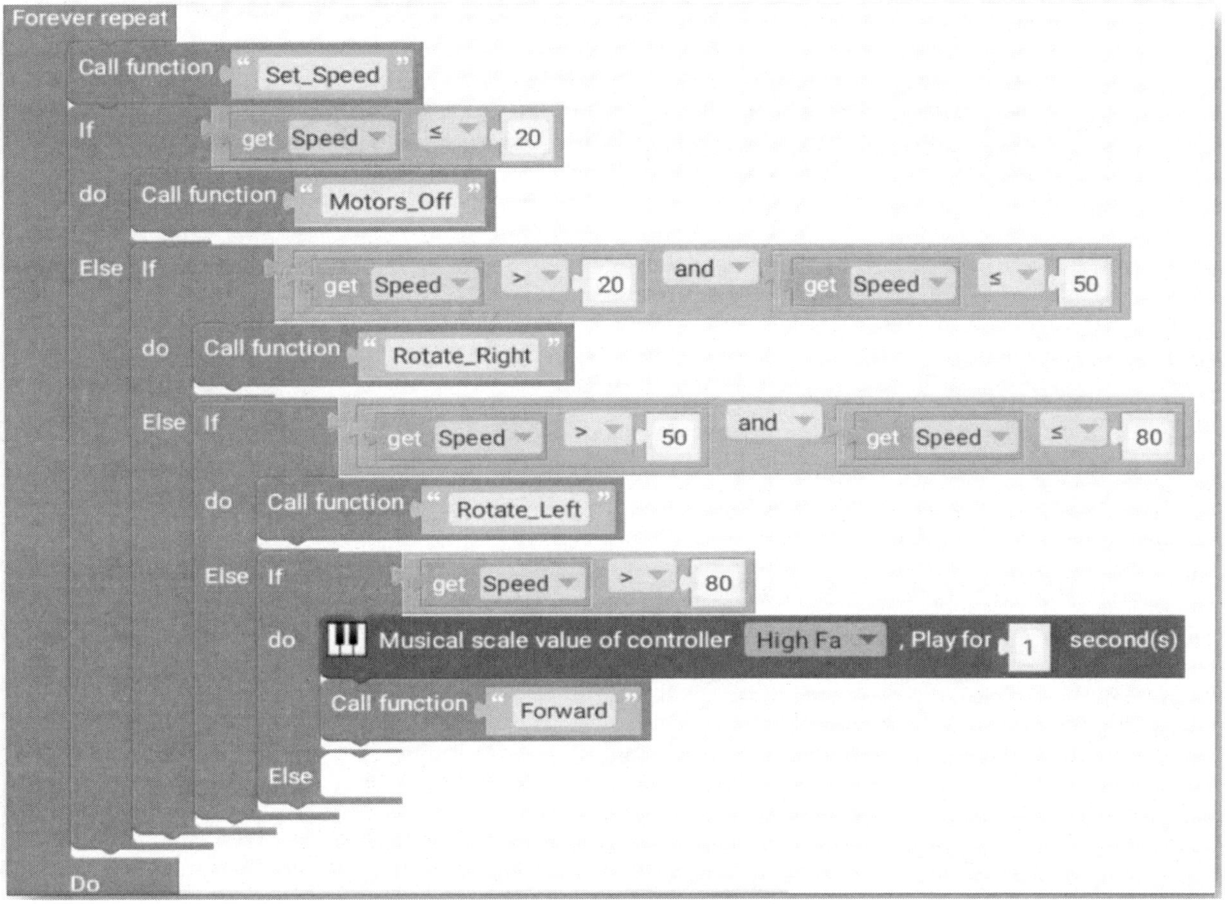

Fig. 2.30 Main Endless Loop used in "Spinning_Top_3_IF_ELSE_IF_1.rpb".

This step would correspond to the "Abstraction" Phase in "Computational Thinking" practice.

The reader should notice that the Conditional Expressions used are the same ones used in "Spinning_Top_3_Par_IFs.rpb", and the last ELSE is "empty" – which is equivalent to a "do nothing" action for the Computing Component. Thus, this program has no Logic issue and should run OK as is, and the reader is encouraged to try it out.

This step would correspond to the "Algorithm" Phase in "Computational Thinking" practice.

However, this solution can be further refined by replacing the Fourth IF-ELSE-DO with an IF-DO (to get rid of the "empty slot"), and all Conditional Statements used can be "simplified":

- Let's consider the First IF-ELSE-DO structure which uses the Conditional Expression **(Speed ≤ 20)**. When/If this expression evaluates to FALSE - meaning **(Speed > 20)**, the Second IF-ELSE-DO will be evaluated next, where this Conditional Expression **((Speed > 20) and (Speed ≤ 50))** is used. The reader can readily see that the Sub-

Expression **(Speed > 20)** is superfluous, so it can be deleted while the original Algorithm is preserved.

- The Third IF-ELSE-DO can also be "simplified" in a similar way.
- The Fourth IF-ELSE-DO is changed to an IF-DO, but the author kept the original Conditional Expression **(Speed > 80)** in Fig. 2.31. This Construct/Conditional Expression can be further simplified but it is left as a "challenge" to the reader.

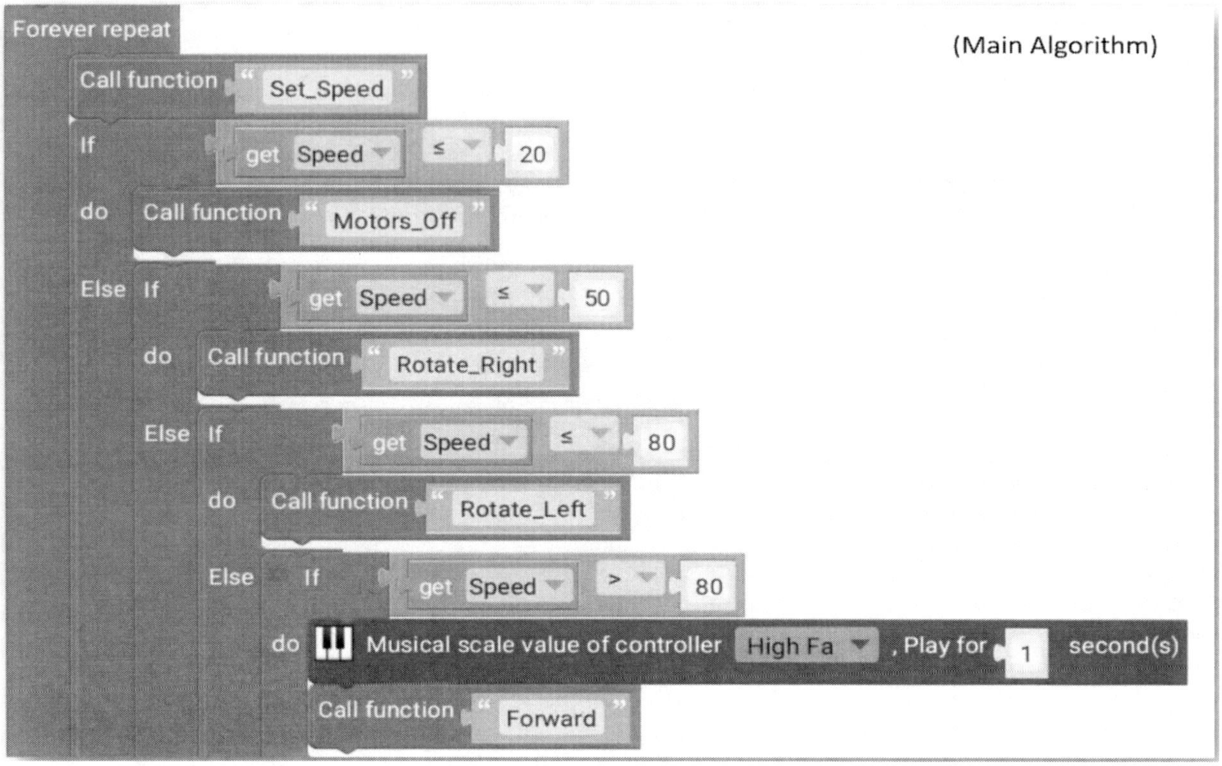

Fig. 2.31 Main Endless Loop used in "Spinning_Top_3_IF_ELSE_IF_2.rpb".

These steps would correspond to the "Algorithm" (Refinement) Phase in "Computational Thinking" practice.

In practice, diligent users would soon realize that backtracking various steps of the Abstraction and Algorithm Phases for improving one's R+BLOCK program will be the "norm" for any meaningful robotic project.

2.5.7 Application to Clap-based Remote Control

First, let's spell out the goals/rules for a new project "Clapping Remote-Control" that can lead to several example codes, labeled as "Clapping_RC_XXX.rpb":

- The CM-150/151 Buzzer will be used to detect the number of hand claps (or loud noises) originating from the human operator to serve as a Remote-Control signal for the DREAM II robot.

- If no hand clap is detected, the robot should stay still.
- If ONE hand clap is detected, the robot should Go Forward.
- If TWO hand claps are detected, the robot should Go Backward.
- If THREE hand claps are detected, the robot should Turn Left.
- If FOUR hand claps are detected, the robot should Turn Right.

Once again, using the "Reactive Control" (RC) approach, the **first step RC1** is to translate each project goal from the above list into a specific Condition/Event matched with an appropriate Action for the robot to do (see Fig. 2.32).

Reactive Control Approach

Given Condition (Event) >> Appropriate Robot Action

Condition/Event	Action
No Hand Clap	Robot Stays Still
1 Hand Clap	Robot Goes Forward
2 Hand Claps	Robot Goes Backward
3 Hand Claps	Robot Spins Left
4 Hand Claps	Robot Spins Right

Fig. 2.32 Event-Action Pairs for "Clapping Remote-Control" Project.

The **second step RC2** is to build upon the results of the previous step **RC1** by mirroring its Event-Action pair into a matching Sensor-Actuator pair as shown in Fig. 2.33.

Input-Output Table

Given Input Sensor(s) >> Activate Appro. Actuator(s)

Input Sensors	Output Actuators
Sound Counter = 0	Both Motors Off
Sound Counter = 1	Motor 1 CCW & Motor 2 CW
Sound Counter = 2	Motor 1 CW & Motor 2 CCW
Sound Counter = 3	Motor 1 CW & Motor 2 CW
Sound Counter = 4	Motor 1 CCW & Motor 2 CCW

Fig. 2.33 Sensor-Actuator Pairs for "Clapping Remote-Control" Project.

Fig. 2.33 shows that all 5 Sensor-Actuator Pairs are mutually exclusive, thus an IF-ELSE-IF construct will be used to represent this Input-Output Table in the first solution named "Clapping_RC_1.rpb". Of course, an equivalent 2nd solution using PARALLEL-IFs is also provided as "Clapping_RC_2.rpb" for the reader convenience.

As this is a Remote-Control application, we need to additionally specify a Speed Control scheme for the robot – which is not specified in the current goals/rules for this project. Let's say that we want to set the robot's maneuver speed **once,** as part of the Init Action for this project. Hence, let's see how we can modify Function **Set_Speed()** to fulfill this new requirement (see Fig. 2.34):

- The First Block of this Function shows that a "Low Do" Musical Note is played for 1 second with the Buzzer, to let the human operator be aware that the robot is ready to accept his/her Speed-Setting Gesture, which is to extend a blocking object over the Middle IR Sensor LOS.
- Of course, the robot has no idea when the operator is going to perform this action, thus a **Wait Until** Loop is used as the Second Block. This Loop uses the Conditional Expression **(Middle IR Sensor Value ≠ 0)**, meaning that if the Middle IR Sensor (MIRS) reads 0 (i.e., no object detected), the Computing Component (CC) re-executes this **Wait Until** Loop as required by Logic. It is only when the operator inserts a blocking object within the LOS of the MIRS that the CC would quit this loop and executes the Third Block.
- This 3rd Block reads in the current value of the MIRS and assigns it to Variable **Speed**, meaning that the operator can "control" the robot's maneuver speed by putting the blocking object closer or farther away from the MIRS: "closer" corresponds to "faster" and vice versa.
- Next, the usual transformation steps to keep the final value of **Speed** between 0 and 100 are performed (Blocks 4-6).
- Block 7 is new and provides visual feedback to the operator by printing out the "set" maneuver speed for the current "Play" session.

- Additionally, Block 8 plays a High Ti for 1 second to let the operator knows that the robot is "done" with setting its maneuver speed.

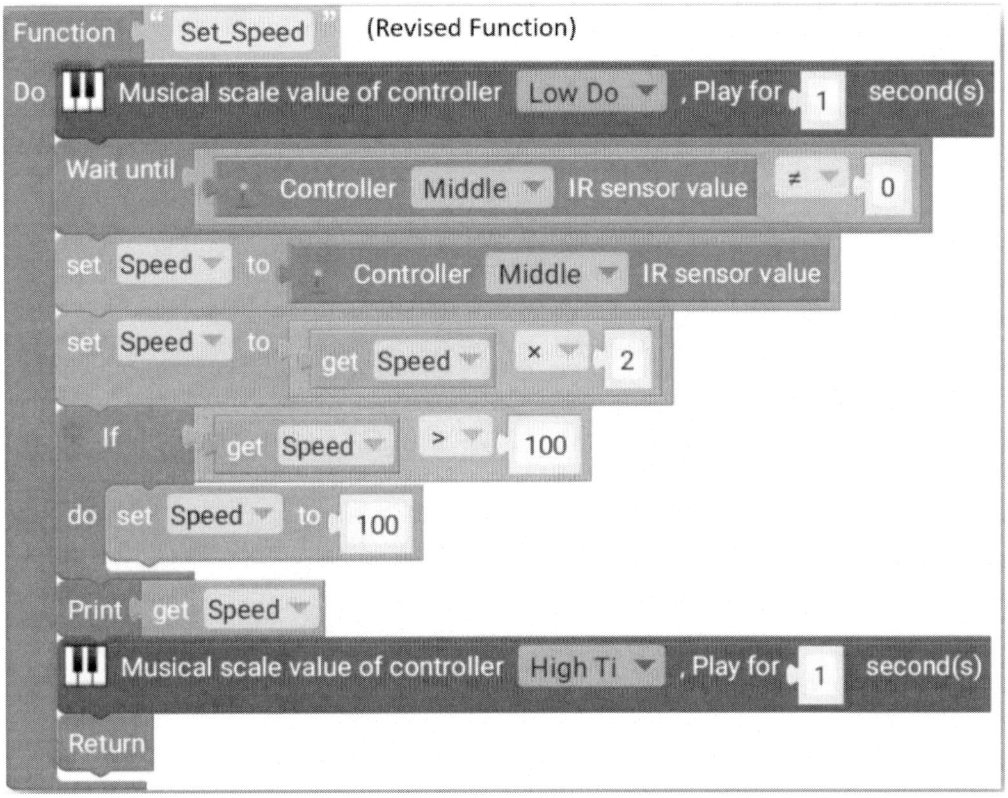

Fig. 2.34 Revised Function **Set_Speed()** in "Clapping_RC_1.rpb".

Fig. 2.35 shows Part 1 of the Main Function of "Clapping_RC_1.rpb":

- Variable **Speed** is set to its final value, as described above, in the First Block.
- The CC waits for 1 second, then plays Melody 3 to signal that the robot is ready for the operator to do Remote-Control using hand claps.
- As usual, an Endless/Forever Loop is used to represent the STA Interactions Cycle:
  - First, the "Result Of Sound Counter" (ROSC) is reset to 0, then a Wait of 1 second is used to make sure that the previous step is accomplished properly.
  - Then a **Wait Until Loop** is activated and remains in effect **as long as** the robot detects NO hand clap. Please note that the Conditional Expression used is **(Result of Sound Counter ≠ 0)** which seems to be the reverse of what we try to accomplish, but the reader needs to remember that this is a **WAIT UNTIL** structure, not a **WAIT AS LONG AS** structure (which is preferred by the author). When the operator generates "detectable" hand claps, this **Wait Until Loop** is exited, and the Final ROSC value is saved into Variable **Claps** by the next Block.

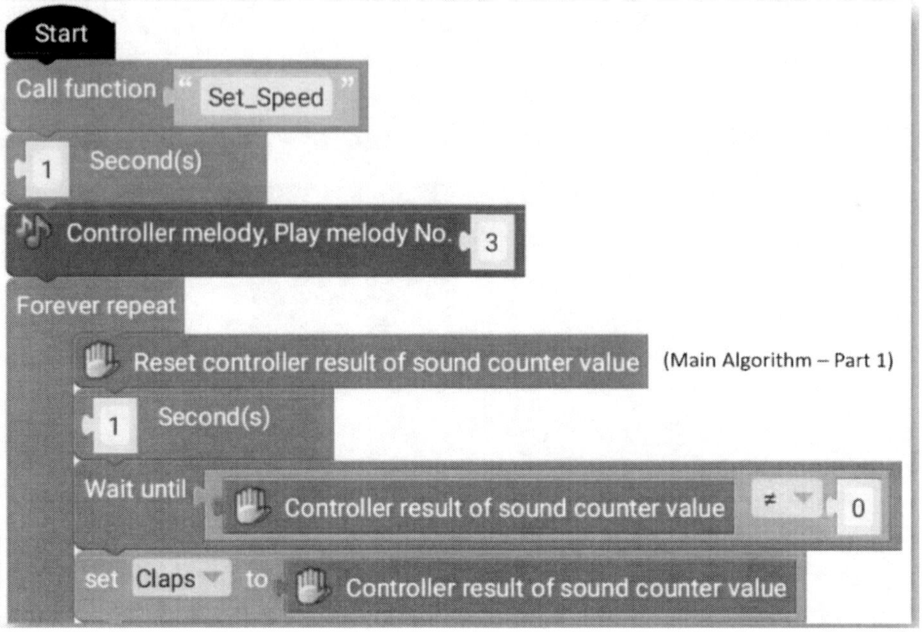

Fig. 2.35 Part 1 of Main Function in "Clapping_RC_1.rpb".

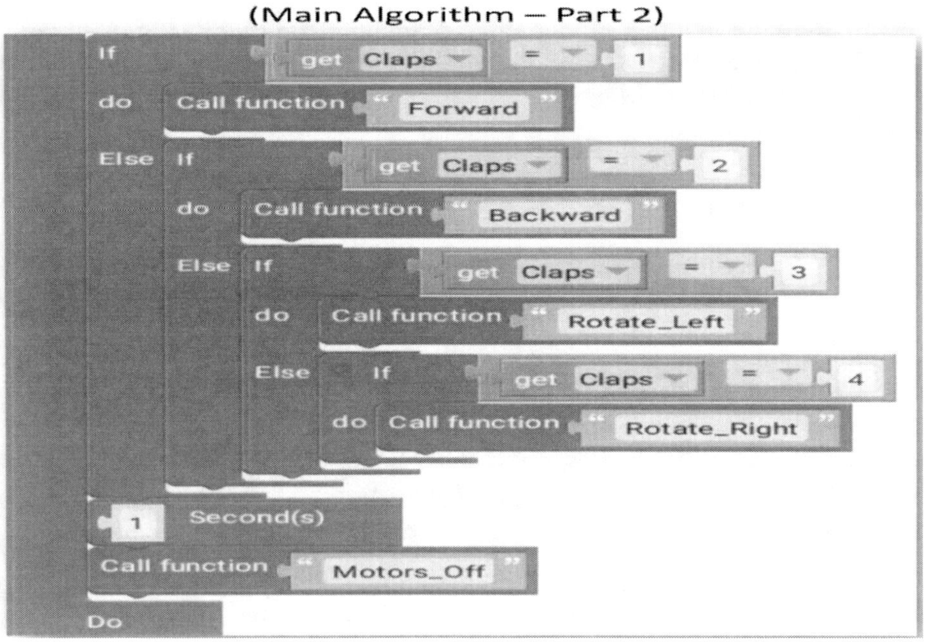

Fig. 2.36 Part 2 of Main Function in "Clapping_RC_1.rpb".

Part 2 of the Main Function of "Clapping_RC_1.rpb" continues in Fig. 2.36:

- An IF-ELSE-IF structure is used to handle, at run time, ONE and ONLY ONE possible Event/Action Pair that may happen, depending on the number of hand claps detected via

Variable **Claps**. Algorithmically, this is most efficient structure to use for mutually exclusive Event/Action Pairs.

- The reader may wonder about the next Blocks used: **Wait for 1 Second** and **Call Function "Motors_Off()"**. They were not specified in the original Goals/Rules. The reason for their use is for "Robot Safety":
    - Previously, we knew that when Motors are set into motion, they will continue to turn until they are commanded to stop. The previous IF-ELSE-IF structure just set them into a given maneuver, and as it **takes time** for the operator to produce a **new set of hand claps**, the author had found it necessary to limit each maneuver to a fixed time interval, if not the robot may run into an obstacle or fall off a table. The reader can of course use a different maneuver time interval.
    - Another side benefit of the **Motors_Off** action is that the robot can only listen for any hand clap "in silence", without interferences from running motors noises.
    - In Chapter 3, when we program in TASK which includes a complete Remote-Control module, the author will show a different "tack" to handle "Robot Safety".

Fig. 2.37 illustrates an alternative Algorithm whereas a PARALLEL-IFs construct is used. This Project is named "Clapping_RC_2.rpb" and is included in this book's Example Codes Package.

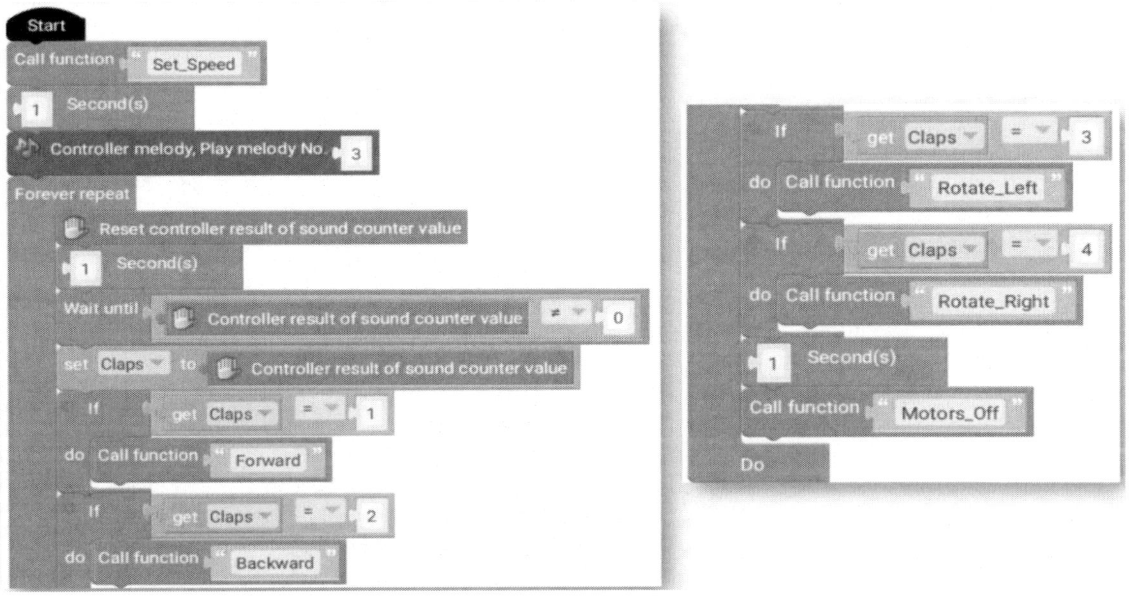

Fig. 2.37 Main Algorithm used in "Clapping_RC_2.rpb".

If the reader needs to use a Delay Time that is shorter than 1 second, he/she is referred to **Function Delay()** which essentially is a **"Do Nothing" Counting Loop**. This function is included in both Projects "Clapping_RC_1.rpb" and "Clapping_RC_2.rpb".

2.6 Object Detection & Line Tracking with NIR Sensors

This Section showcases R+BLOCK projects that require **Fast Sensor Inputs** and **Responsive Motor Actions** which are inherently "weak features" for the "Chromebook/R+Block/Dream II" Computing System, due its ISTD and OSTD issues. But their implementations can be useful reference points when we switch to the "Chromebook/TASK/Dream II" Computing System or the "Chromebook/Python/TASK/Dream II" Computing System, in Chapters 3 and 4.

In this section, we'll use the same "Avoider" robot shown in Fig 2.2, but it will be using the IRSS-10 Sensors connected to Ports 3 and 4, because they have a longer detection range than the Internal IR Sensors of the CM-150/151.

The **Set_Speed()** Function, as shown in Fig. 2.34, was revised to set the Avoider's Maneuvering Speed to be **equal only** to the **Initial Reading of the Middle IR Sensor** (not its doubled value as before) due to ISTD and OSTD issues. In simpler terms, the Avoider robot is not as "speedy" as the Spinning Top robot used in Section 2.5.

A new **Read_Port_IR_Sensor()** Function was created as shown in Fig. 2.38. The key "feature" to notice is the Port IR Sensors readings had to be multiplied by 5, before they are used in the rest of the Avoider project codes in this Section 2.6 (Blocks 1-2 and 5-6 inside the Function's body). The user likely would need a different multiplication factor to match his/her own IRSS-10s.

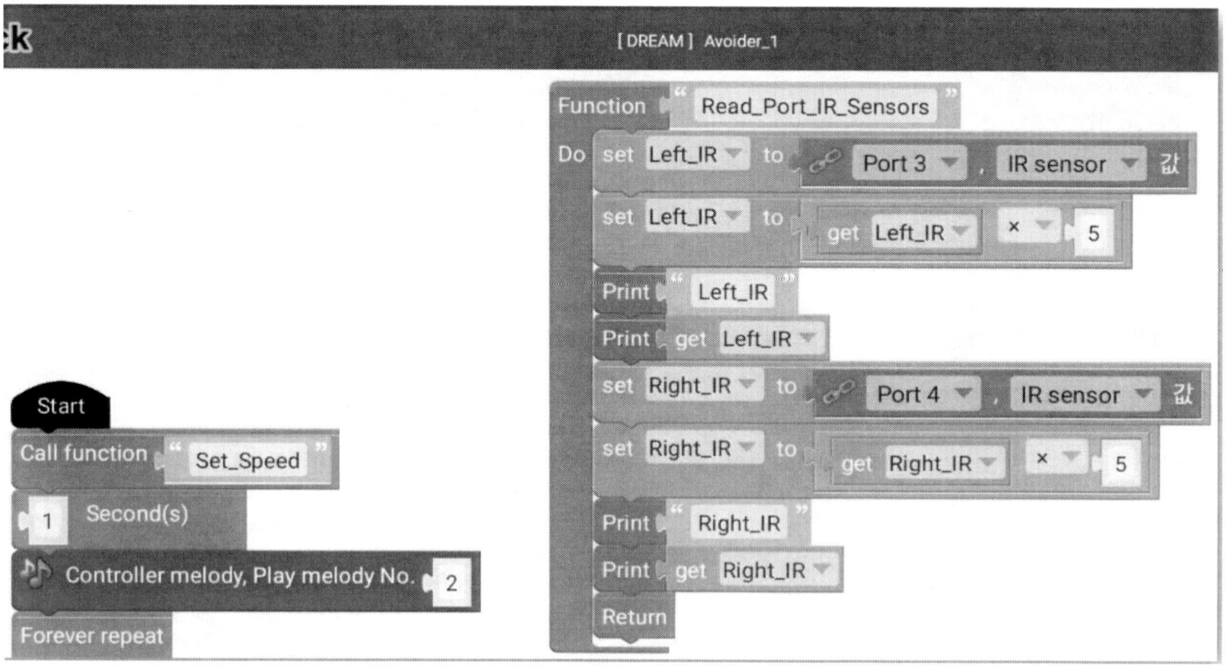

Fig. 2.38 Function **Read_Port_IR_Sensors()** used in Project "Avoider_1.rpb".

2.6.1 Avoider

The "Avoider" will be using its Left (Port 3) and Right (Port 4) IRSS-10 Sensors to detect a single object and to accomplish the tasks described in the following Event-Action list:

- If there is no obstacle in front of Avoider, it can go forward.

- If there is an object on Avoider's Left, it should back away and turn to the Right to avoid the obstacle.
- If there is an object on Avoider's Right, it should back away and turn to the Left to avoid the obstacle.
- If there is an object in front of Avoider, it should go straight Back to avoid the obstacle.

At first analysis, this seems to be a familiar situation of choosing ONE and ONLY ONE Event-Action (or Sensor/Actuator) Pair among several mutually exclusive Event-Action (or Sensor/Actuator) Pairs, thus an IF-ELSE-IF structure should do an adequate job when combined with an Outer Endless Loop (see Fig. 2.39):

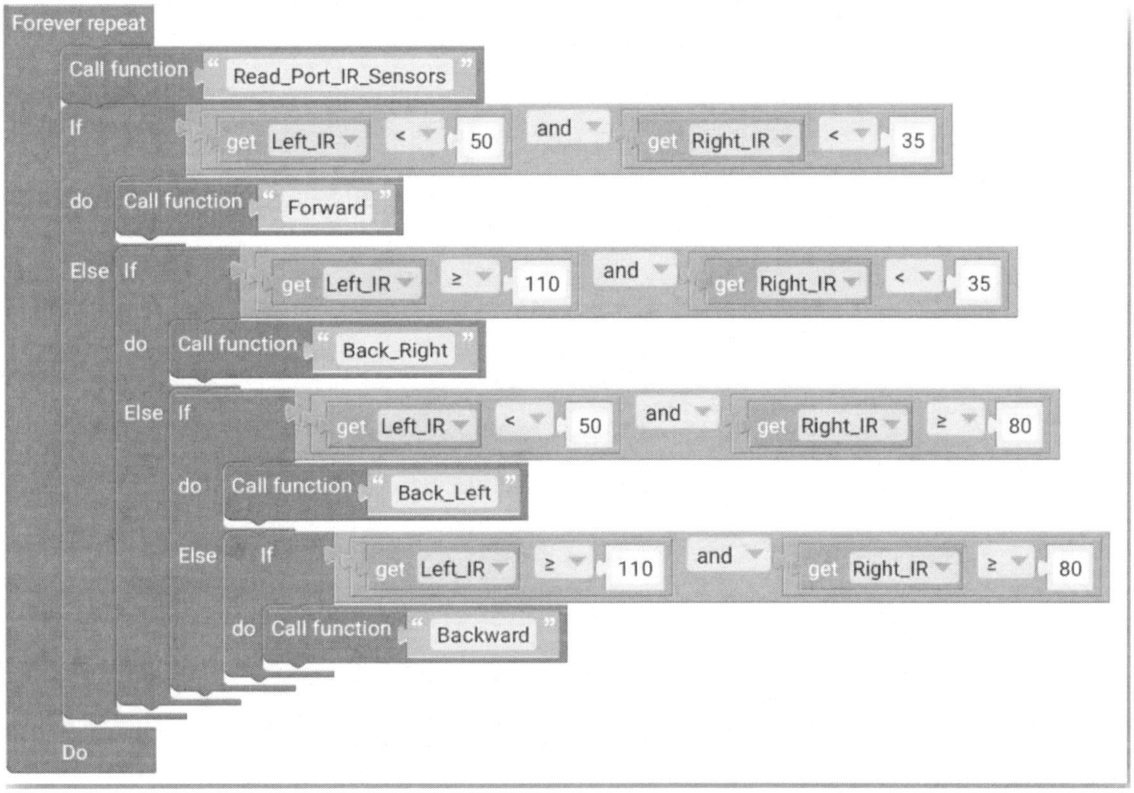

Fig. 2.39 Main Endless Loop used in Project "Avoider_1.rpb".

- For each iteration of the Main Endless Loop, the latest IRSS-10 Sensor readings are saved into Variables **Left_IR** and **Right_IR**, by calling Function **Read_Port_IR_Sensors()**.
- Next, the First Sensor/Actuator Pair is tested to see if the Function **Forward()** needs to be called by evaluating the 1st Conditional Expression **((Left_IR < 50) and (Right_IR < 35))**. The reader can see that different Threshold values (50 and 35) are used for the two IRSS-10 Sensors. This is "normal" as it is very rare to find two IRSS-10 sensors with identical responses to the same environment. If this Conditional Expression evaluates to TRUE, meaning that there is no obstacle detected in front of the robot, therefore it can go forward. If this Conditional Expression evaluates to FALSE, meaning that there is an

obstacle detected somewhere in front of the robot, then the Second Sensor/Actuator Pair is tested.

- The Second Sensor/Actuator Pair uses the 2nd Conditional Expression **((Left_IR $\geq$ 110) and (Right_IR < 35))** to check if the blocking object is on the robot's Left. If this turns out to be TRUE, then the robot would perform/call Function **Back_Right()** to veer Back and Right from the obstacle. If this Conditional Expression evaluates to FALSE, meaning that there is NO obstacle detected on the Left of the robot, then the Third Sensor/Actuator Pair is tested.
- The Third Sensor/Actuator Pair uses the 3rd Conditional Expression **((Left_IR < 50) and (Right_IR $\geq$ 80))** to check if the blocking object is on the robot's Right. If this turns out to be TRUE, then the robot would perform/call Function **Back_Left** to veer Back and Left from the object. If this Conditional Expression evaluates to FALSE, meaning that there is NO obstacle detected on the Right of the robot, then the Fourth Sensor/Actuator Pair is tested. At this point, the robot has determined that there is no obstacle on either side of itself. Therefore, the Fourth Sensor/Actuator Pair is checked next.
- The Fourth Sensor/Actuator Pair uses the 4th Conditional Expression **((Left_IR $\geq$ 110) and (Right_IR $\geq$ 80))** to check if the blocking object is center front of the robot. If this turns out to be TRUE, then the robot would perform/call Function **Backward()** to back away straight from the object. Fig. 2.40 shows the difference between Functions **Backward()** and **Back_Right()**.

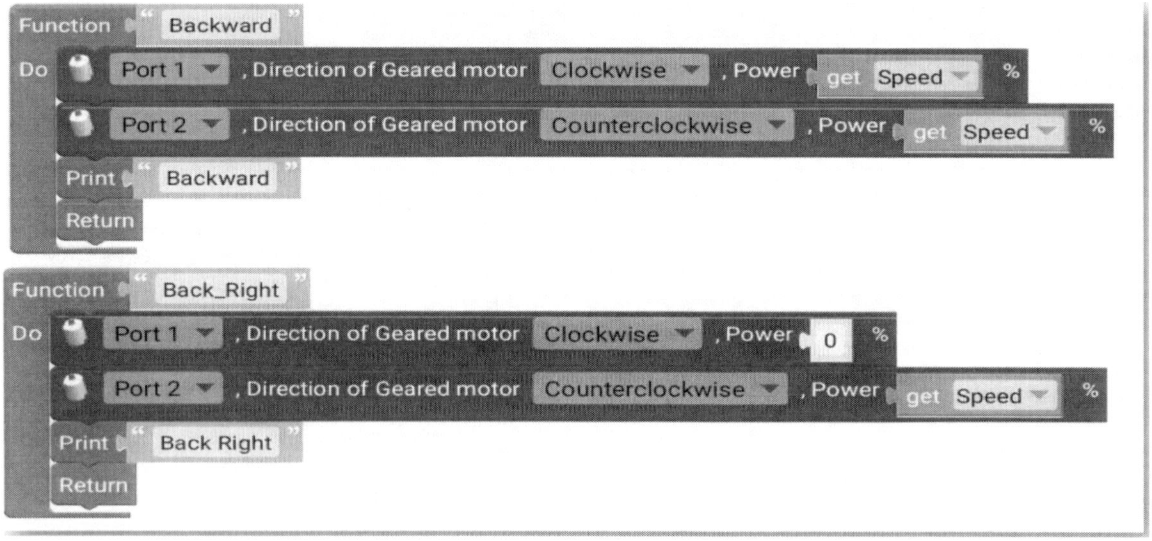

Fig. 2.40 Functions **Backward()** and **Back_Right()** used in Project "Avoider_1.rpb".

Please note that an IF-DO structure is used for the 4th Sensor/Actuator Pair, while the other Pairs use an IF-DO-ELSE structure. Logically, the 4th Condition Expression could evaluate to FALSE, and physically this would mean that there is an obstacle in front of the robot, but it is not close enough to evade yet, therefore it is OK for the robot to continue what it was doing, which most likely was "going forward". Anyhow, this situation would start a new iteration of the Main Endless Loop and a new recheck on the 1st Sensor/Actuator Pair to see whether the robot should next go forward or not.

An interesting question arises: can the order of the Event/Action Pairs be rearranged in the Main Endless Loop? Would the robot's behavior be any different? The author leaves that "exploration" task to the readers.

At run time, the "Avoider_1.rpb" code handled well the situations when the object was on either side of the robot. But it did not handle well the situation when a "flat" object was right in front of it, as it kept on going back and forth at the same spot, i.e., it seemed to be "stuck" in a fixed pattern of behaviors. **This is a side-effect of the current algorithm that resulted in the robot "bouncing" back and forth between the conditions represented by the 1st and 4th Sensor/Actuator Pairs as shown in Fig. 2.39.**

This problem can be easily fixed by adding a Turning maneuver to the current Backward maneuver for the 4th Sensor/Actuator Pair. The turn direction does not matter, the author happens to use a Back Left maneuver in the revised solution "Avoider_2.rpb" (see Fig. 2.41). The user would see much improved performance for frontal objects and corners when using "Avoider_2.rpb".

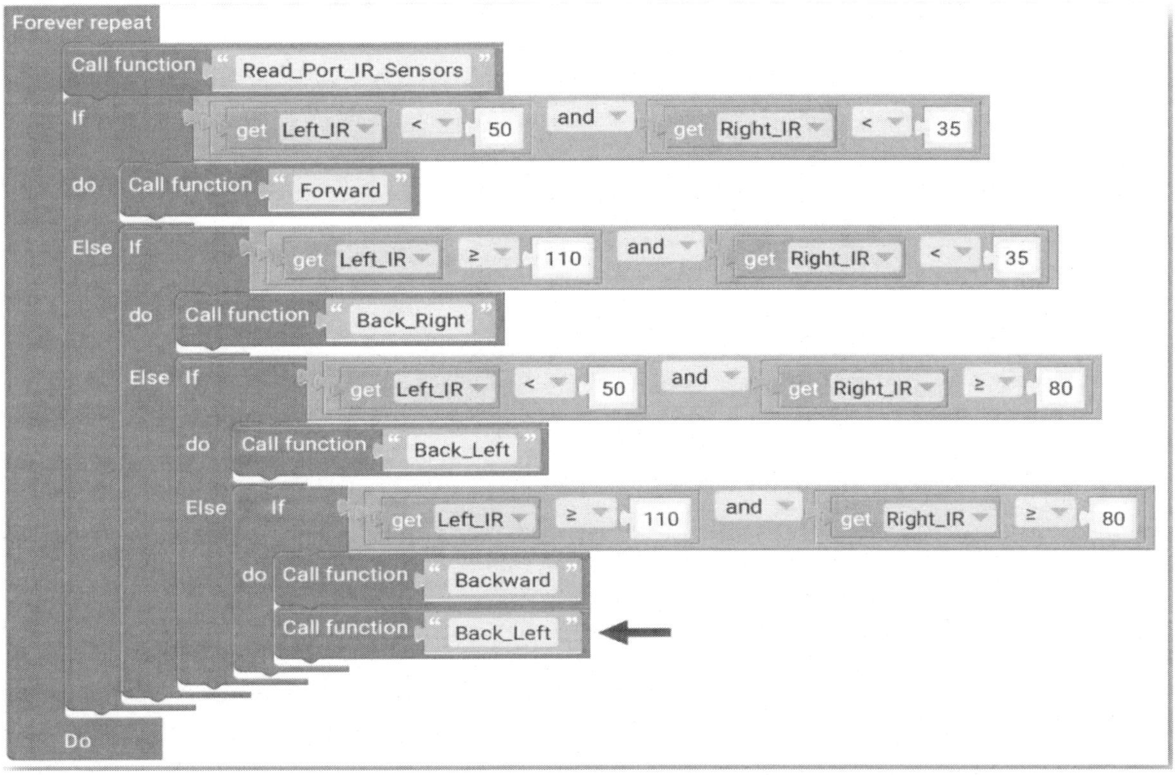

Fig. 2.41 Main Endless Loop used in Project "Avoider_2.rpb".

The next Project "Avoider_3.rpb" offers a different solution technique to these obstacle avoidance tasks. This new approach still relies on the same Event/Action list as described at the beginning of this Sub-Section 2.6.1, but it infers different behavior patterns from this list:

- A "default" action is identified which in this case is "Going Forward".
- A much simpler Main Endless Loop is achieved using several PARALLEL IFs instead of a single big IF-ELSE-IF structure like before. This change came from performing a

deeper analysis of the Event/Action list, and to retain only the "Essential" Event/Action Pairs which turned out to be **Back_Right(), Back_Left()** and **Forward()**.

- However, it was found that a Delay Time was needed for each action made (this part was not needed when the IF-ELSE-IF structure was used). This Delay Time can be tuned for overall optimal performance.

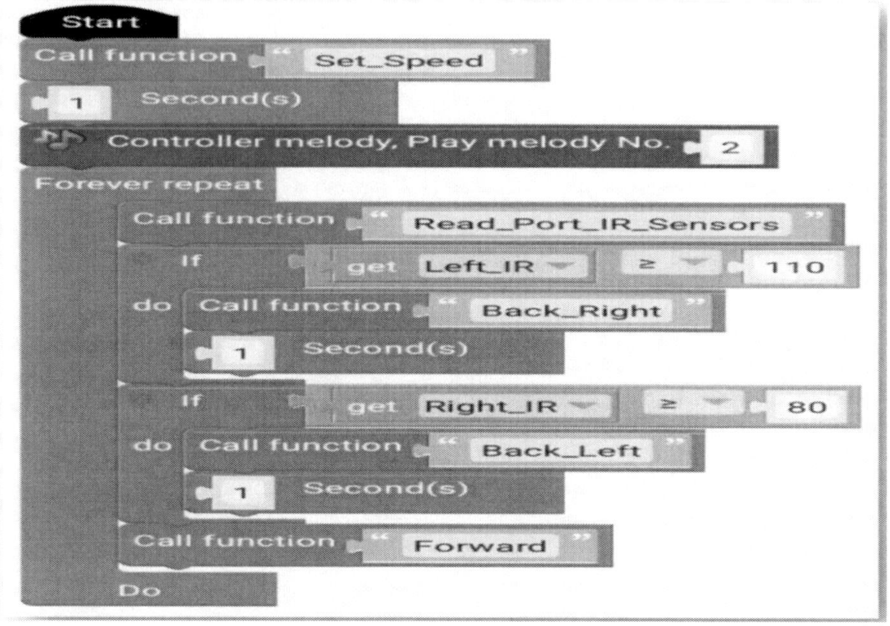

Fig. 2.42 Main Function used in Project "Avoider_3.rpb".

Fig. 2.42 illustrates the Main Function used in Project "Avoider_3.rpb":

- After acquiring the robot's maneuver **Speed** via Function **Set_Speed()** as before, the robot plays Melody 2 to signal the operator that it is ready for "Play".
- The Main Endless Loop starts with the usual Call to Function **Read_Port_IR_Sensors()** to obtain the current Values for Variables **Left_IR** and **Right_IR**.
- The 1st IF-DO Block represents the Event/Action Pair when there is an obstacle on the robot's Left, then the robot should **Back Right** (as before) to steer away from the obstacle. Next, a short delay (1 second) is necessary for the robot to stay in this "Action" for a little while to make this algorithm work properly.
- The 2nd IF-DO Block represents the Event/Action Pair when there is an obstacle on the robot's Right, then the robot should **Back Left** (as before) to steer away from the obstacle. A short delay of 1 second is also necessary for the robot to stay in this "Action" to make this algorithm work properly.
- Please note that the "**default**" maneuver (**Forward**) is always programmed as the "**Last Action**", but with no time delay as for the previous Event/Action Pairs.

It is important to perform a "thought experiment" here to "preview/understand" the possible results of this rather compact Endless Loop at run time:

- Every time that this Endless Loop is executed, the default maneuver **Forward** will be executed, meaning that the robot will always go forward (a little bit), regardless of everything else that may happen - in a way, this is a rather "aggressive" robot.
- Regarding the two PARALLEL IFs, during some execution cycle(s), none of them could be activated (when there is no obstacle), or one of them could be triggered (when the obstacle is on either side of robot), or both could be triggered (for a frontal obstacle). The alert reader may have noticed that seemingly the case for "a frontal obstacle" was not dealt with in this algorithm, so how would the robot behave at run time when encountering a frontal obstacle? The robot would "wiggle" backward right and left because the Functions **Back_Right()** and **Back_Left()** would be called consecutively. This "wiggling" effect also explains why a **DelayTime=0** would not work for the robot, as it would be switching between Functions **Back_Right()**, **Back_Left()** and **Forward()** (within the Endless Loop) **so fast** that the robot would look/act "paralyzed" and therefore unable to deal with a frontal obstacle. The reader is encouraged to test out this situation.

At this point, some readers may have wondered whether the two IF-DO Blocks used in "Avoider_3.rpb" could be changed to REPEAT-IF/WHILE Loops? And would the robot still perform its obstacle avoidance maneuvers as before?

The Project "Avoider-4a.rpb" implements this idea (see Fig. 2.43) and the reader is going to find out that this idea did not work at run time! What went wrong? Was it wrong to use REPEAT-IF/WHILE Loops for this challenge?

As it turns out, the error was not because WHILE Loops were used per se, but it was because these WHILE Loops were implemented incorrectly, as we forgot to provide them with the latest IR Sensor data. We were using the Variables **Left_IR** and **Right_IR** to control the Loops and we forgot to update their values for every new iteration of these Loops! It was a simple change to get to the revised Project "Avoider_4b.rpb" (see Fig. 2.44) which does work properly. Please look at the Conditional Expressions used in the REPEAT IF Blocks in Fig. 2.44, where real-time data from IR Sensor Ports 3 and 4 are now accessed and used to control the WHILE Loops, not copies of their values (**Left_IR** and **Right_IR**). The reader also should try to set the Time Delays to zero and observe any changes in the robot obstacle-avoidance behaviors.

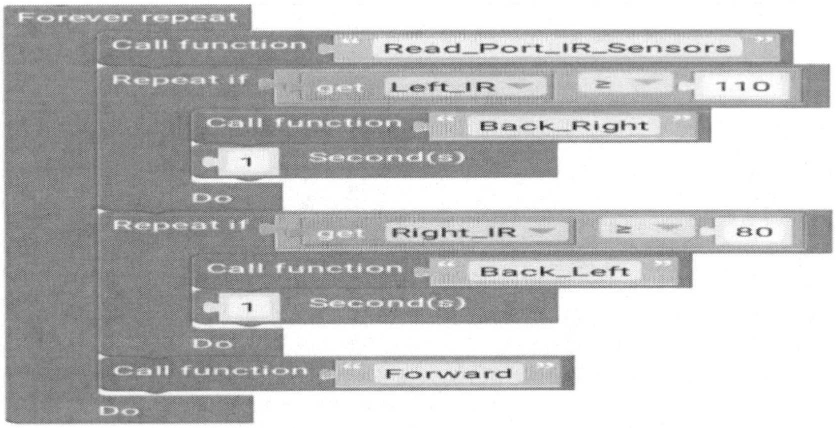

Fig. 2.43 Main Endless Loop used in Project "Avoider_4a.rpb".

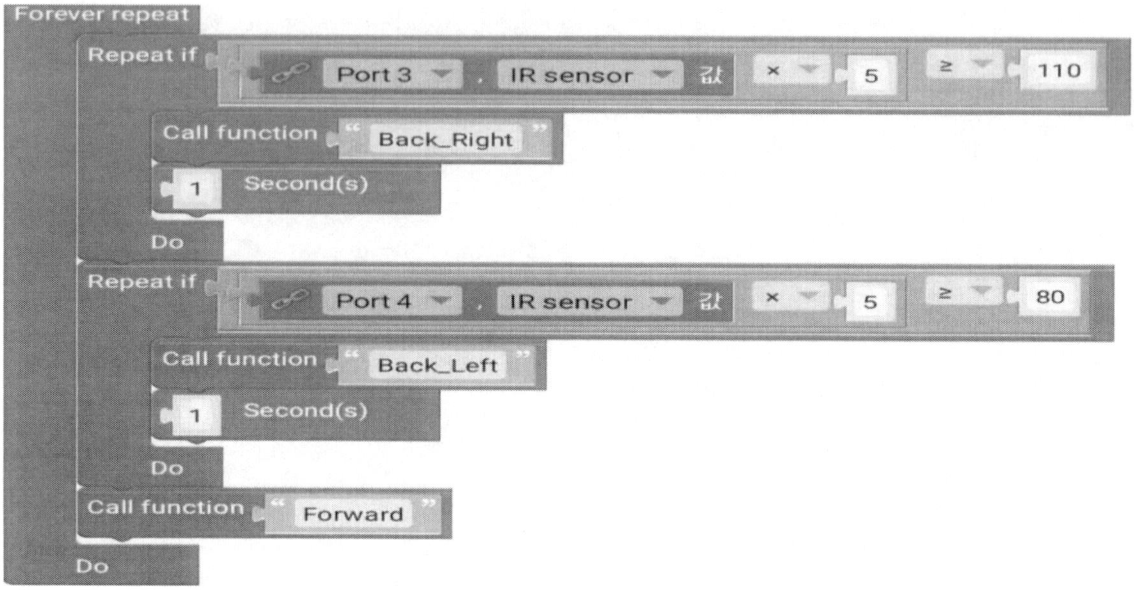

Fig. 2.44 Main Endless Loop used in Project "Avoider_4b.rpb".

The last Avoider Project "Avoider_4i.rpb" builds on Project "Avoider_4b.rpb" and introduces a new concept/procedure using the Initial Values of the IR Sensors at Ports 3 and 4. Fig. 2.45 shows the new Function **Ports_Init()** used to obtain and save the Initial Readings of Ports 3 and 4:

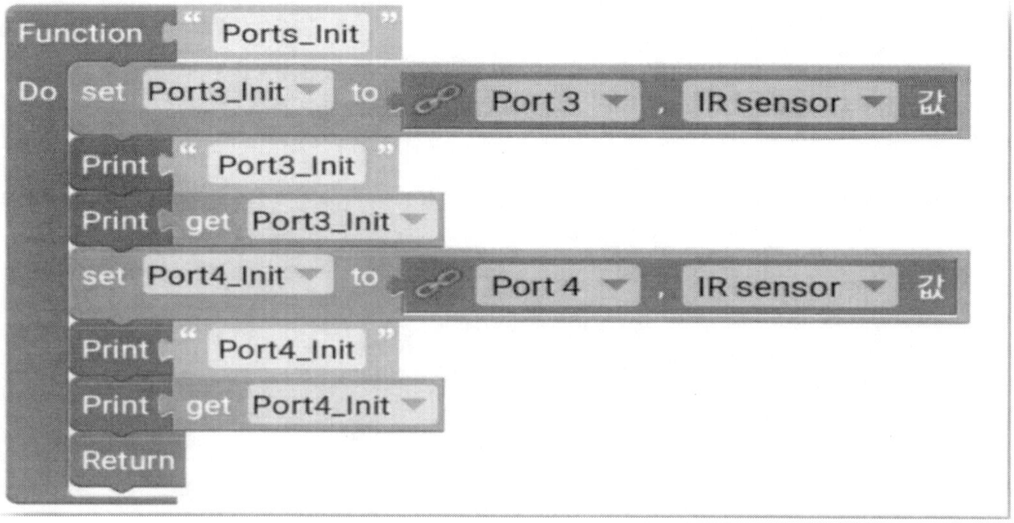

Fig. 2.45 Function **Ports_Init** used in Project "Avoider_4i.rpb".

Fig. 2.46 shows how these new Constants **Port3_Init** and **Port4_Init** are used to "normalize" the Conditional Expressions controlling the REPEAT-IF Loops used in "Avoider_4i.rpb", as both now use the **same Threshold Value of 50**, as compared to Fig. 2.44.

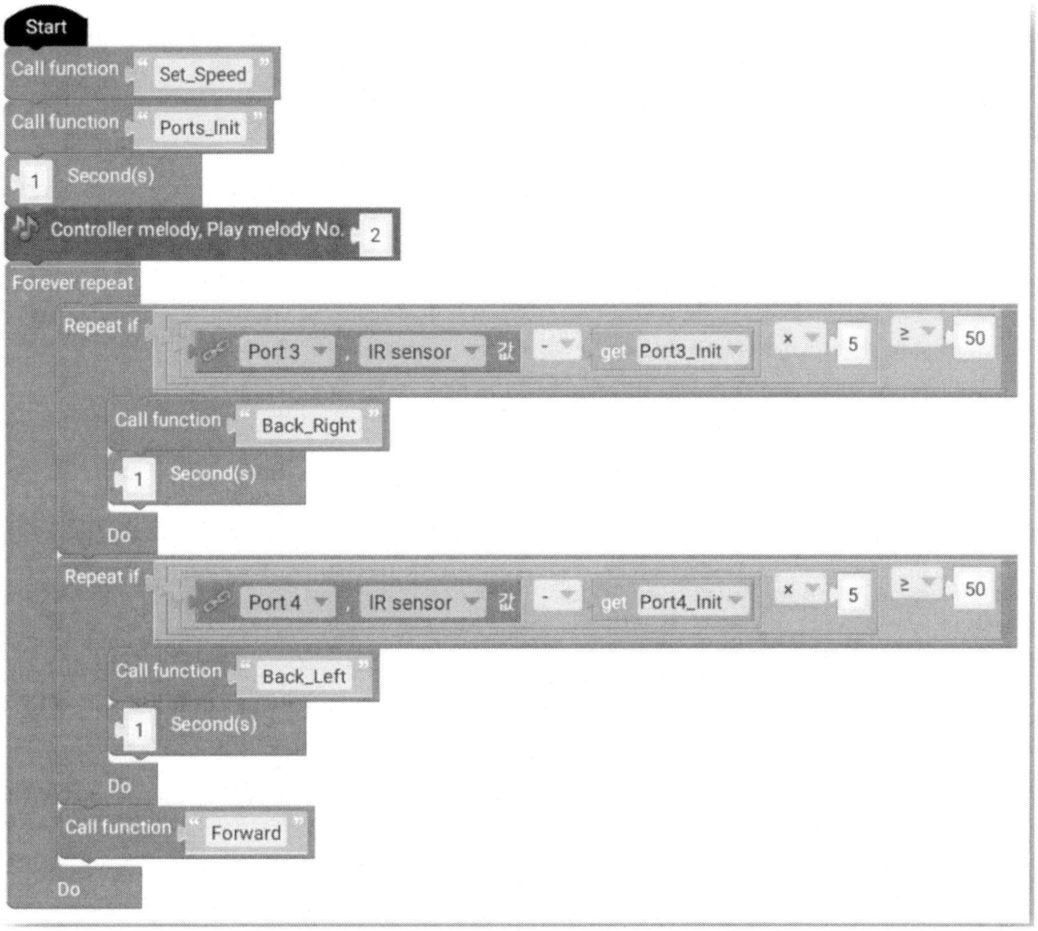

Fig. 2.46 Main Function used in Project "Avoider_4i.rpb".

Thus far, we had been using the IRSS-10s mounted in a "vertical" configuration as shown in Fig. 2.2, but they can also be mounted in an "horizontal" configuration (by rotating them 90 degrees). Interested readers can then rerun/adapt the previous Avoider codes and obtain (slightly) different behaviors/performances of the reconfigured robot.

The author is recommending another interesting experiment that the reader can make with these different versions of the Avoider codes: create a maze using books or boxes and let the Avoider robot roam freely through it. The reader will be surprised with its behaviors depending on which R+BLOCK code is used, especially when the maze has a cul-de-sac.

2.6.2 Follower

By its namesake, the new Project Follower will be using the "Reverse Logic" of Project Avoider, but still using the same robot physical configuration. This time, we'll go back to using the Internal IR Sensors of the CM-150/151.

Essentially, when the Follower robot detects an object via its Left and Right IR Sensors, it should follow/track it. The corresponding Event-Action list can be as follows:

- If there is no object detected, the robot should stay still (default behavior).
- If there is an object detected by the Right IR Sensor, the robot should swing right.
- If there is an object detected by the Left IR Sensor, the robot should swing left.

Should we also plan for the case when an object is detected and located right in front of the robot? Should it "stop", or should it show that it has "locked in" on the object in some other way? From the preceding Avoider coding work, we should be able to let the robot bounce between its two actions of swinging left and right to show that it got the object in its "sight".

To round out the reader's experience with using the Left and Right IR Sensors, three Follower Projects were developed: "Follower_1.rpb", "Follower_2.rpb" and "Follower_3.rpb".

Project "Follower_1.rpb" adopts the same overall Logical Structure of "Avoider_4b.rpb" but with the appropriate Event/Action Pairs, of course (see Fig. 2.47):

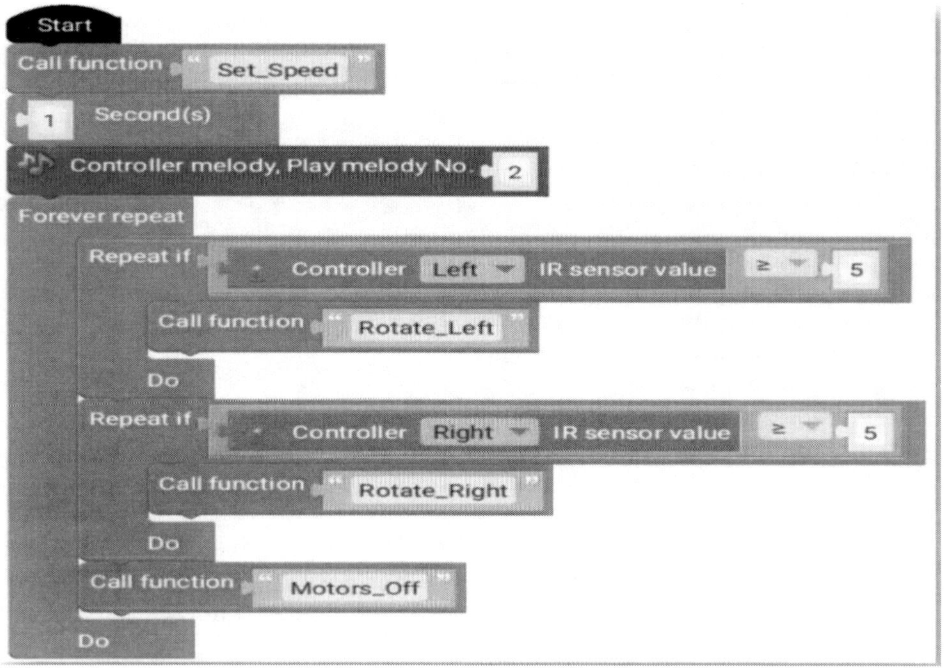

Fig. 2.47 Main Function used in Project "Follower_1.rpb".

- The default behavior is now to stay still, that is to Call Function **Motors_Off()**.
- Two REPEAT-IF/WHILE Loops are used like for "Avoider_4b.rpb", but the Conditional Expressions are now written to allow the detection of an object "as far as possible" using the small threshold value of **5**. This is because the robot's goal is no longer to avoid an object, but to **track** it, that is to **turn into the same direction as the IR Sensor** that detected the object: for example, **Rotate Left** if the **Left IR Sensor** reads in a Value greater than 5.

When the reader tries out "Follower_1.rpb", it is recommended that he/she uses different sizes for the object to be tracked and to note the changes in the robot's tracking behaviors.

The reader is also challenged to come up with another algorithm using PARALLEL-IFs instead, but to yield the same object tracking result. Don't forget to use a small Time Delay to make this approach work!

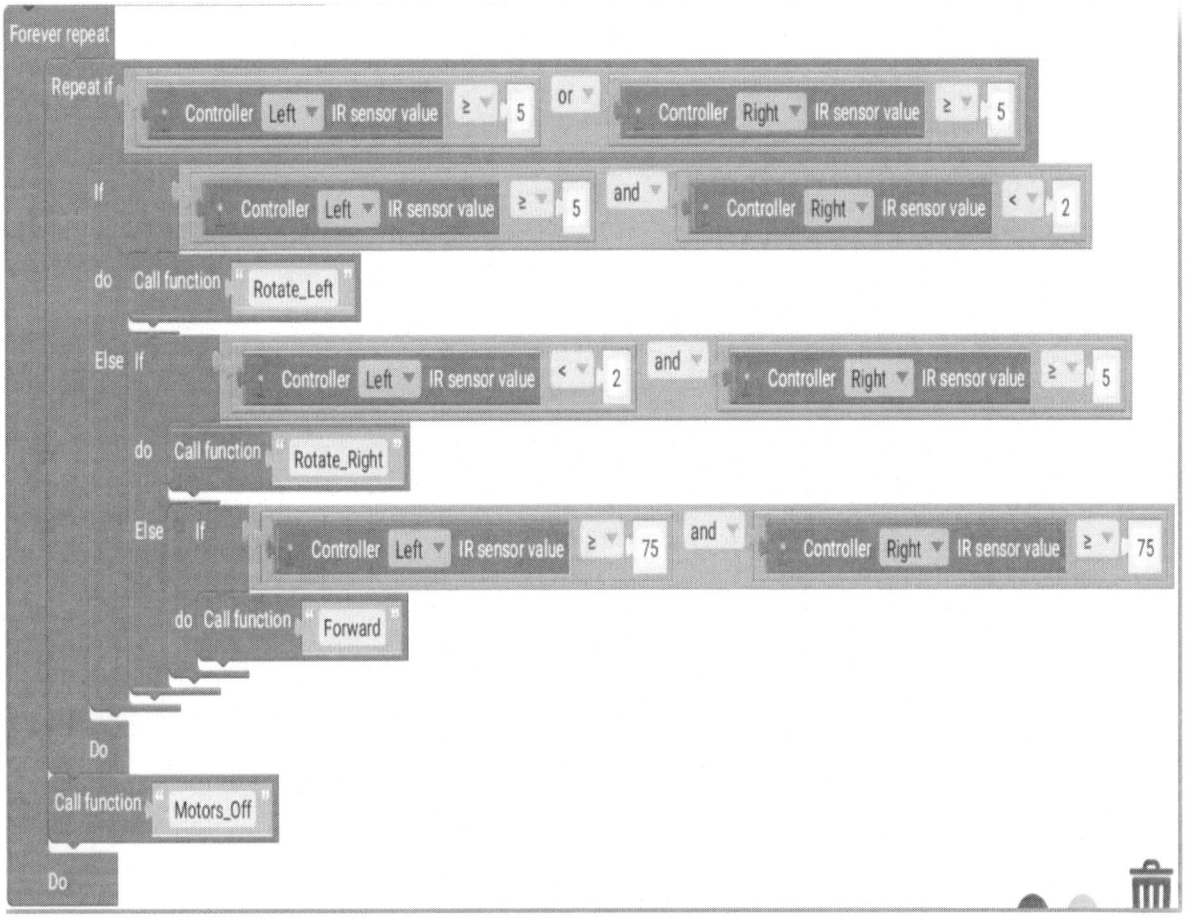

Fig. 2.48 Main Endless Loop used in Project "Follower_2.rpb".

For the next Follower Project (Fig. 2.48), a new Event/Action Pair is added to the previous list:

- When the object gets a bit closer to the robot (e.g., a threshold value of 75 applied to both IR Sensors), the robot should get even closer to the object by Going Forward. At this point during run time, if the user then moves the object away from the robot at a **proper speed**, he/she should see the effect of the robot "following" the object at a short distance away, as if there was an "invisible tractor beam" at work!

Surprisingly, there are two possible solutions: one using a combination of REPEAT-IF/WHILE Loop and an IF-ELSE-IF structure (Follower_2.rpb), and another one using only an IF-ELSE-IF structure (Follower_3.rpb). It is very possible that the readers can find other solutions too!

The Main Endless Loop for "Follower_2.rpb" is illustrated by Fig. 2.48. It uses a Default Maneuver (CALL FUNCTION "Motors_Off()") and an OUTER REPEAT-IF Loop which encompasses an INNER IF-ELSE-IF structure that determines the ONE appropriate action among 3 possible and mutually exclusive Actions "Rotate_Left", "Rotate_Right" or "Forward":

- When no obstacle is detected, the REPEAT-IF Loop would be skipped and Function **Motors_Off()** is called, thus the robot would stay still as its default behavior.
- When **either Left or Right IR Sensor** reads in a Value **greater or equal to 5** (i.e., detects an object), the REPEAT-IF Loop is entered, and the Controller checks further with the IF-ELSE-IF structure to see which IR Sensor was the one detecting the object.
- The First IF-DO Block is used to check if the Left IR Left Sensor is the **only one** detecting the object (threshold "5" exceeded). If this First Event/Condition turns out to be TRUE, then the robot **Rotates Left**.
- But if this First Event/Condition turns out to be FALSE, the Controller next checks on the Second IF-DO Block to see if the Right IR Sensor is the **only one** detecting the object (threshold "5" exceeded). If this Second Event/Condition turns out to be TRUE, then the robot **Rotates Right**.
- However, if this Second Event/Condition turns out to be FALSE also, then the Controller next checks to see if the object is close to both IR Sensors with the Third IF-DO structure. If both IR Sensors read in Values that are greater or equal to "75", meaning that the object is located in front of the robot, it is then commanded to **Go Forward**. This corresponds to the "invisible tractor beam" situation at run time.
- If this Third Event/Condition turns out to be FALSE, the controller reaches the end of the IF-ELSE-IF structure without being able to determine a concrete action to do, so the controller would just loop back to check on the Conditional Expression use in the REPEAT-IF Loop "((Left_IR_Sensor >= 5) OR (Right_IR_Sensor >= 5))". If this condition is still TRUE, it would search through the IF-ELSE-IF structure again. If this condition becomes FALSE, the controller finally quits the REPEAT-IF Loop, next calls upon Function **Motors_Off()** to stop the robot, and then reiterates the Outermost Main Endless Loop all over again.

Please note that when using the REPEAT-IF Loop in "Follower_2.rpb", we need to use "current" IR Sensor Data inputs in all our Conditional Expressions, so that most recent data from the IR Sensors are used. Thus, this solution is quite "Input-Intensive", straining the BT communications link between the Chromebook and the CM-150/151.

The third Follower Project "Follower_3.rpb" offers a "compromise" by using Function **Read_Sensors(),** only once per iteration of the Main Endless Loop, to update the Variables **Left_IR** and **Right_IR** (see Fig. 2.49):

- Next, the familiar IF-ELSE-IF structure is used to check on the Four Event/Action Pairs listed previously for the Follower_2 Project: "Rotate Left", "Rotate Right", "Forward" and "Stay Still".
- Please note that the Third IF-DO is now an IF-DO-ELSE (as compared to the IF-DO used in "Follower_2.rpb", offering a more streamlined syntax for "Follower_3.rpb" over "Follower_2.rpb".

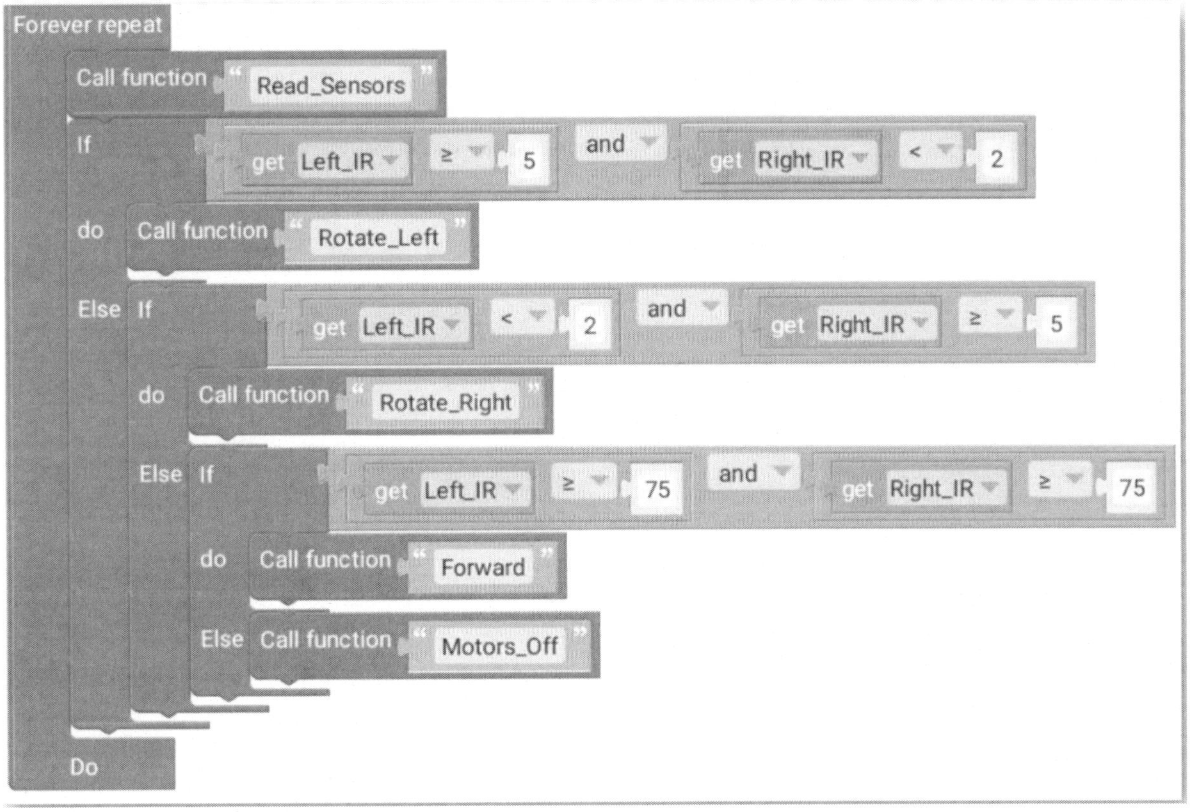

Fig. 2.49 Main Endless Loop used in Project "Follower_3.rpb".

At run-time, which solution, "Follower_2.rpb" or "Follower_3.rpb", provides the better performance for the reader's environment? There was no discernable difference in performance between these two solutions that the author could see during his own testing. In theory, "Follower_3.rpb" should be "faster" because it accesses the IR Sensors only twice per each iteration of the Main Endless Loop, i.e., it uses less resources from the CM-150/151 and the BT communications link, as compared to "Follower_2.rpb" which can use from 4 to 8 accesses to the IR Sensors per each iteration of its Main Endless Loop. However, "Follower_2.rpb" can be more "accurate" in its actions as it always uses "fresh" data inputs.

Recapping Sections 2.5 and 2.6, the author's goal was to show that programming robot behaviors can be fun and challenging at the same time! Starting from the (now familiar) Event/Action Pairs analysis, three techniques can be used to create R+BLOCK codes and after some initial actions, they all used an outermost ENDLESS LOOP for the main algorithm:

- In Technique 1, for each iteration of the Main Endless Loop, latest readings from the used sensors were first recorded, then an IF-ELSE-IF structure was used to represent all possible but mutually exclusive Event/Action Pairs that corresponded to robot behaviors required to fulfill a given challenge. Example programs using Technique 1 are "Spinning_Top_3_IF_ELSE_IF_1&2.rpb", "Clapping_RC_1.rpb", "Avoider_2.rpb", and "Follower_3.rpb".
- In Technique 2, the sensors reading task is still required first, but the IF-ELSE-IF structure is replaced by PARALLEL-IFs. This technique requires a deeper analysis of the original Event/Action Pairs list to identify a "default" behavior, and other Event/Action Pairs

that can work “on their own” or “in concert”. This technique also requires a short Time Delay incorporated into each robot action. Compared to Technique 1, Technique 2 usually results in a smaller number of Blocks used in the R+BLOCK code, meaning that the Computing Component would take less time to do each iteration of the Endless Loop, and thus the robot may become more responsive to changes in the “World” (review Fig. 2.25). Example programs using this technique are “Spinning_Top_3_Par_IFs.rpb”, “Clapping_RC_2.rpb” and “Avoider_3.rpb”.

- Technique 3 is a variation of Technique 2 and it uses “parallel” LOOPs instead of “parallel” IFs. It combines “Data Acquisition” tasks in its Conditional Statements, thus it uses the least number of statements, as compared to the other two Techniques. Example programs using Technique 3 are “Avoider_4b.rpb”, “Avoider_4i.rpb” and “Follower_1.rpb”. “Follower_2.rpb” is a hybrid solution combining Techniques 3 and 1.

2.6.3 Smart Avoider

The “Smart Avoider” Project seeks to combine Autonomous-Behavior and Remote-Control techniques so that the Avoider robot can “save itself” when, on purpose, the operator drives the robot into an obstacle. This project combines concepts and programming techniques from “Clapping_RC_2.rpb” and “Avoider_2.rpb” with additional algorithm changes.

The overall Control Algorithm can be expressed as a pseudo-code shown below:

Main Endless Loop {

If (NIR Sensors detect no obstacle) then ***{*** *// Take “Clapping RC” Path*

When new Sound Counter Result comes in, save it in ***Claps***

Depending on the Actual value of ***Claps****, activate the appropriate Robot Maneuver.*

A short delay to perform the required Maneuver, then ***Stop*** *the Robot.*

} *// End of Clapping RC Path*

Else ***{*** *// Obstacle within NIR detection ranges so take “Escape” Path*

Stop the robot

Figure out whether Obstacle is Frontal, on Left or on Right of robot and activate appropriate robot’s avoidance moves <u>*until obstacle out of NIR detection ranges*</u>

Use Musical Melody to signal robot’s successful obstacle avoidance

} *// End of Escape Maneuver*

} (End of Main Endless Loop)

Figs. 2.50 through 2.53 show details of the actual R+BLOCK code “Smart_Avoider.rpb”. Fig. 2.50 shows the Main Function and three Utility Functions named **Set_Speed()**, **Init()** and **Print_IR()**:

- Function **Set_Speed()** sets Variable **Speed** to 25% power.

- Function **Init** sets three NIR Threshold Variables **T1**, **T2** and **T3** to their respective values of 30, 20 and 5. **T1**, **T2** and **T3** will be used in the detection of an obstacle and the performing of appropriate escape maneuvers when this obstacle is detected.
- Function **Print_IR()** is used to display the current values for Variables **Left_IR** and **Right_IR** for the programmer to match the current robot's maneuvering action with these IR Sensor data inputs.
- The Main Function starts out by Calling Functions **Set_Speed()** and **Init()**, and waits for 1 second before Playing Melody 3.
- Next, Function **Stop()** is called, mainly for safety reasons.
- Then the Main Endless Loop (which follows the logic flow described in the previous pseudo-code) is entered:
    - Function **Print_IR()** displays current values for **Left_IR** and **Right_IR**.
    - Next, an IF-DO-ELSE structure is used to check the current Left and Right IR Sensors Values against **T1(=30)** to determine whether an obstacle is "detected" or not. If an obstacle is **not detected**, meaning that **(Left_IR_Sensor < T1) AND (Right_ IR_Sensor < T1)**, Function **Clap_RC** is called. Otherwise, if an obstacle is **detected**, i.e., **(Left_IR_Sensor ≥ T1) OR (Right_ IR_Sensor ≥ T1)** – *via De-Morgan's Theorem*, Function **Escape()** is called. Please note that the actual complexity of programming the appropriate RC and Escape Maneuvers is hidden at this coding level.

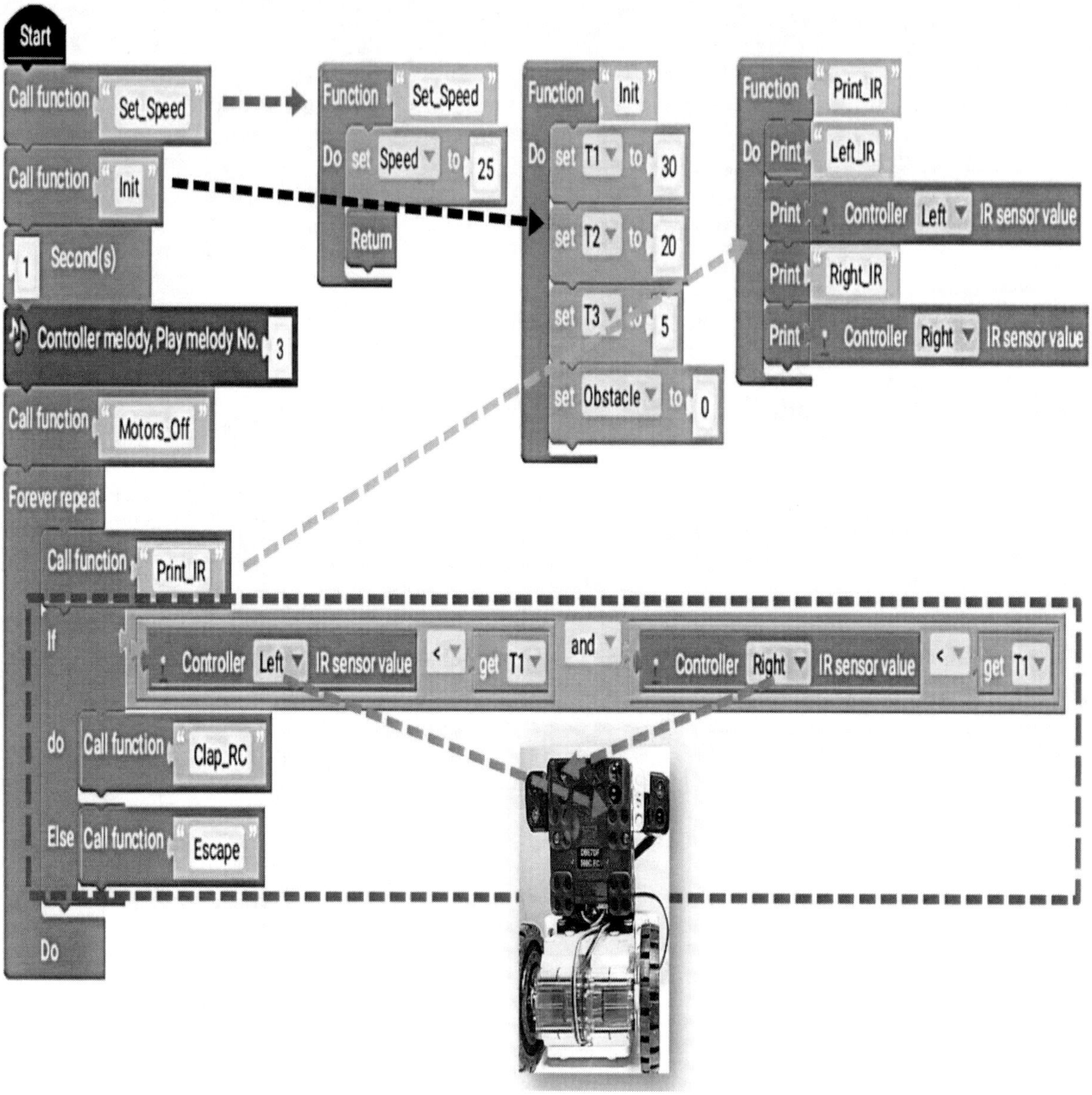

Fig. 2.50 Main Function and Functions **Set_Speed()**, **Init()** and **Print_IR()** used in Project “Smart_Avoider.rpb”.

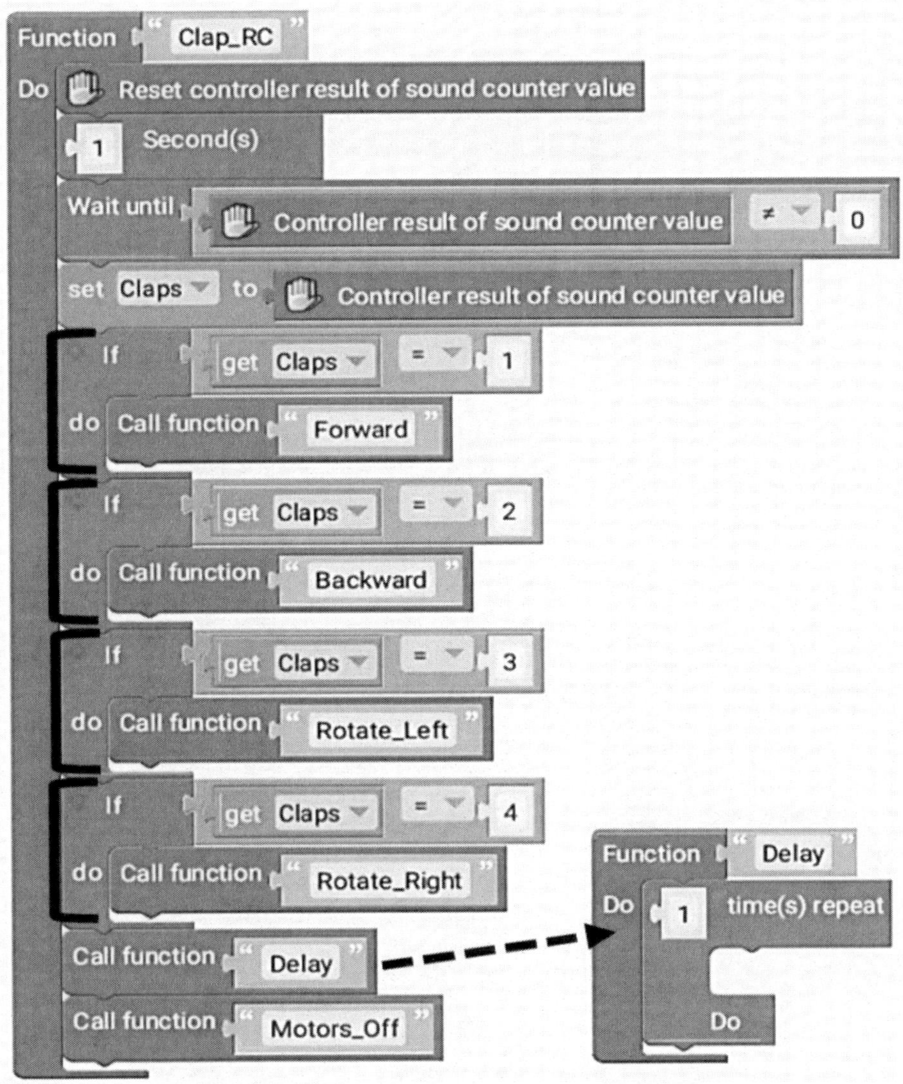

Fig. 2.51 Function **Clap_RC()** used in Project "Smart_Avoider.rpb".

Fig. 2.51 illustrates the details of Function **Clap_RC()**:

- First, it resets the **Sound Counter Result** to 0 and waits for 1 second for this process to finish up on the CM-150/151 side.
- It then waits until the Operator claps his/her hands loud enough to result in a non-zero value for the **Sound Counter Result**. When this event happens, it saves the actual Hand Claps "detected" into Variable **Claps**.
- Next, 4 Parallel IF-DO structures are used to match up 4 possible Events: 1 clap, 2 claps, 3 claps or 4 claps, into the appropriate Robot Maneuvers "Forward", "Backward", "Rotate_Left" and "Rotate_Right".
- Previously, in Fig. 2.37, we used a Time Delay of 1 second to let the robot perform its commanded maneuver, before stopping the robot. But in this project, a shorter Time Delay is needed to provide a "timelier" Escape Response from the robot when an obstacle

is detected. Thus, Function **Delay()** was developed to create a Time Delay that is adjustable to fractions of a 1 second at run time. Essentially, it uses a Counting Loop that does "Nothing" as its body contains no actionable Block. The shortest Time Delay possible is obtained using a count of 1 and the reader is encouraged to use other counter values to see how the robot's performance changes at run time.

Figs. 2.52 and 2.53 describe the steps needed for the robot to "escape" a detected obstacle:

- Function **Escape()** starts out by assigning 1 to Variable **Obstacle** and stops the robot by using Function **Motors_Off()**.
- Essentially, the "Escape" algorithm needs to figure out which one among three possible Event/Action Pairs that may have happened at run time:
    - If the obstacle happens to be "Frontal", Left and Right IR Sensors would provide Values that are "about the same". Then the robot should roll Backward a bit, next it should keep on Turning Right or Left until the obstacle is no longer detected.
    - If the obstacle is on the Left of the robot, the Left IR Sensor would yield a Higher Value than the Right IR Sensor. Then the robot should Turn Right until the obstacle is no longer detected.
    - If the obstacle is on the Right of the robot, the Right IR Sensor would give a Higher Value than the Left IR Sensor. Then the robot should Turn Left until the obstacle is no longer detected.
- In Fig. 2.52, this "Escape" algorithm is implemented as an Outer REPEAT-IF-DO Loop that contains one IF structure to deal with the case of a "Frontal Obstacle", and two Inner REPEAT-IF-DO Loops to deal with the cases of "Left Obstacle" and "Right Obstacle" respectively (see Fig. 2.53 also).
- However, to understand the design of the Outer REPEAT-IF-DO Loop, we need to refer back to the Conditional Expression used in the IF-DO-ELSE structure found in the Main Endless Loop (see Fig. 2.50). From previous discussions, when an obstacle is detected, this Conditional Expression becomes **(Left_IR_Sensor $\geq$ T1) OR (Right_IR_Sensor $\geq$ T1)**, where T1=30. Thus, the Conditional Expression used for the Outer REPEAT-IF-DO follows the **same structure**, except now **T2=20** is used (see Fig. 2.52), which is "stronger" logic-wise to "guarantee" that the Outer REPEAT-IF-DO will be activated when an obstacle is detected (from a previous section of code in the Main Endless Loop).
- Once this Outer REPEAT-IF-DO Loop is entered, Variable **NIR_Dif** is computed and next tested to see if it is between -10 and 10. If this turns out to be TRUE (i.e., "frontal" obstacle), then the robot rolls Backward for 1 second - otherwise no action is done, and the First Inner REPEAT-IF-DO Loop is activated next.

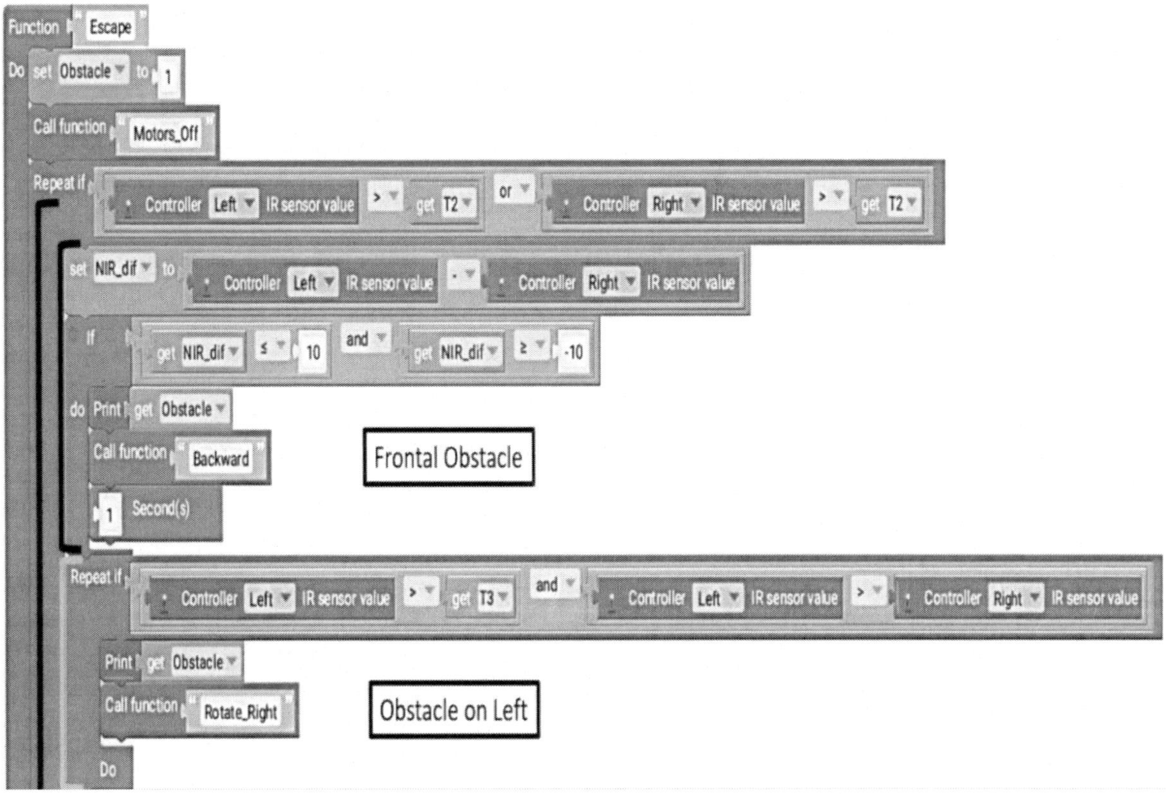

Fig. 2.52 Part 1 of Function **Escape()** used in Project "Smart_Avoider.rpb".

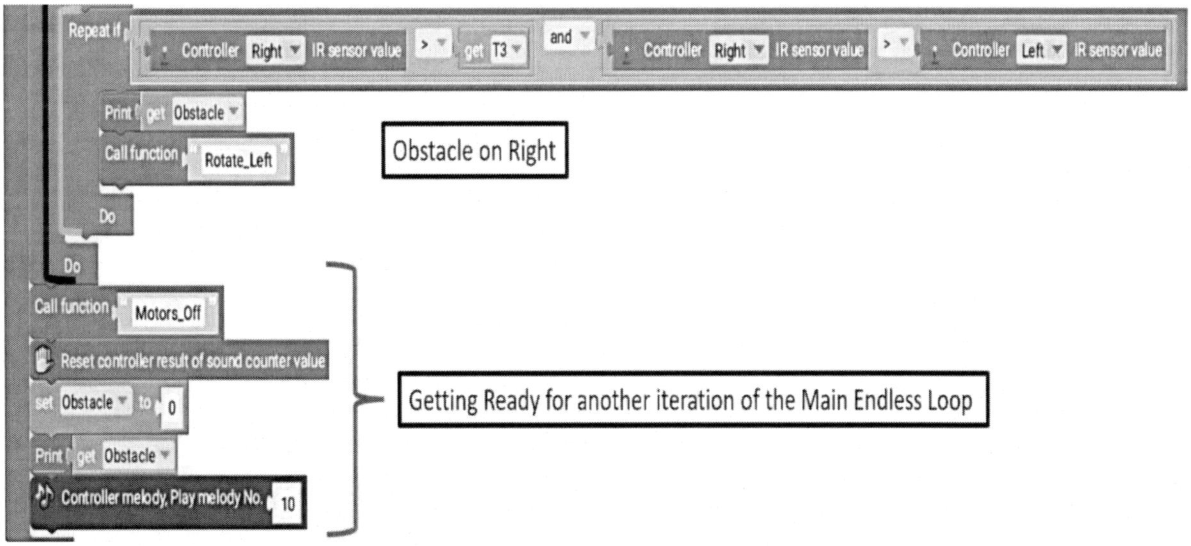

Fig. 2.53 Part 2 of Function **Escape()** used in Project "Smart_Avoider.rpb".

- The First Inner REPEAT-IF-DO Loop (Fig. 2.52) is designed to check if the obstacle is found "only" on the Left of the robot, meaning that **(Left_IR > Right_IR)**. The "extra" condition

(Left_IR > T3), where T3=5 which practically corresponds to "empty space", is used to make sure that the robot would rotate itself "well away" from the obstacle. If this situation turns out to be true at run time, then the robot is instructed to keep on Rotating Right until the obstacle is no longer detectable (i.e., **Left_IR ≤ T3**).

- Similarly, the Second Inner REPEAT-IF-DO Loop (Fig. 2.53) is designed to check if the obstacle is found "only" on the Right of the robot, meaning that **(Right_IR > Left_IR)**, and the extra condition **(Right_IR > T3)** is used for the same reason as explained above. If this turns out to be the case at run time, then the robot is instructed to keep on Rotating Left until the obstacle is no longer detectable (i.e., when **Right_IR ≤ T3**).
- Please note that for a single obstacle, once the First Inner REPEAT-IF-DO Loop is activated, the Second Inner REPEAT-IF-DO Loop would not apply at all (i.e., no chance for the robot to rotate back into the obstacle).
- Please note that for a "Single Frontal Obstacle", the robot would roll backwards first, then would likely rotate right to evade further from the obstacle, because of the current ordering of the two Inner Loops.
- Once the obstacle is evaded, i.e., break-out from the Outer REPEAT-IF-DO Loop, the robot would Stop, reset the Sound Counter Result, and play Melody 10 to get ready for another iteration of the Main Endless Loop.

This "Escape" algorithm was designed to respond to a single obstacle such as a flat wall or fence, but the reader can try to send this robot into a wall/fence corner to see how the robot would maneuver itself then. The author had found that most of time this algorithm would handle a corner well, but there are instances where it would get "stuck" bouncing back and forth between the two walls/fences.

The reader is also encouraged to rearrange the IRSS-10 sensors into a "horizontal" configuration and revise the "Smart_Avoider.rpb" code to use Ports 3 and 4 IR Sensors instead.

As a comparison, the reader may check out the equivalent "Smart Avoider" solution using MIT SCRATCH-2 running on a Desktop PC (https://www.youtube.com/watch?v=o8w_YVocCpA)

The "Smart Avoider" project showcased a mixed "RC/Autonomous" approach to control the Avoider robot, in the last two projects of Section 2.6, we will go "full-autonomous".

2.6.4 Simple Line Tracker

For this project, the Avoider is rotated 90 degrees "down" and a small frontal wheel is attached (see Fig. 2.54) to keep the four IR Sensors within a good working range for surveying the ground surface.

Fig. 2.54 also shows a small R+BLOCK program, handy for collecting IR Sensors data of the White Background or the Black Track which will be needed in the "Simple Line Tracker" and "Yin-Yang Line Tracker" Projects.

In the "Simple Line Tracker" Project, we will be using the CM-150/151 Internal IR Sensors and we are going to assume that the Operator has already lined up the robot on the Black Track as shown in Fig. 2.54 before activating the robot "Line Tracking" algorithm. This assumption would reduce the Event/Action Pairs to monitor to three Pairs that are mutually exclusive:

- If the Left IR Sensor sees the Black Track and the Right IR Sensor sees the White Background, then the robot should Turn Left in an attempt to re-align itself with the Black Track.

- If the Right IR Sensor sees the Black Track and the Left IR Sensor sees the White Background, then the robot should Turn Right in an attempt to re-align itself with the Black Track.
- Otherwise, the robot should continue to go Forward.

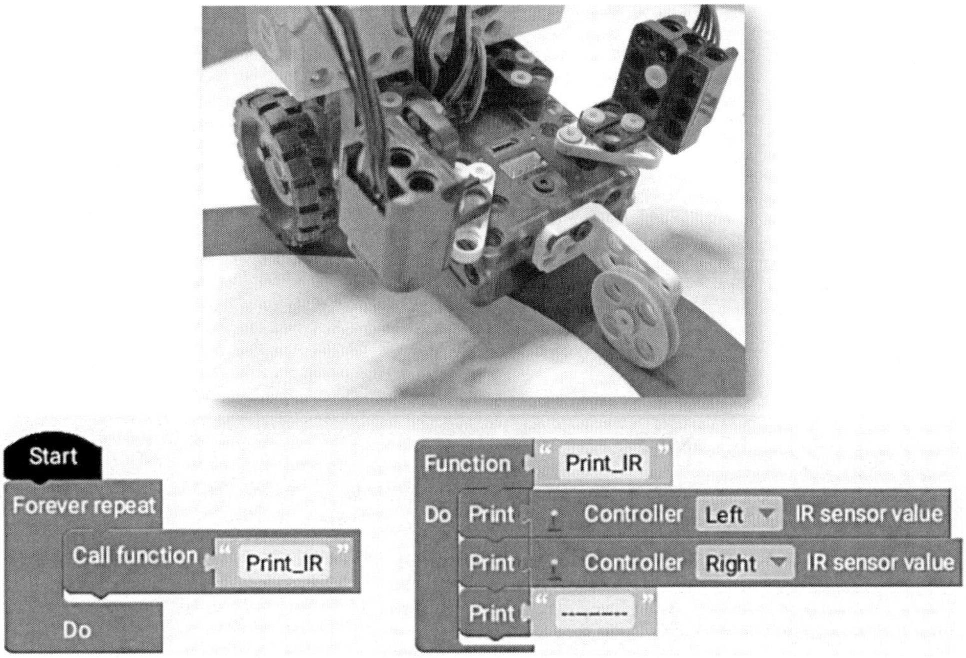

Fig. 2.54 Reconfiguration of "Avoider" into "Line-Tracker".

From previous examples, readers probably already guess that we will be using an IF-ELSE-IF structure to implement the handling of these 3 Event/Action Pairs (see Fig. 2.55):

- First, Variable **Speed** is set to 22% power and Melody No. 2 is played.
- The Main Endless Loop implements an IF-ELSE-IF structure to handle the 3 Event/Action Pairs described previously.
- The Threshold Values for switching between "Black" and "White" were 60 and 90 respectively for Left IR and Right IR, and they were determined for the author's set up only, using the simple R+BLOCK program as shown in Fig. 2.54. The reader can follow the same procedure to determine his/her appropriate Threshold Values. Please note that it is quite "normal" for the IR Sensors to respond differently when being tested exactly on the same "color" patch (i.e., Black Track or White Background)

At first glance, the algorithm used for the "Line-Tracker" robot is not that different from the one used for the "Avoider" and "Follower" robots (see Figs. 2.39 and 2.48). However, the "Line-Tracker" robot requires higher "precision" maneuvers which require a high rate of Data Input and Actuator Output coordination (which was not necessary for the "Avoider" and "Follower" robots). This is the issue of **Input/Output Systemic Time Delays** previously discussed in Sub-Sections 2.4.3 and 2.4.4. In practice, the author had not been able to go beyond a Speed Power setting of 22% when running the "Line_Tracker.rpb" code, using the Chromebook and R+BLOCK environment, without the robot running "Off-Track" at tight corners of the Black Track. Looking forward to Chapters 3 and 4, where the same Line-Tracker project is implemented in TASK code on CM-150-151 or Python code on the Chromebook, without any "Off-Track" issue and at a much higher Speed/Power.

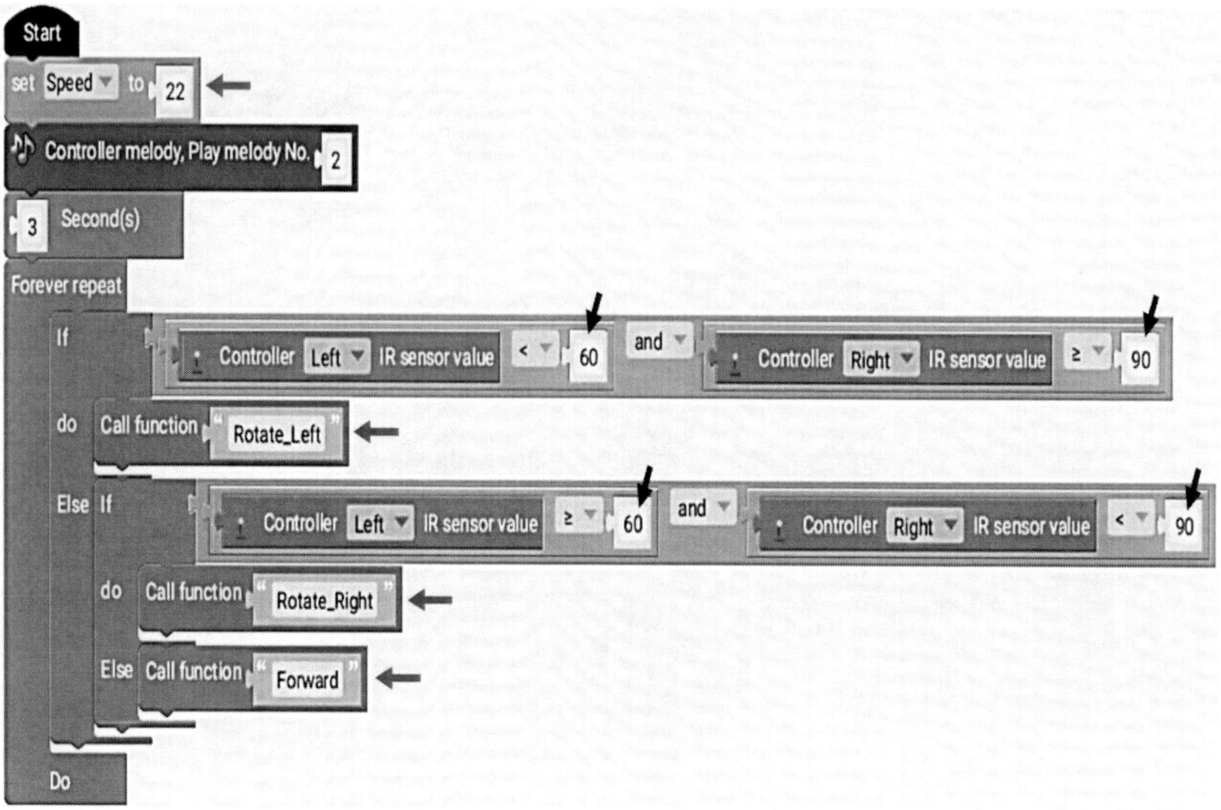

Fig. 2.55 Main Function used in "Line_Tracker.rpb".

As a comparison, the reader may check out the equivalent "Line Tracker" solution using MIT SCRATCH-2 running on a Desktop PC (https://www.youtube.com/watch?v=MNlFmCwqP8I).

2.6.5 Yin-Yang Line Tracker

For the "Yin-Yang Line Tracker" (YYLT) Project, we will be using all 4 IR Sensors mounted on the Line-Tracker robot (see Fig. 2.56): the two Internal IR Sensors, and the two IRSS-10s mounted to Ports 3 (Left) and 4 (Right).

In the "Yin-Yang Line Tracker" Project, the Event/Action Pairs to monitor are based on the ones used for the "Simple Line Tracker" Project as the robot is still required to follow a "track", but this track can now flip from Black to White. Additionally, we need to use the IRSS-10s to figure out which type of Background (Black or White) is the robot running on at any point in time.

Again, we are going to assume that the Operator has already lined up the robot on the Yin-Yang Track as shown in Fig. 2.56 before activating the robot "Yin-Yang Line Tracking" algorithm. This assumption reduces the Event/Action Pairs to monitor to five Pairs that are mutually exclusive:

- If the Left Internal IR Sensor sees the Black Track, and both IRSS-10s report seeing a White Background, then the robot should Turn Left in an attempt to re-align the robot with the Black Track.
- If the Right Internal IR Sensor sees the Black Track, and both IRSS-10s see a White Background, then the robot should Turn Right to re-align itself with the Black Track.
- If the Left Internal IR Sensor sees the White Track, and both IRSS-10s see a Black Background, then the robot should Turn Left to re-align itself with the White Track.
- If the Right Internal IR Sensor sees the White Track, and both IRSS-10s see a Black Background, then the robot should Turn Right to re-align itself with the White Track.
- Otherwise, the robot should continue to go Forward.

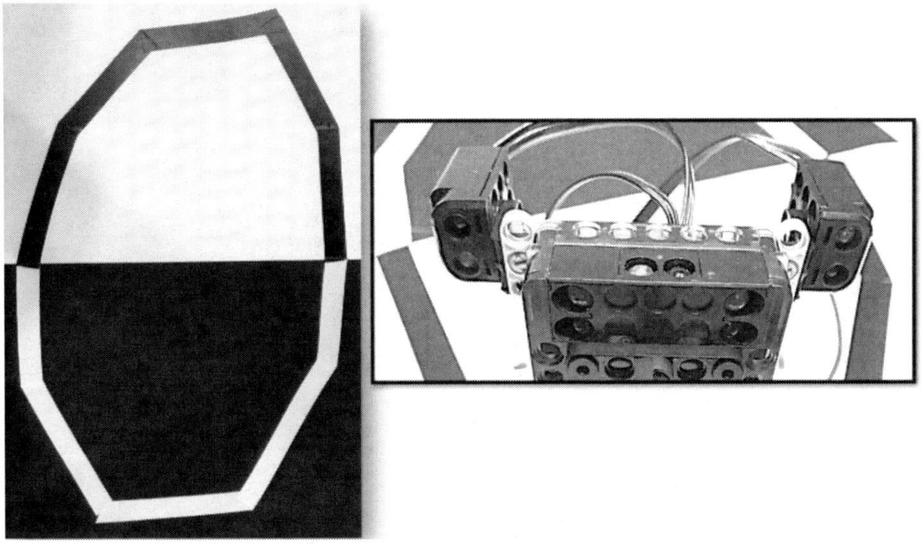

Fig. 2.56 Yin-Yang Line Track and the 4 IR Sensors used in "Yin_Yang_Line_Tracker.rpb".

Fig. 2.57 displays the Main Function used in the "Yin_Yang_Line_Tracker.rpb" program:

- Once again, the familiar IF-ELSE-IF structure is used, but the Conditional Expressions are more complex, as for each Conditional Expression 3 IR Sensors would need to be checked against their respective Threshold Values, which were determined using a similar code as shown in Fig. 2.54. The reader can follow the same procedure to determine his/her appropriate Threshold Values.

As the reader might have guessed, the run time performance for the "Yin_Yang_Line_Tracker.rpb" program was worse as compared to the "Line_Tracker.rpb" program as the Data Input requirements had doubled. At the Speed setting of 22%, the YYLT robot did not stay "On-Track" for very long. When the author reduced the Speed Power setting to a value below 22% in an attempt to reduce the required Data Input rate, but then this lower setting was not enough to let the robot roll forward or perform tracking maneuvers!

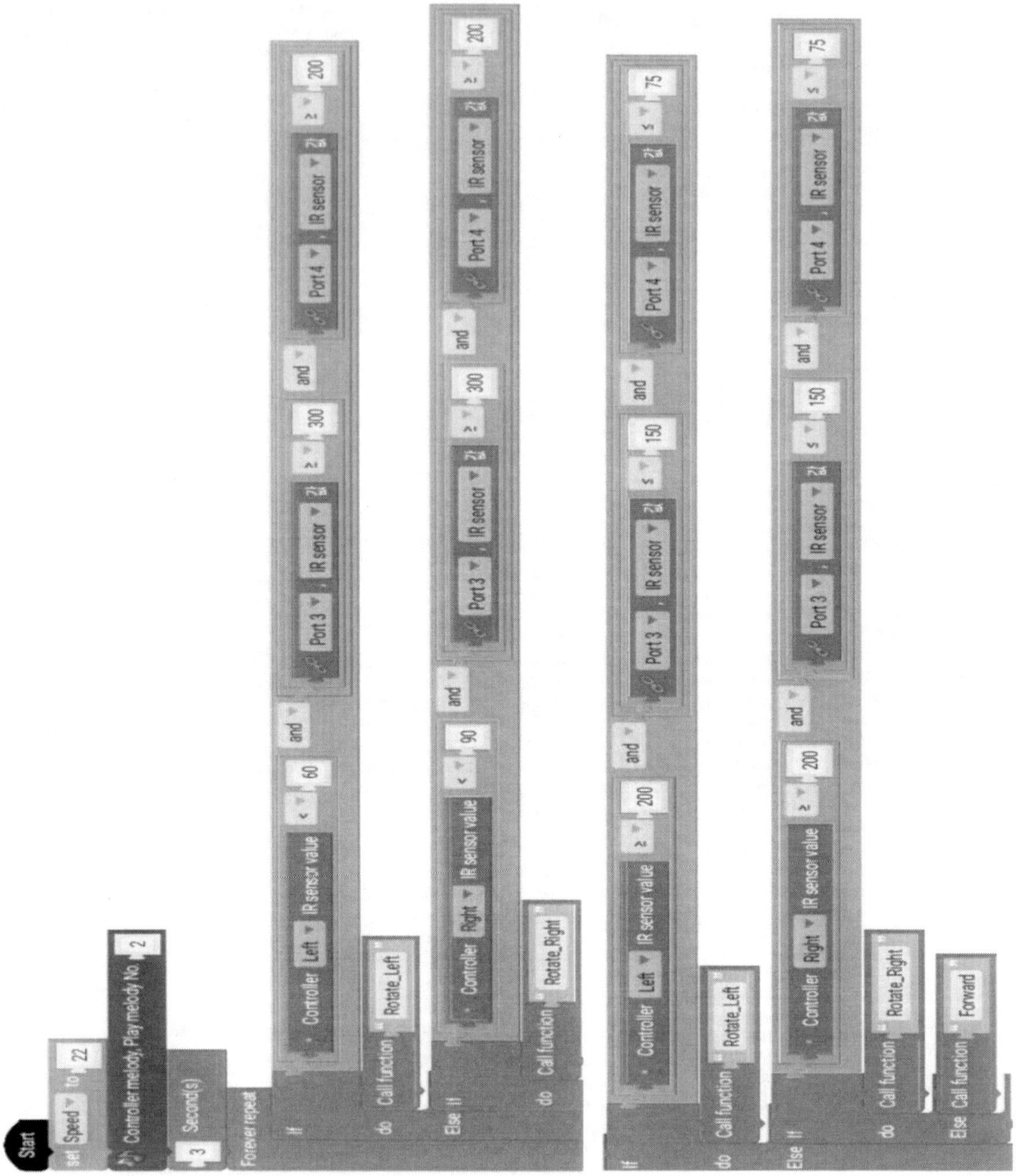

Fig. 2.57 Main Function used in "Yin_Yang_Line_Tracker.rpb".

As a comparison, the reader may check out the equivalent "Yin-Yang Line Tracker" solution using MIT SCRATCH-2 + EDBOT running on a Desktop PC (https://www.youtube.com/watch?v=HEZqA5PvYxs).

In summary for Section 2.6, for the "Chromebook-R+Block-Dream II" Computing System is not currently well suited for robotic projects that require Fast Sensor Inputs and Responsive Motor Actions due its ISTD and OSTD issues.

2.7 Position Control of Servo Motors

The DREAM II School Set comes with 1 Servo Motor SM-10 (https://emanual.robotis.com/docs/en/parts/motor/servo_motor/) which can be programmed to **hold its horn** at a given **Goal Position** between **-150** and **+150** degrees (see Fig. 2.58) – this is called **Position Control Mode**:

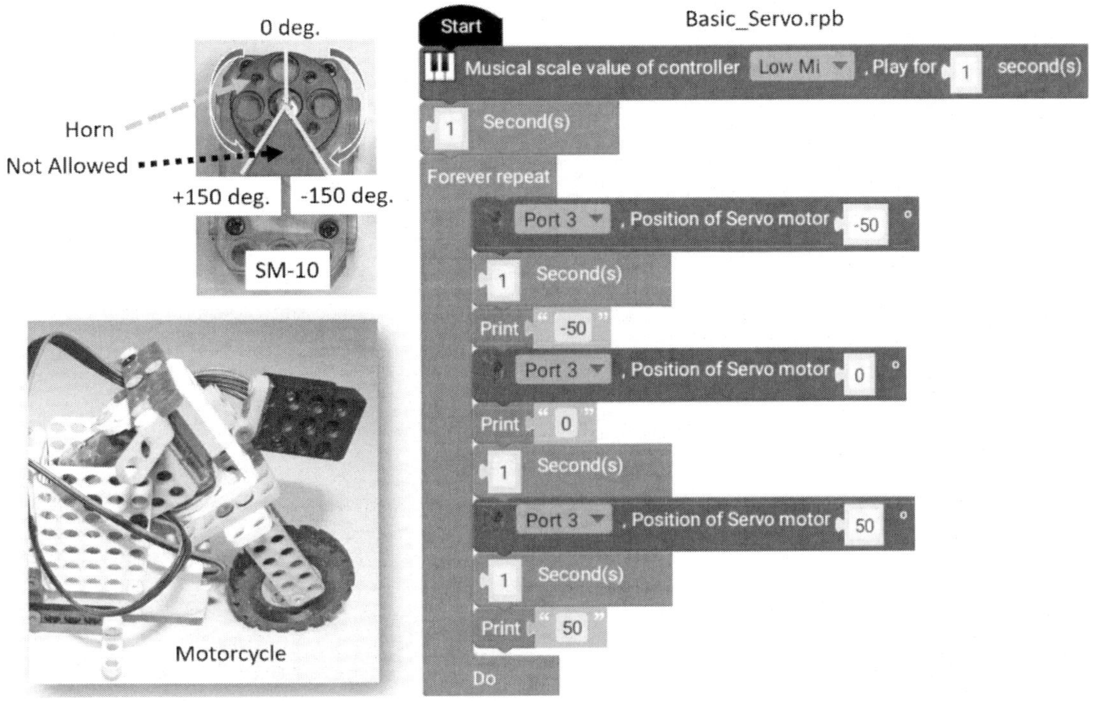

Fig. 2.58 Servo Motor SM-10: Features and Use Example as Steering Column for Motorcycle.

- Please note that there is a "Forbidden Zone" spanning 60 degrees that cannot be used by the SM-10. The reader needs to take this issue into account when designing his/her own robot.
- On the CM-150/151, the SM-10 can be used out of Ports 3 and 4, and it can be used in Position Control Mode as previously described, or in Continuous Turn Mode just like the GM-10 motors out of Ports 1 and 2, but in this Book, we will work only with the Position Control Mode.
- The current version of R+BLOCK only supports one type of command for the SM-10 in Position Control Mode, which is to **Set** its **Goal Position** in positive or negative degrees (see insert "Basic_Servo.rpb" in Fig. 2.58).
- The SM-10 has two more Position Control Features that are currently accessible **only** via TASK or PYTHON programming (see Chapters 3 and 4):
    - Its Goal Velocity, i.e., rotational speed (0-100%).
    - Its Present Position [-150 to 150 degrees].

The reader can use the example code "Basic_Servo.rpb" on an SM-10 just by itself to see how the Position Control Mode works. Please note that the SM-10 works only at **100% speed** when used in Position Control Mode with R+BLOCK. Later to be shown in Chapters 3 and 4, this rotational speed can be changed in TASK and PYTHON programs.

2.7.1 Smart TriCycle

The author chose the **TriCycle** robot to illustrate the use of the SM-10 to control its Steering Wheel (see Fig. 2.59 – more construction details are shown in Appendix B). Essentially, the Tricycle's backend is the same as for the Avoider. The SM-10 is connected to the Controller's front, and then a Steering Wheel unit is mounted onto its horn. The Steering Wheel unit uses a Passive Rubber Tire and also has an IRSS-10 Sensor mounted on it.

Fig. 2.59 TriCycle Robot.

The author developed the "Smart_Tricycle.rpb" Project, which has the same overall Operational Goals as the "Smart_Avoider.rpb" Project (see Sub-Section 2.6.3), to illustrate the adaptations/solutions in Mechanical Engineering and Robotics Programming issues that differ between the two robots:

- The Avoider robot used two IR Sensors to determine the Presence and Location of an Obstacle in Front, on the Left or on the Right sides of the robot. The Tricycle robot will need to provide the same operational capabilities, but with a Servo Motor and only 1 IR Sensor.
- The reader probably had noticed that the Avoider robot was quite maneuverable due to its small/compact shape of its 3 Ground Contact Points (see top-right picture insert in Fig. 2.60). On the other hand, the TriCycle has a more elongated shape for its 3 Ground

Contact Points, especially its Front Contact Point is now a Rubber Tire with a larger ground contact surface than the small plastic "knob" that was used on the Avoider. Thus, the TriCycle will have to deal with a larger Ground Frictional Force (due to the Rubber Tire) which is also located with a longer Moment Arm from the Powering Motors located in its backend. Thus, the author has to use two Speed Variables in Function **Set_Speed()** (top left in Fig. 2.60):

- **Speed** (=25) which is used for going Forward and Backward while the Steering Wheel is set Straight Forward, which yields smaller rolling resistance with the ground surface.
- **Turning Speed** (=40) which is larger to compensate for the larger rolling and sliding resistances between the Steering Wheel and the ground surface, due to its being set at an angle with respect to the longitudinal axis of the robot.

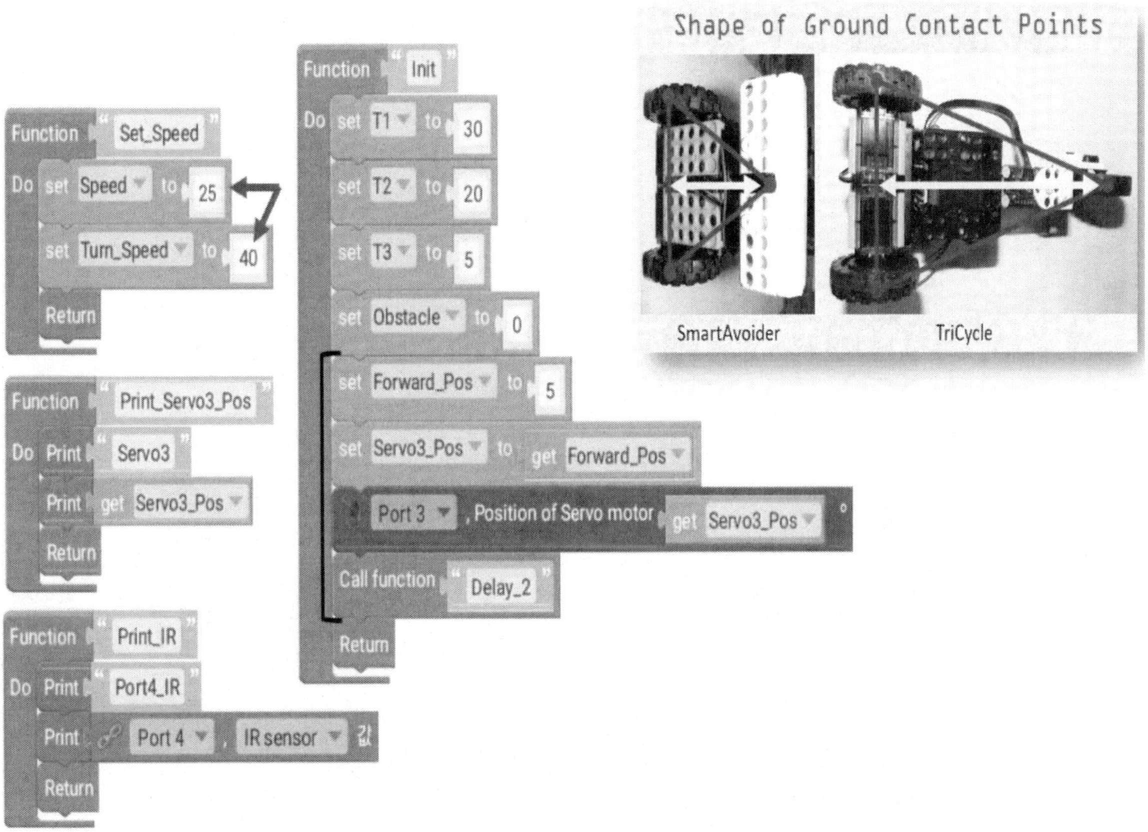

Fig. 2.60 Selected Utilities Functions used in "Smart_Tricycle.rpb".

- Function **Init()** in Fig. 2.60 shows previously used Variables such as **T1**, **T2**, **T3** and **Obstacle**. It also shows two new Variables:
  - **Forward_Pos** set to 5, **when theoretically it should be set to 0**, but due to Steering Wheel's construction inaccuracy and "wear and tear" on the SM-10 used by the author, this variable was needed. For the reader's robot, he/she can use the "Basic_Servo.rpb" code to find the appropriate value, but please make sure

that the TriCycle is placed on the **actual practice ground surface** when performing this determination.
  - Because R+BLOCK does not currently provide a direct Block/Command to read the **Current Present Position** of the SM-10, the author developed a "work-around" solution by using Variable **Servo3_Pos** which serves as a **Goal Position** value to be set to the SM-10's horn, but it can also serve as an "approximate" **Present Position** of the SM-10's horn (which is an information much needed in developing the RC Maneuvering and Obstacle Evading algorithms later).
  - Please also note the **required Delay()** Function used, each time that a Goal Position is set to the SM-10 on Port 3.

Fig. 2.61 displays the Main Function used in "Smart_Tricycle.rpb", which practically is the same as for the Avoider (see Fig. 2.50), except for the Conditional Expression used for the detection of an obstacle, which now depends only on one IRSS-10 Sensor connected to Port 4 of the Controller.

Fig. 2.61 Main Function used in "Smart_Tricycle.rpb".

Let us have a closer look at Function **Clap_RC()** (see Fig. 2.62):

- It still uses the same pattern of 1/2/3/4 clap-events to correspond to Forward/Backward/Steer_Left/Steer_Right actions, but the actual robot maneuvers are now quite different:

- “3-claps” would now only rotate the Steering Wheel Left (+10 degrees) for each “3-claps” detection and the rear GM-10 motors are not activated during this maneuver (see Function **Steer_Left()**).
- Likewise, “4-claps” would only rotate the Steering Wheel Right (-10 degrees) for each “4-claps” detection and with no activation of the rear GM-10 motors (see Function **Steer_Right()**).

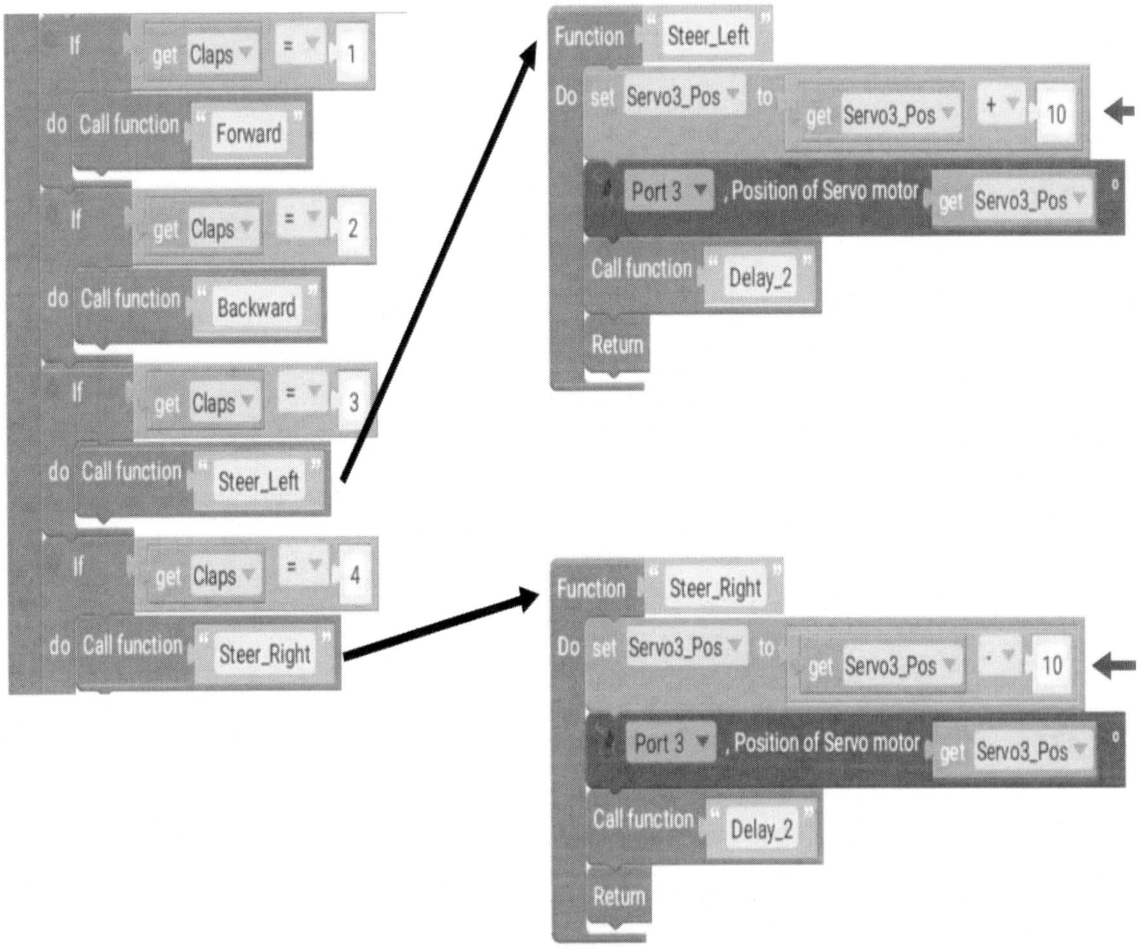

Fig. 2.62 Clap Control Pattern used for Function **Clap_RC()** in “Smart_Tricycle.rpb”.

Functions **Forward()** and **Backward()** are now much more complex, Figs. 2.63 and 2.64 will be used to explain the programming steps used by Function **Forward()**. The reader can review similar steps used in Function **Backward()** by reading the provided example code “Smart_Tricycle.rpb”.

Fig. 2.63 lists the complete Function **Forward()**:

- It contains an IF-ELSE-IF structure to implement 3 Event/Action Pairs that depend on the Current Value of Variable **Servo3_Pos**:

- When the Steering Wheel is found to be set "more or less" Straight Forward, that is within ± 5 degrees along the robot's longitudinal axis, the rear GM-10 motors are activated using **Speed** to make the robot roll Forward (i.e., same as for the Avoider).
- When the Steering Wheel is found to set at some Angle (more than 5 deg.) to the Left of the robot's longitudinal axis, the robot should Rotate Left (see Fig. 2.64).
- And when the Steering Wheel is found to set at some Angle (more than 5 deg.) to the Right of the robot's longitudinal axis, the robot should Rotate Right (also see Fig. 2.64).

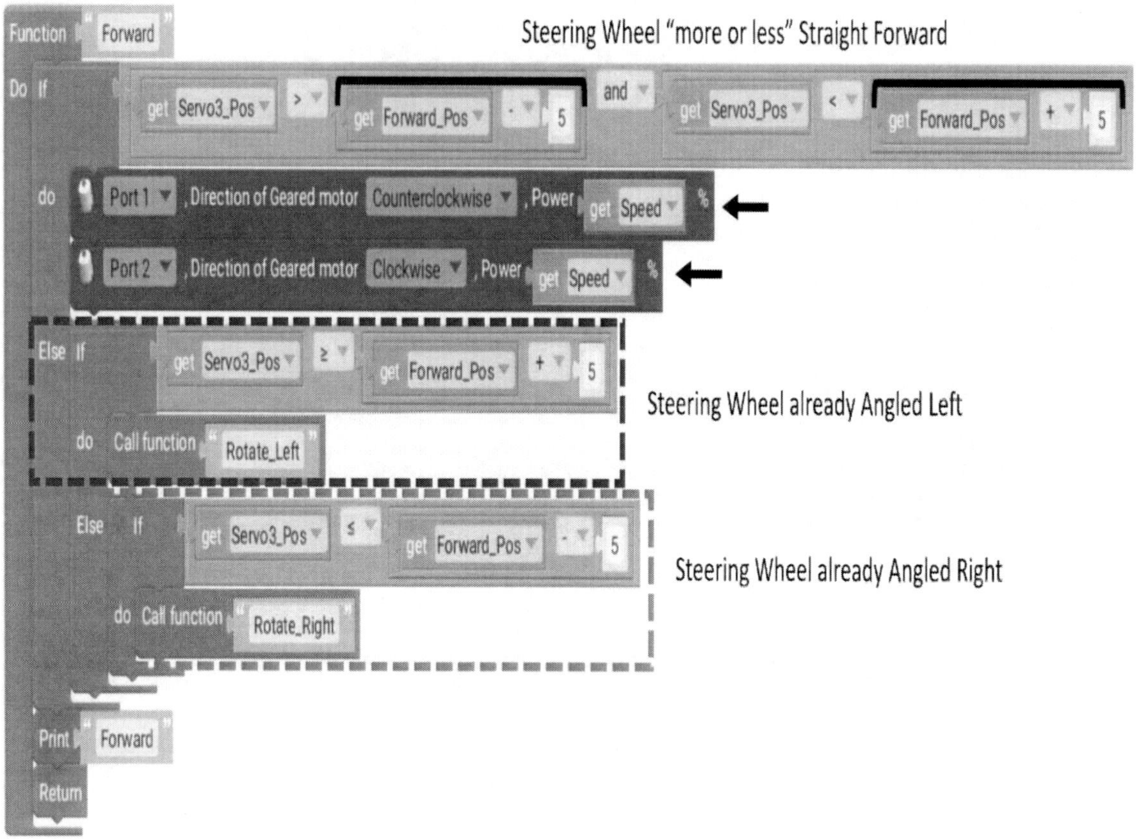

Fig. 2.63 Function **Forward()** used in "Smart_Tricycle.rpb".

Fig. 2.64 lists more details for Functions **Rotate_Left()** and **Rotate_Right()**:

- These Functions are constructed similarly to the ones used for the Avoider (see Fig. 2.20), except now Variable **Turn_Speed** is used to handle the higher frictional forces at play due to the use of a Steering Wheel.

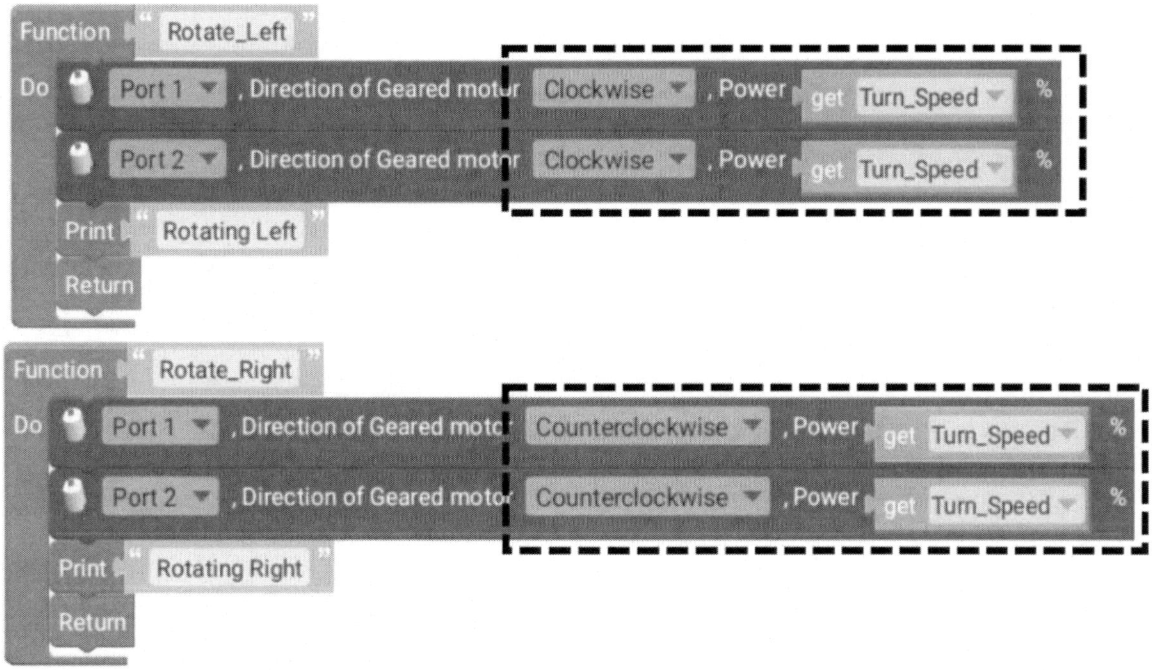

Fig. 2.64 Functions **Rotate_Left()** and **Rotate_Right()** in "Smart_Tricycle.rpb".

Let us next have a closer look at Function **Escape()** (see Fig. 2.65 for Part 1) which starts by setting Variable **Obstacle** to 1, next **stops** the robot, and then prints out the Current Value of Variable **Servo3_Pos**. Variable **Servo3_Pos** serves as the **Key** information needed to determine the appropriate robot maneuvers to be implemented next:

- Let us recall that Function **Escape()** is activated when Port 4 IR Sensor reports a Value that is greater or equal to **T3** (=30), i.e., **when an obstacle is detected** (see Fig. 2.61). However, because of the use of a Single IR Sensor to do the Obstacle Detection work, much is still "**unknown**" about the "**geometrical**" relationship between the robot and the obstacle at this point in time, such as:
    - Is the Steering Wheel angled to the Right or to the Left of the robot's Longitudinal Axis?
    - Or is the Steering Wheel set Straight Forward? In this case, is the robot Longitudinal Axis "perpendicular" to the Obstacle Plane? Or is this Axis at an angle to this plane, and then which way? To the Left or to the Right?
    - All these possible Event/Action Pairs will of course be implemented with a big IF-ELSE-IF structure, with Part 1 of it shown in Fig. 2.65.
- Fig. 2.65 illustrates the first two Event/Action Pairs dealing with the Steering Wheel's angle with respect to the robot's Longitudinal Axis:
    - If **Servo3_Pos** reports a Value more than 5 degrees to the Right of its **Forward_Pos** then the robot might as well keep on "Escaping to the Right".
    - Conversely, if **Servo3_Pos** reports a Value more than 5 degrees to the Left of its **Forward_Pos** then the robot might as well keep on "Escaping to the Left".

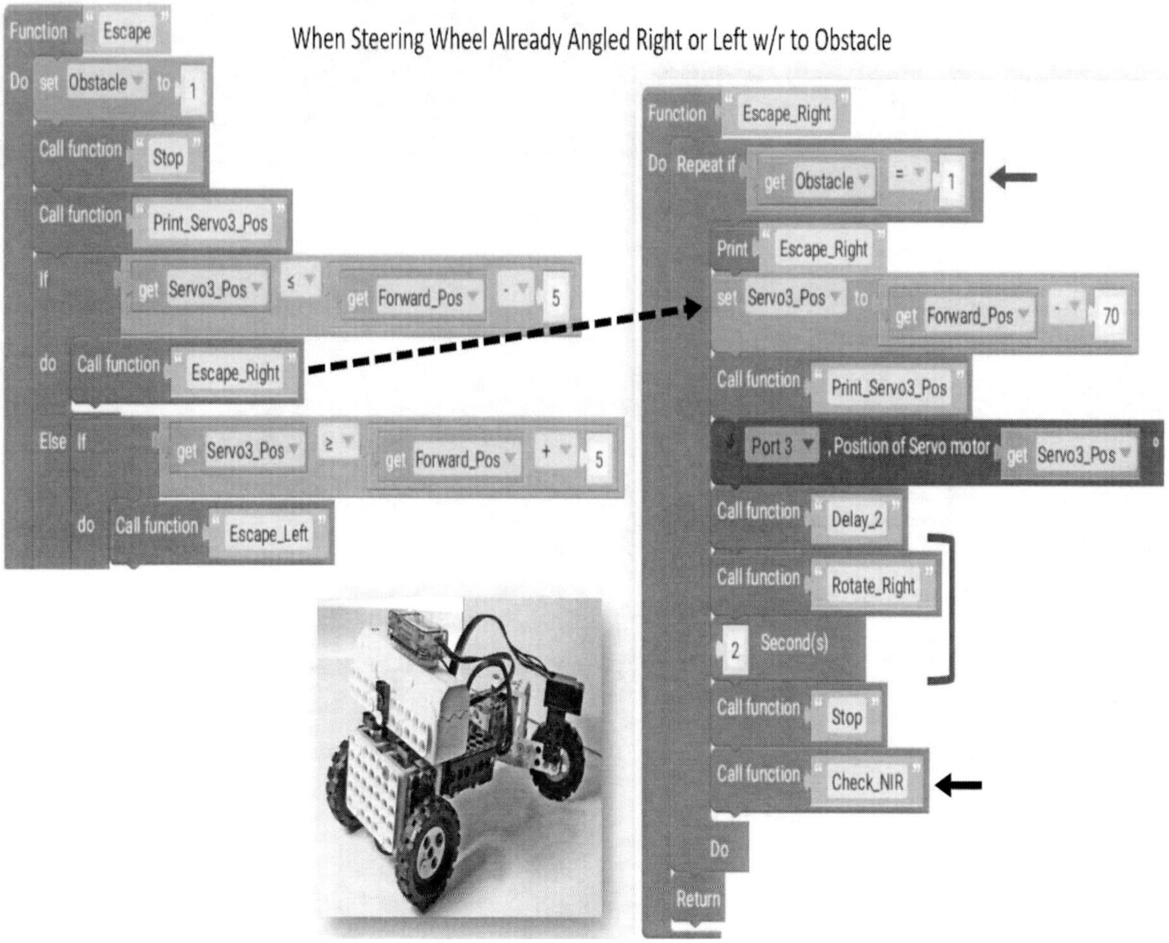

Fig. 2.65 Part 1 of Function **Escape** used in "Smart_Tricycle.rpb".

- For example, Fig. 2.65 also shows the details for Function **Escape_Right()**:
  - This Function is a REPEAT-IF-DO Loop that depends on Variable **Obstacle** (which is already set to 1 previously when an obstacle is first detected). Therefore, this Loop would be activated at this point in time.
  - Next, it cranks the Steering Wheel further to the Right (70 degrees to the Right of its **Forward_Pos** angle, i.e., to the Right of the robot's longitudinal axis).
  - After a short wait/delay to let Servo 3 get to this Goal Position, the robot Rotates Right for 2 seconds (the reader can also use shorter or longer time delays depending on the Value used for Variable **Turn_Speed**). The robot then stops.
  - Next, Function **Check_NIR()** is used (see Fig. 2.66) to determine if the obstacle is still somewhere in the robot Line Of Sight (LOS):
    - First, the robot straightens the Steering Wheel to its **Forward_Pos** value and waits a bit for Servo 3 to get there, before reading the Current Value reported by Port 4 IR Sensor.
    - If Port 4 IR Sensor reports a Value greater than **T3** (=5), the obstacle is "still there", so **Obstacle** is set to 1, otherwise it is set to 0.

- At this point, Function **Check_NIR()** returns (with **Obstacle** set to either 1 or 0) to the End of the REPEAT-IF-DO Loop of Function **Escape_Right()**, which then loops back to its beginning and checks on the Current Value of **Obstacle**. If (**Obstacle == 1**), then the robot would Rotate Right for another 2 seconds and check upon Port 4 IR Sensor once more. Eventually, the robot should be able to evade the obstacle, then **Obstacle** is set to 0, then the robot would get out of Function **Escape()** to be back in the Main Endless Loop, ready to be Remote-Controlled by the Operator once again.

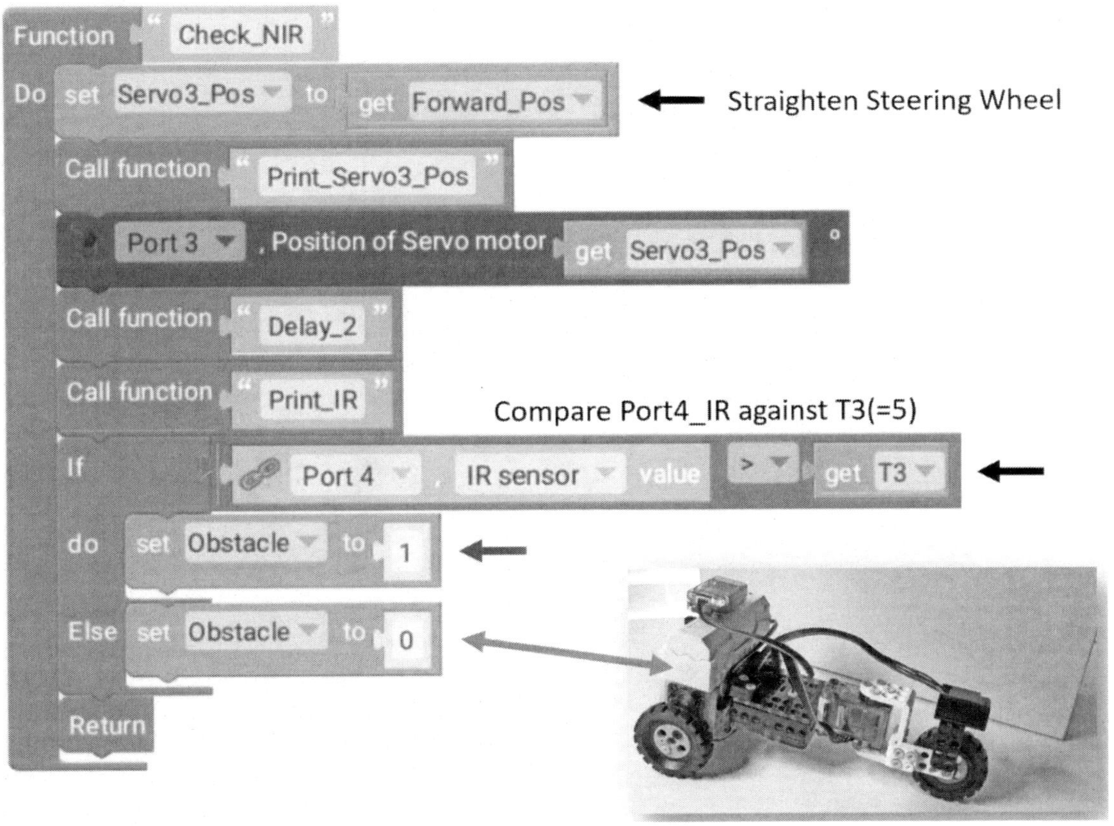

Fig. 2.66 Function **Check_NIR()** used in "Smart_Tricycle.rpb".

Figs. 2.67, 2.68 and 2.69 describe the programming steps used when the robot encounters the obstacle with its Steering Wheel set Straight Forward (i.e., Part 2 of Function **Escape()**), that is when the robot is aligned with its Longitudinal Axis, which can form 3 possible angles with the Obstacle Plane:

- Is the robot's Axis at an acute angle to the Obstacle Plane with respect to its Left side? (Case A).
- Or is it at an acute angle to the Obstacle Plane with respect to its Right side? (Case B).
- Or is it pretty much perpendicular to the Obstacle Plane? (Case C).

These three "mutually exclusive" Event/Action Pairs will thus be implemented as a Nested IF-ELSE-IF structure within the Outer IF-ELSE-IF structure already described in Fig. 2.65.

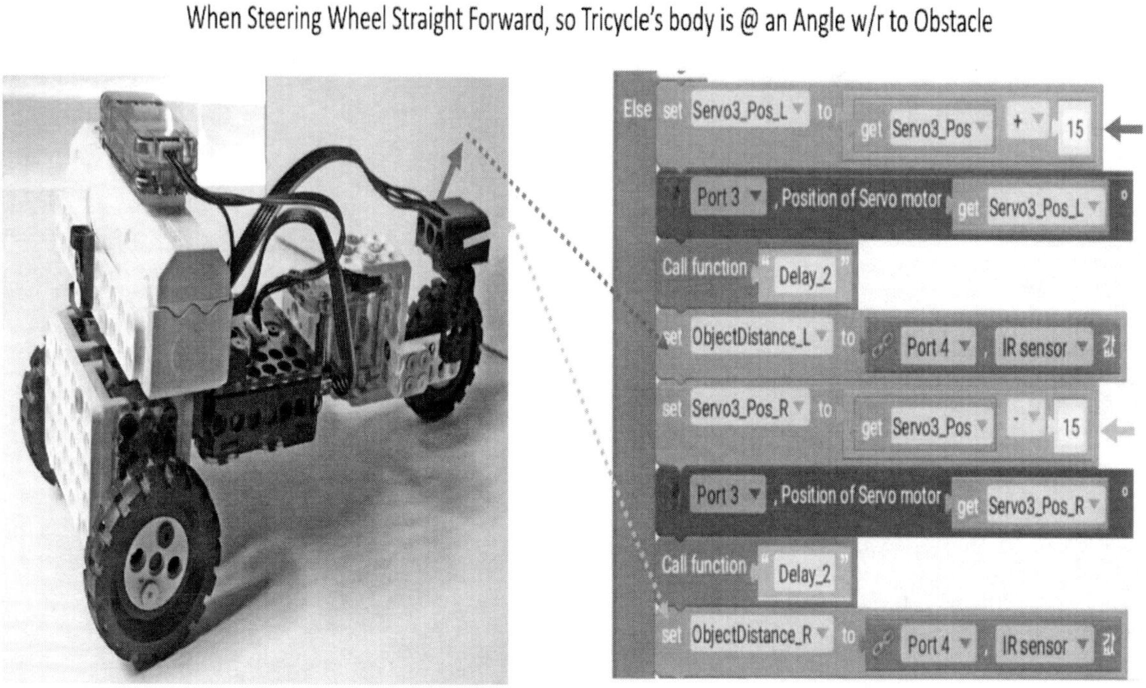

Fig. 2.67 Part 2-A of Function **Escape()** used in "Smart_Tricycle.rpb".

Fig. 2.67 shows Part2-A of Function **Escape()** whereas the robot would swing the Steering Wheel 15 degrees to the Left from its current position **Servo3_Pos**, wait a bit and record the Current Value of Port 4 IR Sensor into Variable **ObjectDistance_L.** Next, the robot performs similar operations but for the Right side and records the Current Value of Port 4 IR Sensor into Variable **ObjectDistance_R.** Now, the robot has enough information to figure out which one of the previous three Event/Action Pairs apply at a particular point at run time.

Fig. 2.68 displays Part 2-B of Function **Escape** which lays out the steps used for Case (A) and Case (B):

- If **(ObjectDistance_L > ObjectDistance_R)**, meaning that the Obstacle Plane is closer to the robot's Left than to its Right, the robot should then **Escape Right** (Case A).
- Else, if **(ObjectDistance_R > ObjectDistance_L)**, meaning that the Obstacle Plane is closer to the robot's Right than to its Left, the robot should then **Escape Left** (Case B).

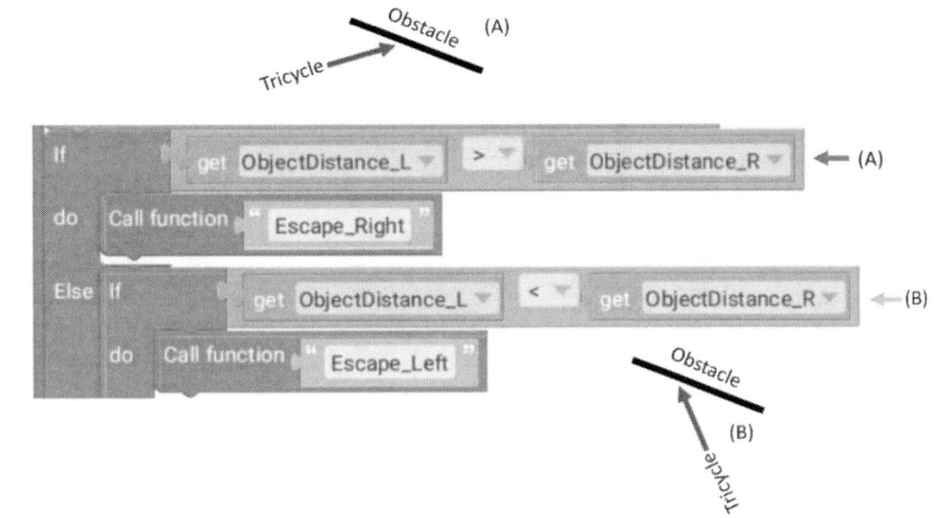

Fig. 2.68 Part 2-B of Function **Escape()** used in "Smart_Tricycle.rpb".

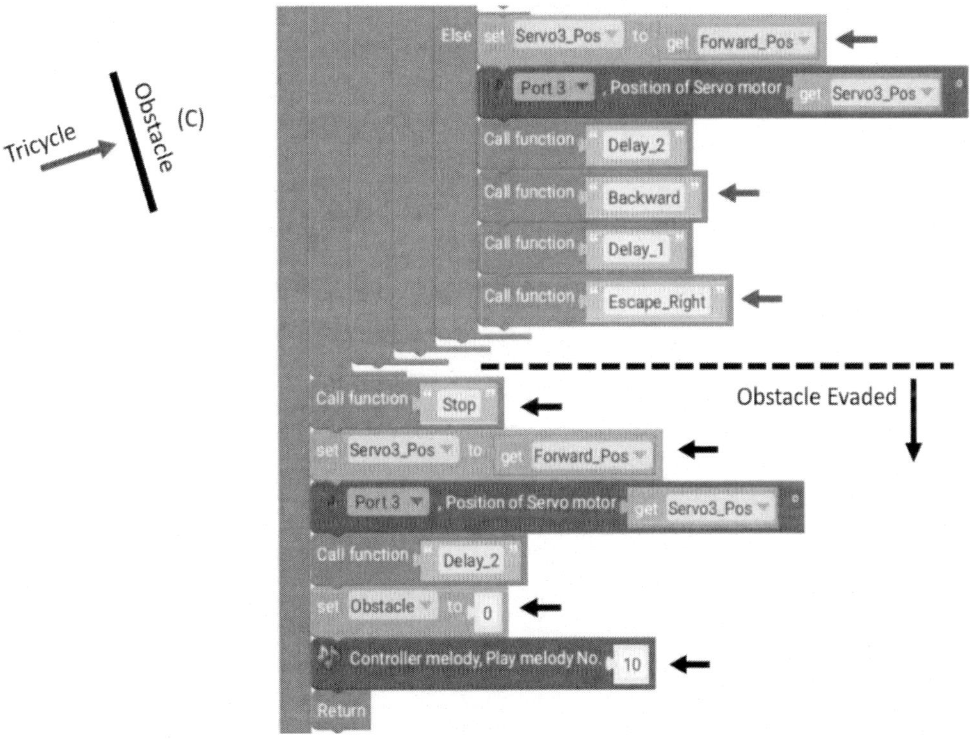

Fig. 2.69 Part 2-C of Function **Escape()** used in "Smart_Tricycle.rpb".

If both previous situations do not apply, meaning that the robot's Axis is mostly perpendicular to the Obstacle Plane (Case C), then the robot should roll **Backward** a bit and **Escape Right** (see Fig. 2.69 for Part 2-C of Function **Escape()**). Of course, for this last step, the robot can **Escape Left** also if the reader prefers.

Once the robot evades the Obstacle, it would stop and straighten it Steering Wheel. Lastly, it sets **Obstacle** to 0, and plays Melody 10. And then back to the Main Endless Loop as shown in Fig. 2.61.

Fig. 2.70 shows a sequence of pictures recording an actual Obstacle Evading Event.

Fig. 2.70 An example Obstacle Evading Event using program "Smart_Tricycle.rpb".

2.7.2 Dowel Scanner

In this last project using R+BLOCK, the author developed a Dowel Scanner robot as shown in Fig. 2.71 by using a different mechanical configuration of the Port 3's SM-10 and the Port 4's IRSS-10 (more construction details are available in Appendix B).

For this project, a primary requirement is for the IRSS-10 to detect the Centerline of a White Dowel while the SM-10 is set to rotate from -50 to 50 degrees. Recalling that at present R+BLOCK only allows rotating the SM-10 at full speed, the author first used the program "Servo_Scan.rpb" to ascertain if this basic task can be accomplished. As shown in the Print-Out on the left side of Fig. 2.72, this "100% Power" scanning procedure does not provide enough time elapsed for reading in even just a few "meaningful" Port 4 IR Sensor data. Thus, this first approach was not feasible at present for R+BLOCK.

Fig. 2.71 Dowel Scanner Robot.

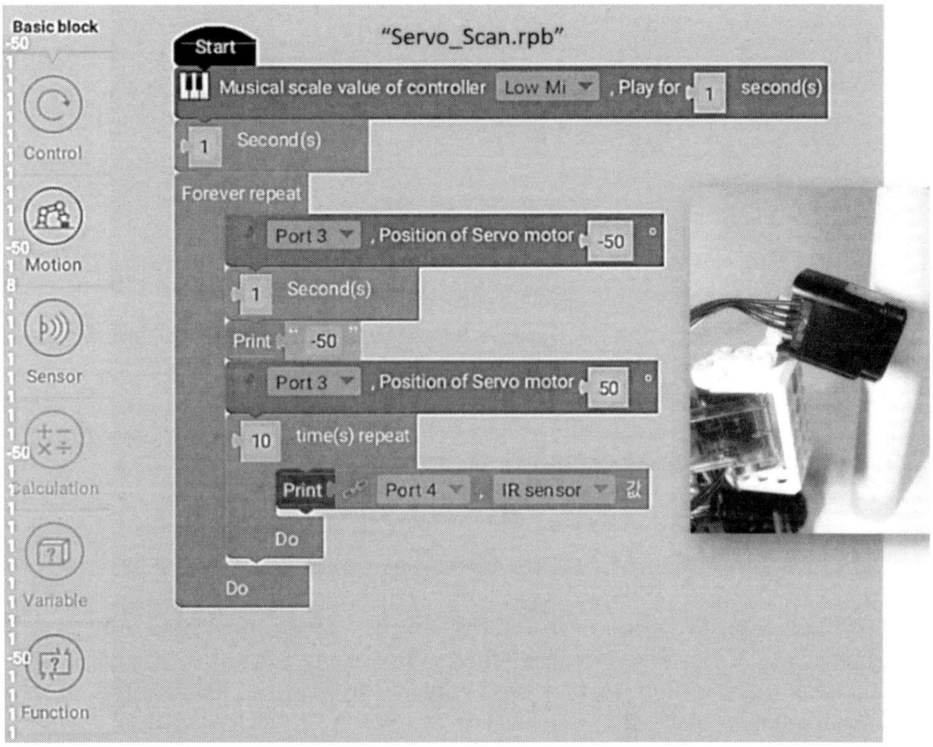

Fig. 2.72 Runtime Performance of "Servo_Scan.rpb".

The author's second approach used a "Go-Stop-Scan" type of procedure implemented as "Discrate_Servo_Scan.rpb" (see Fig. 2.73) whereas:

- Port 3 Servo would still scan from -50 degrees (Right) to 50 degrees (Left), but only by **3 degrees each time**. For each time, it would stop, and let the Port 4 IR Sensor read in and print its Current Value. Then it would rotate 3 more degrees over to the Left.

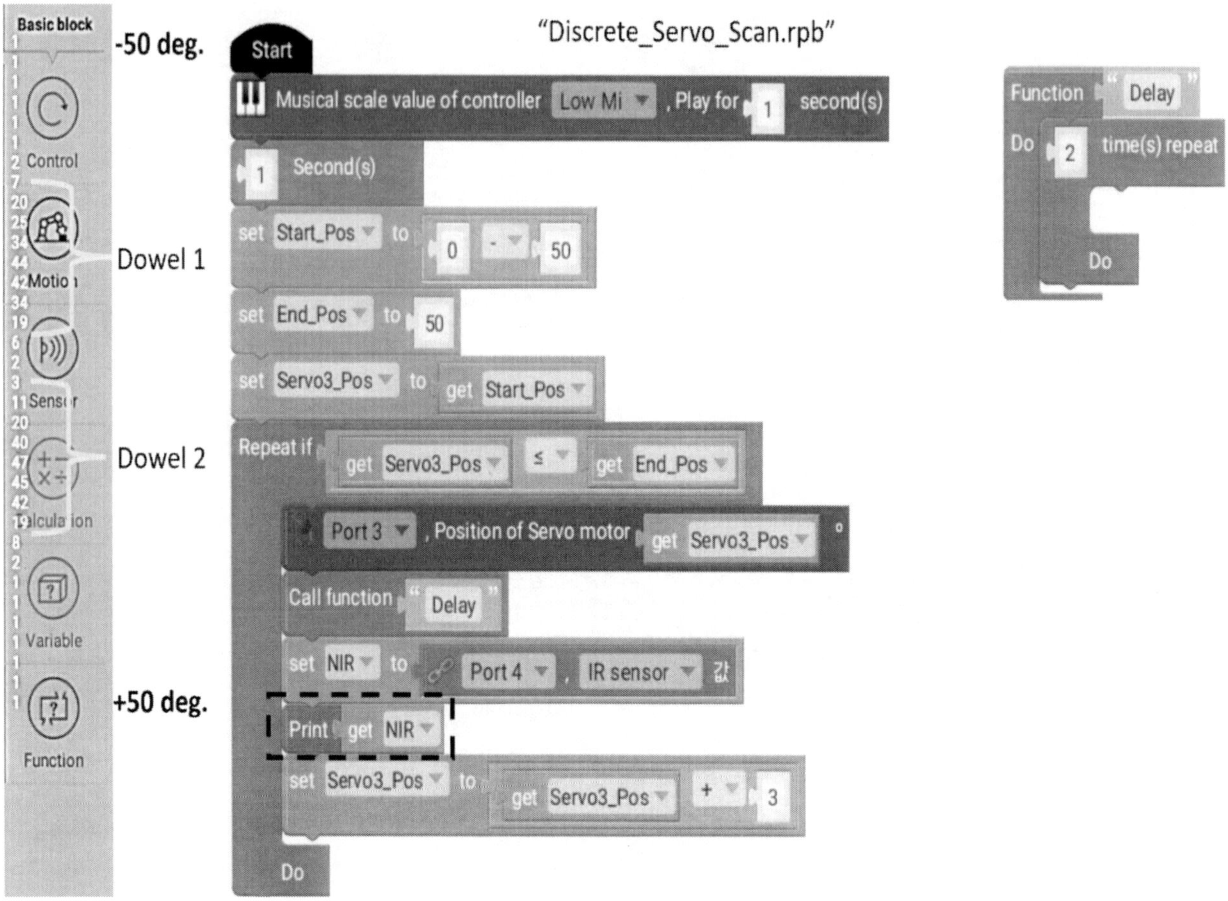

Fig. 2.73 Runtime Performance of "Discrete_Servo_Scan.rpb".

- The left-most column of runtime "numbers" in Fig. 2.73 shows that this approach worked well:
  - The series of "1" shows the "Background" IR Sensor readings.
  - Then the IR Sensor's values started to increase as the "Right Edge" of Dowel 1 is picked up, but once passed its Center Line, these IR values started to decrease.
  - Next, a few "low" readings were recorded, corresponding to the "empty space" between Dowel 1 (just passed) and Dowel 2 (about to be scanned). Please note that the IR "sensing beam" of the IRSS-10 is rather "wide".
  - Once again, the IR Sensor's values started to increase as the "Right Edge" of Dowel 2 is picked up, but once passed its Center Line, these IR values started to decrease back to the "Background" IR readings.

Based on these results, the author developed a Dowel-Scan algorithm based on 5 Zones as depicted in Fig. 2.74:

- First, when Port 3 Servo gets set to the Goal Position of **-50 degrees**, it is assumed that Port 4 IR Sensor would be "staring" at "empty space", thus **BackgroundNIR** is recorded and **DowelThreshold** is next defined based on this value of **BackgroundNIR**. The "Go-Stop-Scan" procedure is then applied to make Servo 3 rotate from -50 degrees to +50 degrees, by 3 degrees each time.

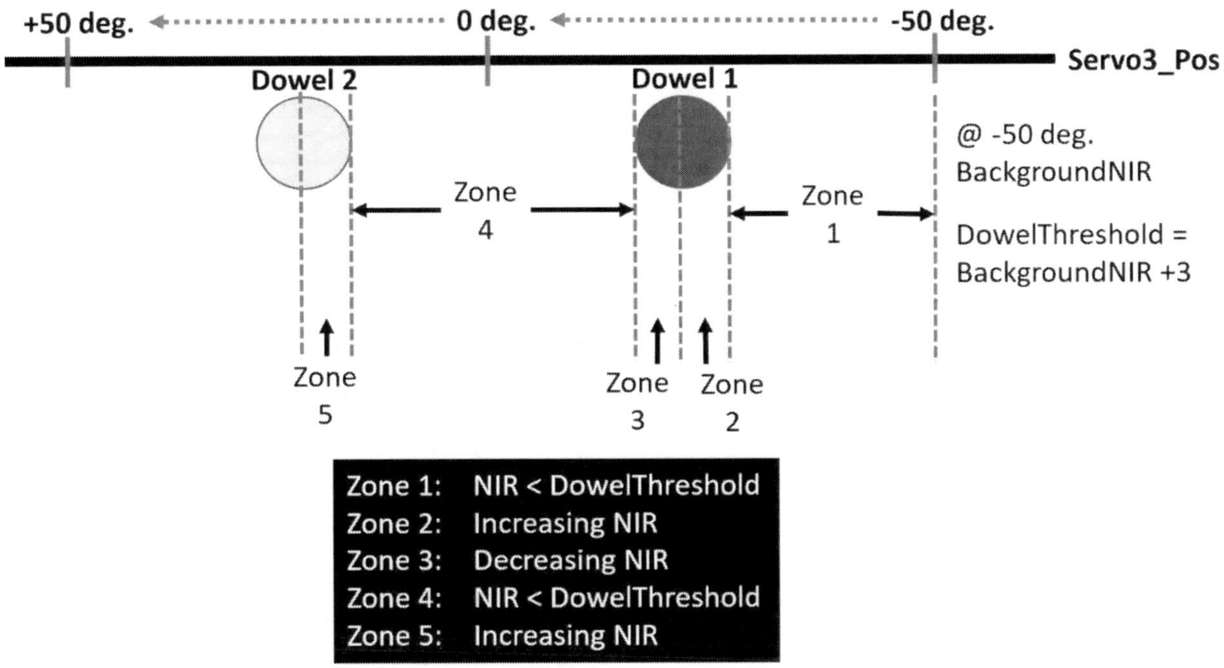

Fig. 2.74 Dowel-Scan Algorithm based on 5 Travel Zones used in "Dowel_Scan.rpd".

- During the Servo 3's travel through Zone 1, Port 4 IR Sensor would see/record "empty space", thus its Current **NIR** readings would be "less than" **DowelThreshold**.
- Next, Zone 2 would produce a pattern of increasing **NIR** values, while Zone 3 would yield a pattern of decreasing **NIR** values. The transition between Zone 2 and Zone 3 would be considered as the "approximate" Center Line of Dowel 1 (via Servo 3's Present Position).
- In Zone 4, the IR Sensor would see/record "empty space" again, thus its Current **NIR** readings would be "less than" **DowelThreshold** (like Zone 1).
- Lastly, Zone 5 would produce a pattern of increasing **NIR** values like Zone 2. The "end" of Zone 5 would be considered as the "approximate" Center Line of Dowel 2 (via Servo 3's Present Position). And the Dowel-Scan algorithm ends here, as the goal of determining the locations of the two Dowels is accomplished.

Figs. 2.75 through 2.82 describe various Functions used in the Project "Dowel_Scan.rpb":

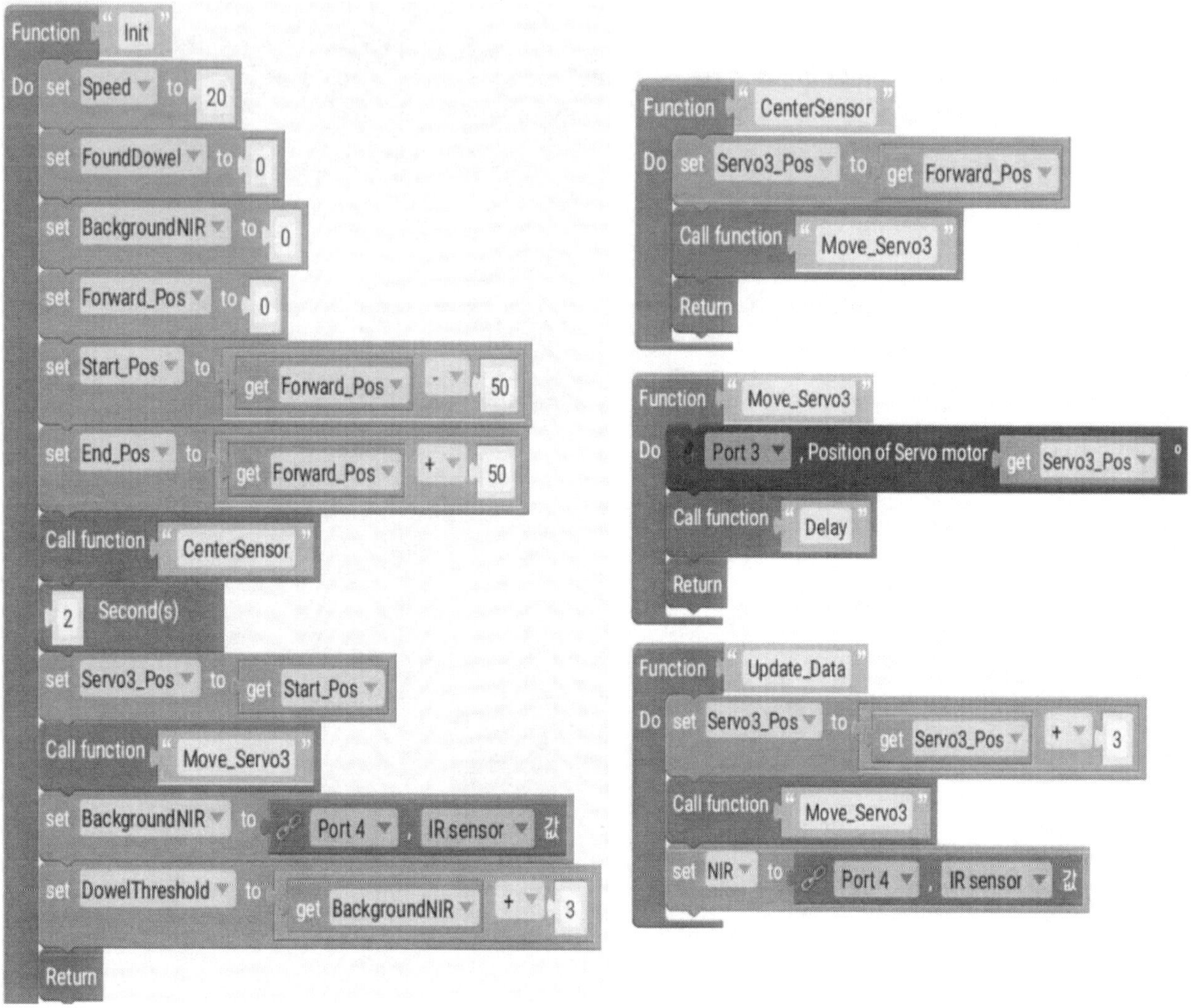

Fig. 2.75 Functions **Init(), CenterSensor(), Move_Servo3()** and **UpdateData()** used in "Dowel_Scan.rpb".

- Function **Init()** sets up initial values for various Variables used such as **Speed, FoundDowel, BackgroundNIR, Forward_Pos, Start_Pos, End_Pos. Servo3_Pos** and **DowelThreshold**.
- Function **CenterSensor()** is used in conjunction with Variable **Forward_Pos** to let the Operator verify that, at this point, the robot has the IR Sensor aimed squarely Forward. The reader's robot may have to use a value different than 0 as used herein by the author.
- Servo 3 is then moved to **Start_Pos** and **BackgroundNIR** is read in at this servo position, next it is used to compute **DowelThreshold** (= **BackgroundNIR + 3)**.
- Function **Move_Servo3()** shows that after setting Servo 3's Goal Position to the Current Value of **Servo3_Pos,** a short time delay is used via Function **Delay()** (already

described back in Fig. 2.51) to wait for Servo 3 to be "settled down". Usually, after this move, Port 4 IR Sensor is used to record its Current Value as Variable **NIR** as shown is Function **Update_Data()**.

Fig. 2.76 lists the steps used in the Main Function of "Dowel_Scan.rpb":

- Play **Low-Mi** for 1 second.
- Call Function **Init()** and play **Melody 1**.
- Then Function **ServoScan()** is used to determine where the two Dowels might be, and also which one is the "closer" one to the robot, which could be either Dowel 1 or Dowel 2 (see Fig. 2.74), as the dowels were placed at random in front of the robot at the start of a typical demonstration run. Function **ServoScan()** would "traverse" through the 5 Zones (see Figs. 2.77 through 2.80) as previously described in Fig. 2.74.
- Lastly, Function **VerifyDowels()** is used to make the robot straighten forward its IR Sensor, which is then used first to find the "closer" dowel and to approach it up to a certain distance, next it switches to the search of the "farther" dowel and also approaches it up to a certain distance (more details in Figs. 2.81 and 2.82).

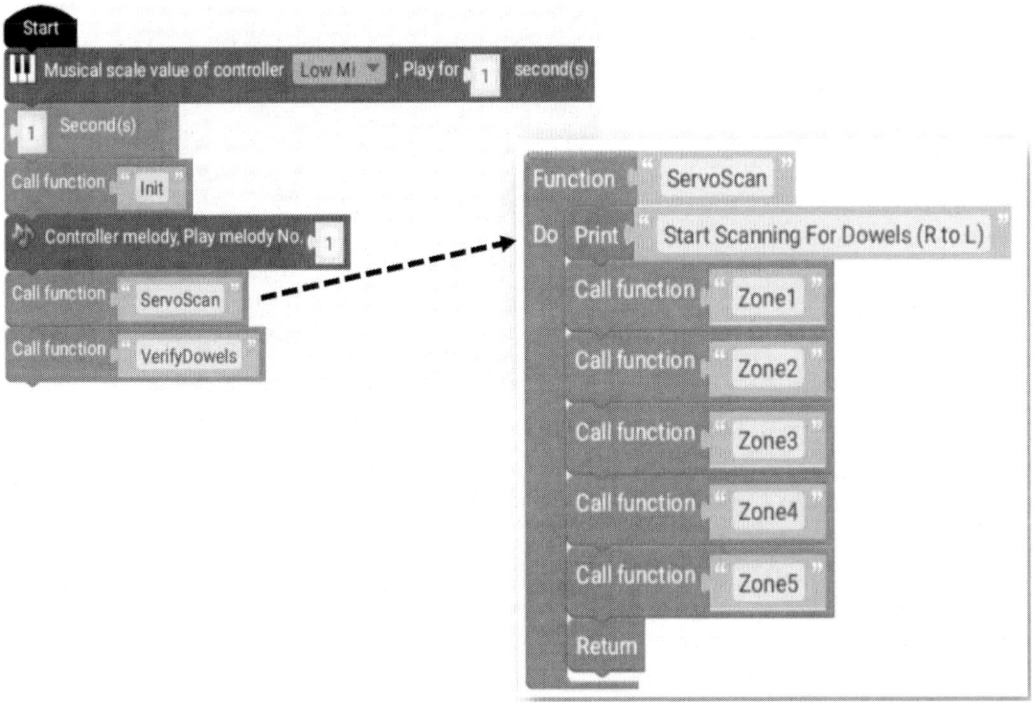

Fig. 2.76 Main Function and Function **ServoScan()** used in "Dowel_Scan.rpb".

Fig. 2.77 illustrates Function **Zone1()** where:

- First, "Zone 1" is printed out to notify the Operator that the robot is currently searching "Zone 1" (see Fig. 2.74).

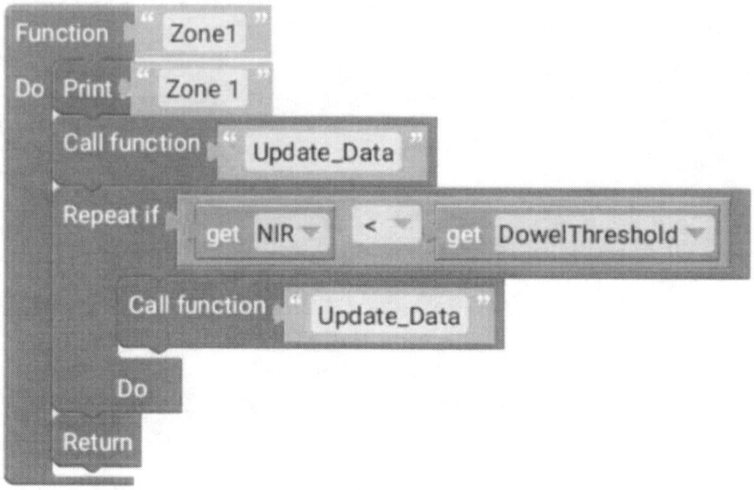

Fig. 2.77 Function Z**one1()** used in "Dowel_Scan.rpb".

- This travel/search sub-procedure is implemented as a REPEAT-IF-DO Loop that is controlled by the Conditional Expression **(NIR < DowelThreshold**), as "physically" Port 4 IR Sensor would see "empty space" within Zone 1. Please note that Function **Update_Data** (see Fig. 2.75) is invoked before entering the REPEAT-IF-DO Loop, and also inside this Loop's body. This is to ensure that, for each loop's iteration, Servo 3 is rotated Left by 3 degrees, so that Port 4 IR Sensor can provide its latest Value to be recorded in Variable **NIR**.
- This REPEAT-IF-DO Loop is designed to "exit", when Port 4 IR Sensor picks up a little bit of the Right Edge of Dowel 1, and this would make **NIR** become larger than **DowelThreshold** and this Conditional Expression would therefore evaluate FALSE.

Fig. 2.78 describes Function **Zone2()**:

- First, "Zone 2" is printed out to notify the Operator that the robot is currently searching "Zone 2" (see Fig. 2.74) with the goal of finding the Center Line of Dowel 1.
- Within Zone 2, physically, **NIR** increases its numerical value to a maximum value obtained "around" the Center Line of Dowel 1. Algorithmically, this search procedure needs to use two Variables: **NIR** to contain the "latest/current" value of Port 4 IR Sensor and **PreviousNIR** to contain the "older" value from the previous iteration of this search procedure.
- This travel/search procedure is implemented as another REPEAT-IF-DO Loop that is controlled by the Conditional Expression **(NIR > PreviousNIR**). Please carefully note that **PreviousNIR** is set to **BackgroundNIR** before entering this Loop's body. As this time, **NIR** is bound to be larger than **DowelThreshold** (see Fig. 2.77) which itself is 3-unit larger than **BackgroundNIR** (see Fig. 2.75). Therefore, the Conditional Expression **(NIR > PreviousNIR)** for sure would evaluate to TRUE at this point, and this Loop is entered.

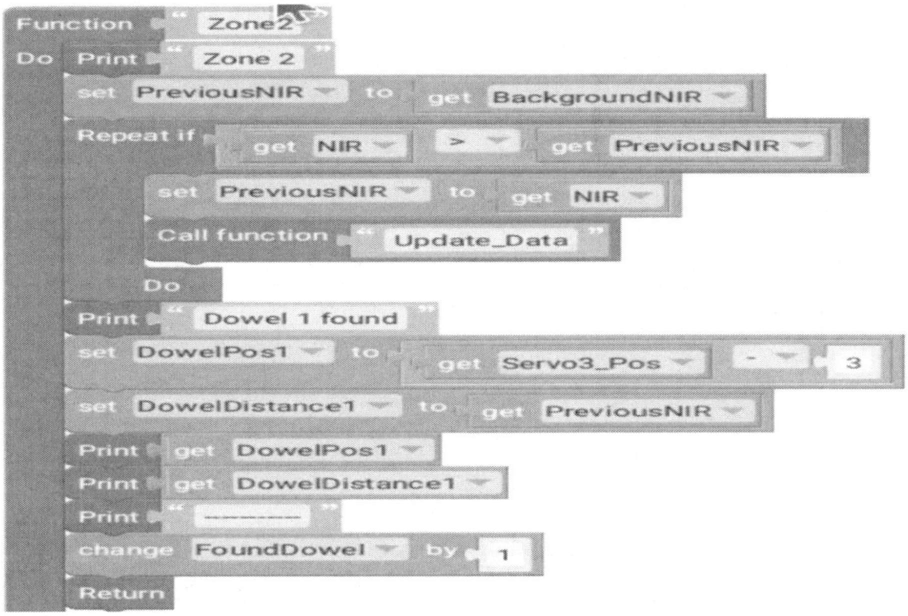

Fig. 2.78 Function Z**one2()** used in "Dowel_Scan.rpb".

- For each Loop's iteration, the current **NIR** value is saved into **PreviousNIR** before **NIR** is updated by calling Function **Update_Data()**. Therefore, **(NIR > PreviousNIR)** evaluates TRUE at the beginning of Zone 2 and Servo 3 is kept turning to the Left by 3 degrees for each Loop's iteration until the Center Line is passed and therefore **(NIR <= PreviousNIR**), and this Loop exits.
- At this point, "Dowel 1 found" is printed out.
- Next, **DowelPos1** is set with the value contained in the current value of **Servo3_Pos** but subtracted by 3 degrees as the Dowel's Center Line had been "passed over" by one Loop iteration earlier.
- Likewise, **DowelDistance1** is set to **PreviousNIR** and not **NIR** as the reader may think.
- These two Variables get printed out to update the Operator and **FoundDowel** is incremented by 1, before Function **Zone2()** returns to Function **ServoScan()**.

Fig. 2.79 lists the details of Functions **Zone3()** and **Zone4():**

- In Zone 3, Port 4 IR Sensor would exhibit a pattern of decreasing **NIR** values as it scanned "away" from Dowel 1's Center Line toward "empty space" (see Fig. 2.75). We could have used a search procedure similar to the one used in Zone 2, but Dowel 1 was already "found", so a simpler procedure was implemented using a REPEAT-IF-DO Loop that is controlled by the Conditional Expression **(NIR > DowelThreshold**). Structurally, it is the same as for Zone 1, except that the "Greater Than" Operator is used instead of the "Less Than" Operator.

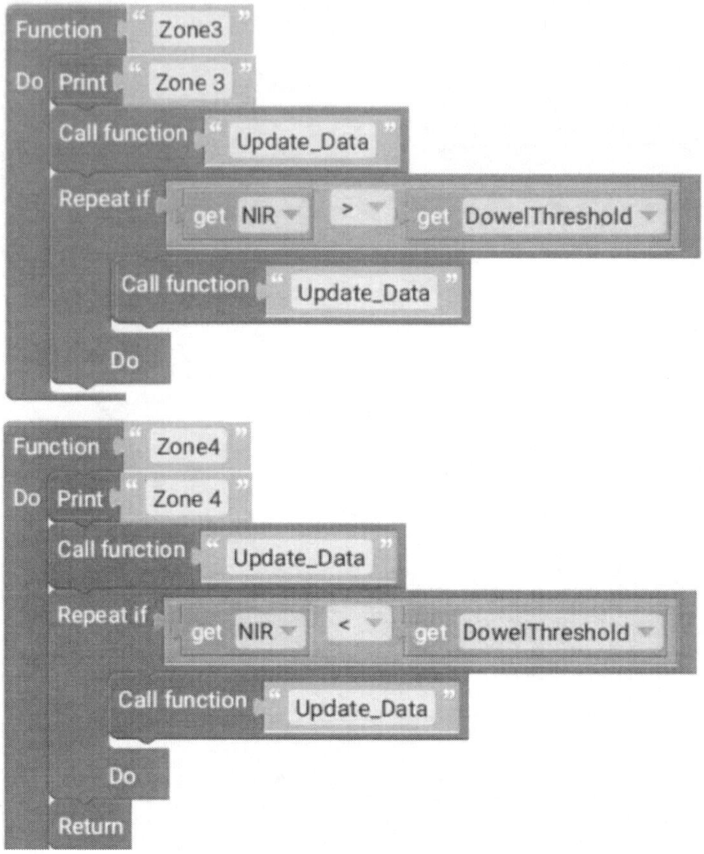

Fig. 2.79 Functions **Zone3()** and **Zone4()** used in "Dowel_Scan.rpb".

- Function **Zone4()** is practically identical to Function **Zone1()** (Fig. 2.77) as both have the same operational goal of detecting the Right Edge of a Dowel.

Fig. 2.79 displays the details of Function **Zone5()** which has the same operational goal of detecting the Center Line of a Dowel as for Function **Zone2()**, but it is now for Dowel 2 instead. Thus, Function **Zone5()** only differs from Function **Zone2()** with the Variables used being **DowelPos2** and **DowelDistance2**.

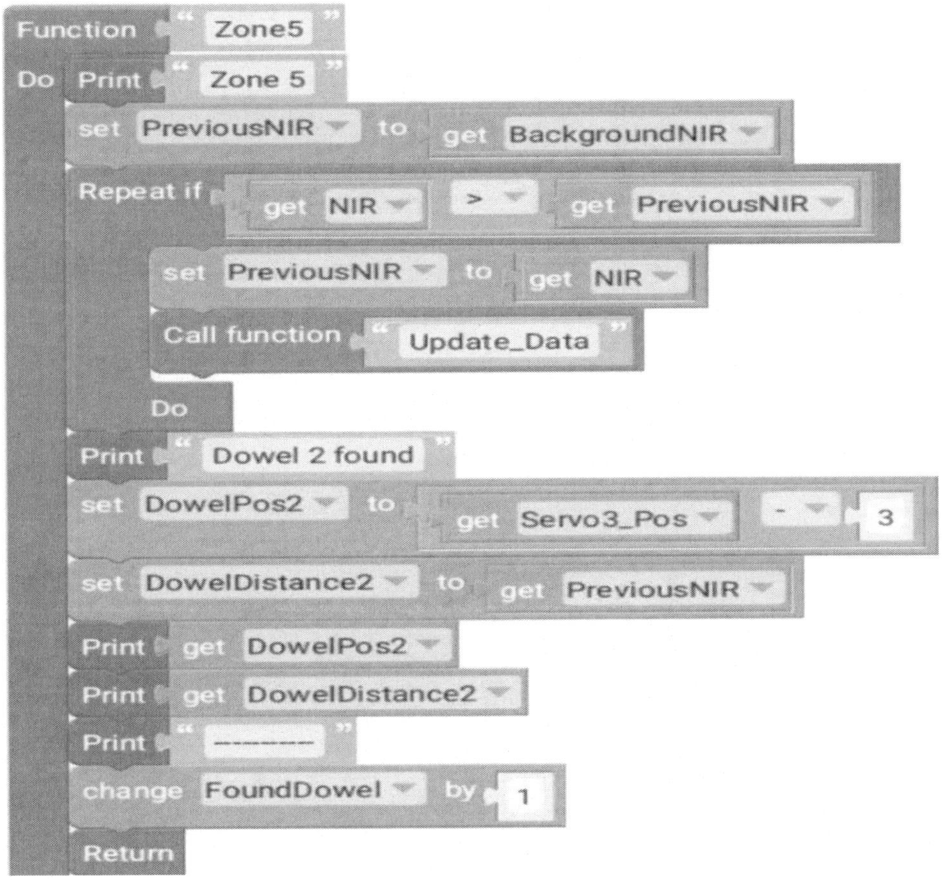

Fig. 2.80 Function **Zone5()** used in "Dowel_Scan.rpb".

Fig. 2.81 illustrates the procedures used in Function **VerifyDowels()** which has the operational goal of "showing" that the robot now "knows" where the two Dowels are located:

- First, Function **CenterSensor()** is called to set Port 4 IR Sensor squarely forward again, i.e., "staring" at the "empty space" between the two Dowels.
- Next, an IF-ELSE structure is implemented where the Conditional Expression **(DowelDistance1 > DowelDistance2)** is used. If this Expression turns out to be TRUE at runtime, then the Right Dowel 1 is closer to the robot than the Left Dowel 2, otherwise, the Left Dowel 2 is the closer one to the robot.
- If the Right Dowel 1 is closer to the robot, it is commanded to Rotate Right "continuously" until it can find Dowel 1 and approaches it within a given distance. This procedure is implemented by invoking Function **Rotate_Right()** before calling Function **FindApproachDowel()** (see Fig. 2.82). Then, the robot Rotates Left to find Dowel 2 and approaches it within a given distance by activating Function **Rotate_Left()** then Function **FindApproachDowel()**. In actual practice, when the robot stops in front of Dowel 1, the Operator is supposed to "promptly" remove Dowel 1 to "help" the robot in scanning for Dowel 2 next.
- If the Left Dowel 2 is closer to the robot, it is commanded to Rotate Left "continuously" until it can find Dowel 2. This procedure is implemented by invoking Function **Rotate_Left()** before calling Function **FindApproachDowel()**. Then, the robot Rotates

Right to find Dowel 1 and approaches it within a given distance by activating Function **Rotate_Right()** then Function **FindApproachDowel()**. In actual practice, when the robot stops in front of Dowel 2, the Operator is supposed to "promptly" remove Dowel 2 to "help" the robot in scanning next for Dowel 1.

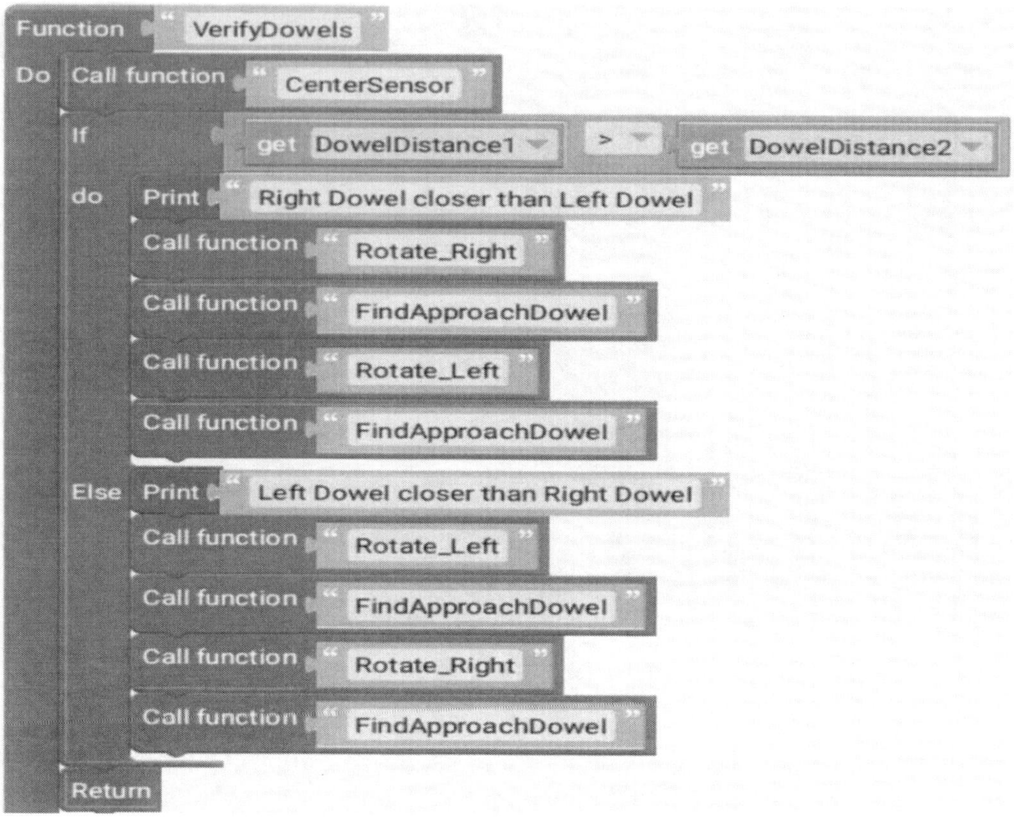

Fig. 2.81 Function **VerifyDowels()** used in "Dowel_Scan.rpb".

Fig. 2.82 shows the details of Function **FindApproachDowel()**:

- At this point in the algorithm, Port 4 IR Sensor was starting from "empty space" and "rotating" into a Dowel, thus a WAIT-UNTIL Loop is used to detect the "on-coming" Dowel Edge (either Right or Left). This Loop only needs to use the simple Conditional Expression **(Port 4 IR Sensor > DowelThreshold)**.
- When this Expression evaluates TRUE, this WAIT-UNTIL Loop exits, and the robot stops. Considering the ISTD and OSTD issues with R+BLOCK, in practice the robot would wind up being further into the Dowel than its Edge (which is a bonus in this case!).
- Next, Port 4 IR Sensor is used to read in its Current Value to which 5 is added before saving the result into Variable **ProximityLimit**.
- Then, Function **Forward()** is used in conjunction with a WAIT-UNTIL Loop to make the robot rolls Forward toward the Intended Dowel, but it would stop at a short distance from this Dowel, controlled by the value of **ProximityLimit**.

- The “last” IR Sensor is then printed out and Melody 3 is played, reminding the Operator to promptly remove the “Verified” Dowel.

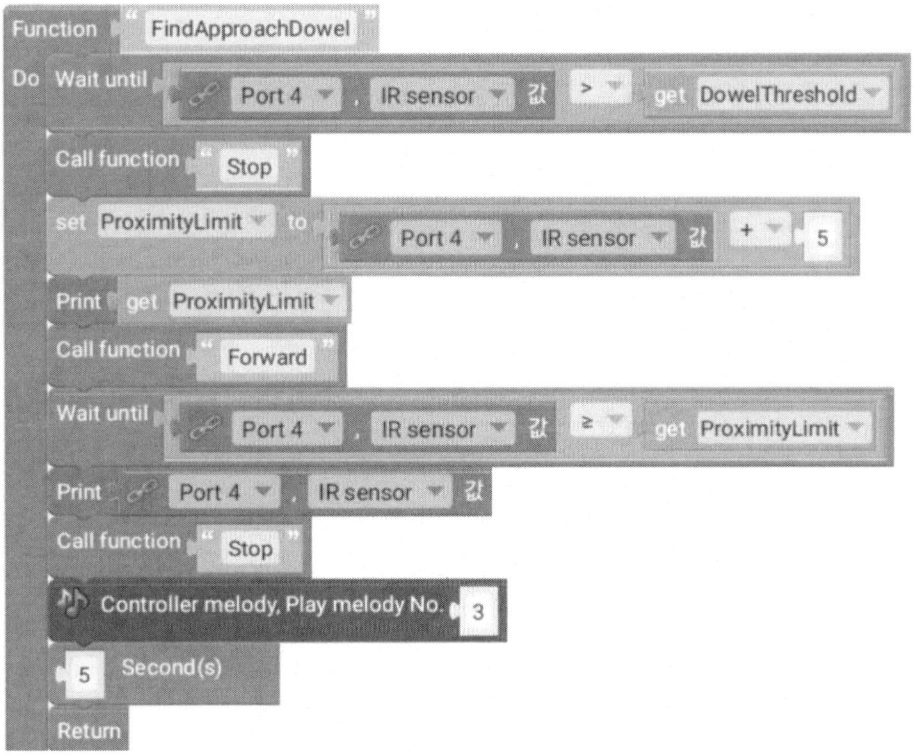

Fig. 2.82 Function **FindApproachDowel()** used in “Dowel_Scan.rpb”.

Fig. 2.83 shows the runtime printouts of two demonstration runs:

- The left one was for a setup where the Right Dowel 1 was closer to the robot. All “numbers” looked “fine”, except for the value of **98** reported for **ProximityLimit** for Right Dowel 1 (not sure what happened there!).
- The right one was for a setup where the Left Dowel 2 was closer to the robot. All “numbers” looked “fine” for this run.

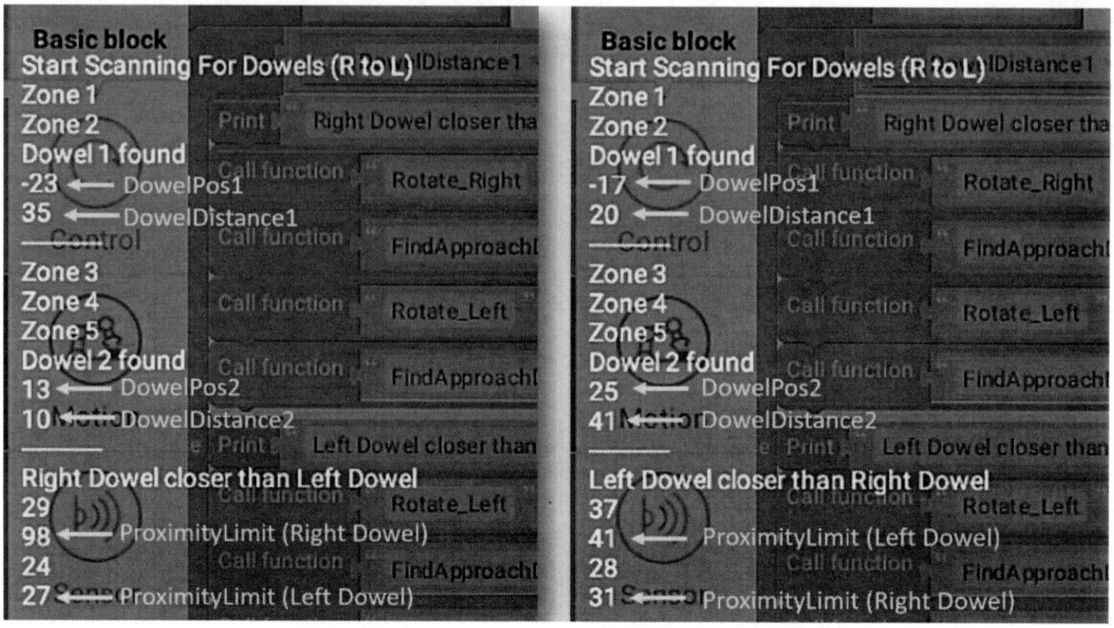

Fig. 2.83 Runtime Performances for Project "Dowel_Scan.rpb".

2.8 Author's View of R+BLOCK

The current R+BLOCK's Version 1.1.3 (from 2020) is incomplete: for example, Remote Control and some Servo functionalities are not yet implemented (as compared to the TASK tool). The issues of Systemic Time Delays are quite significant, but the author had shown several "work-around" approaches to achieve overall operational objectives for all robots considered, although at lower functional speeds. In spirit with the old adage "When there is a will, there's a way".

The author also discovered a "digitization" bug regarding the Integral IR Sensors (Left-Middle-Right) and the IRSS-10 External IR Sensors on Ports 3 and 4. Essentially, R+BLOCK does not report the same numerical values from the above sensors when used against a given obstacle set at a given distance from the sensors, **as compared to the ones reported by ROBOTIS MANAGER**. Interestingly, this issue does not occur when using TASK 2/3 (Chapter 3) or Python (Chapter 4). This issue had been reported to ROBOTIS, but the author does not know when a fix will happen.

The upcoming Chapter 3 will offer TASK equivalent solutions to the same projects to provide the reader with a useful comparative study illustrating concepts and techniques adaptations that can be made when the development environment supports new or different programming features.

2.9 References

Beecher K (2017) Computational Thinking: A Beginner's Guide to Problem-Solving and Programming. BCS Learning & Development Ltd, Swindon, UK.

Ben-Ari M, Mondada F (2018) Elements of Robotics. Springer Publishers International, Cham, Switzerland.

Bergin J (2012) Pedagogical patterns: advice for educators. CreateSpace Independent Publishing Platform, p. 115.

Denning PJ, Tedre M (2019) Computational Thinking. The MIT Press, Cambridge.

Matarić MJ (2007) The robotics primer. The MIT Press, Cambridge.

Thai CN (2018) Learning Robotics with ROBOTIS DREAM System. CNT Robotics LLC Publisher. Printed by Amazon KDP.

Thai CN (2019) Advanced Application Programming with ROBOTIS DREAM System. CNT Robotics LLC Publisher. Printed by Amazon KDP.

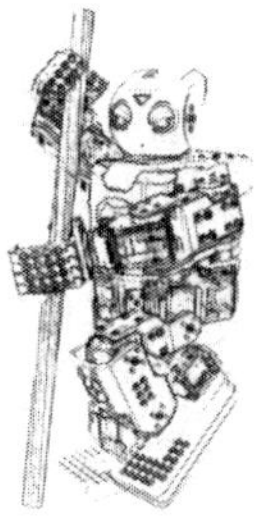

Chapter 3: Using R+TASK

To accommodate the broadest range of readers, this chapter is also written for a complete beginner in robotics and computer programming as for Chapter 2 (in case some readers are not interested in learning how to use R+BLOCK). Consequently, some materials would have to be repeated or referred to, so the author is offering his apologies to readers who have mastered Chapter 2.

One key difference between R+BLOCK and R+TASK is where the robot's code is executed from, **at runtime**:

- R+BLOCK's codes are stored and executed on the Chromebook (CB) where the User and Output Interfaces also reside and coordinate all runtime activities from the robot.
- R+TASK's source codes are edited and stored on the Chromebook (CB), once the source codes are compiled into binary codes which are then downloaded to the CM-150/151 Controllers. These binary codes are executed by the CM Controllers as soon as their Power Buttons are pushed, independently from the TASK's User and Output Interfaces which still reside on the Chromebook. Thus, the TASK programmer will have to do some extra work in coordinating the running code on the CM Controllers and the User/Output Interfaces on CB. This issue is showcased in Sub-Section 3.1.1.

For Chromebooks, the reader has a choice of using R+m.TASK 2 or R+m.TASK 3 to program the CM-150/151, thus the author provides both versions in his example programs package. However, **only TASK 3** programs will be showcased in Chapter 3.

As compared to R+BLOCK, ROBOTIS has published many more on-line resources for using TASK 2 and TASK 3, so the reader is referred to these following links for text information and tutorial videos for the basic use of TASK 2 and TASK 3:

- R+m.TASK 2: https://emanual.robotis.com/docs/en/software/rplus_mobile/mtask20/.
- R+. TASK 3: https://emanual.robotis.com/docs/en/software/rplustask3/.
- Video tutorials from ROBOTIS for R+TASK 2 and 3: https://www.youtube.com/playlist?list=PL2z6VsTHIiDOeSprJXq-Kf04aIQ9tZpgH.
- Basic use of R+m.TASK 2 on Mobile Devices: https://www.youtube.com/watch?v=vUY43bIk2ug.
- Quick tour of R+TASK 3 on Windows PC: https://www.youtube.com/watch?v=YcTObdgL5qU.

If the reader is new to the TASK tools V. 2 or V. 3, please study the previous support materials first.

This book uses V. 2.3.8 for R+m.TASK 2 and V. 3.1.7 of the ROBOTIS TASK V.3 Tools and the reader is referred to this link for ROBOTIS software download (https://emanual.robotis.com/docs/en/software/#roboplus-r). Currently, Spring 2022, TASK 2 V.2.3.8 can handle both CM-150 and CM-151 Controllers fine, but TASK 3 V. 3.1.7 does not properly download TSK3 codes down to CM-150, although it functions "normally" for CM-151.

The "Spiral" learning approach (Bergin, 2012) will again be used for instructional purposes where the author would show how to use the ROBOTIS TASK tool with projects having increasing scopes and sophistications.

3.1 Learning R+TASK Programming Basics

We will be using the Avoider platform in Section 3.1 and if the reader has not yet built this robot, please construct the Avoider robot as per instructions described in Section 2.2 or Appendix B.1.

The R+TASK V.2 and V.3 Tools offer most of the standard programming structures found in other languages such as C/C++. Below is a short description of the most useful features for a beginning programmer (more details on specific items will be provided at appropriate sections in this chapter):

- **ASSIGNMENTS** – TASK provides three types of assignment statements: LOAD, COMPUTE and POLYNOMIAL.

  The syntax for LOAD is "A = B" with the usual meaning of "Operand A is assigned the Value of Operand B".

  The syntax for COMPUTE is "A = B *Op* C" where the RHS represents the Value obtained when performing the "*Op*" operation between Operands B and C. The "*Op*" operations supported are the 5 basic arithmetic operators "+, - , * , / , and % (remainder operation)", and the two bit-level operators such as "&" (AND) and "|" (OR). VARIABLE Variables are supported but arrays are not.

  POLYNOMIAL is "format-free". It is the usual "mathematical expression" in standard computer languages but it can only include the arithmetic and logical operators listed in the previous paragraph and of course the parentheses "(" and ")" are used to group sub-expressions as needed. **POLYNOMIAL is not available on TASK 2.**

  Only integer arithmetic operations are supported in TASK 2 and 3. Depending on the type of Variables, their internal storage sizes can be from 1 to 4 bytes. Negative integers use the Two's Complement format (https://en.wikipedia.org/wiki/Two%27s_complement).

- **LABEL** and **JUMP** statements are supported.
- **CONDITIONS** – For Conditional statements, TASK provides the usual logical operators: "&&" and "||" respectively for logical AND and logical OR operations, and also "==", "!=" , "<", "<=", ">", ">=" for equality and inequality tests. The standard IF, IF-ELSE-IF, ELSE structures are supported, but nested conditional statements using parentheses are not supported (so the user will have to apply DeMorgan's theorems to expand complex logical statements as needed).
- **LOOPS** – for Repetition statements, TASK provides "WAIT WHILE", "LOOP WHILE", "LOOP FOR", "ENDLESS LOOP" and "BREAK LOOP". Standard computer languages normally offer a DO-WHILE loop structure which is not supported in TASK.
- **FUNCTION** definition and calling are provided in TASK. A special **CALLBACK** function can also be used: it is triggered by a hardware timer every 7.81 ms independently of the main TASK program. CALLBACK can only handle up to a maximum of **two** external device calls.

3.1.1 Sequence Control & Variables

In Sub-Section 3.1.1, we are going to learn that by default the CM-150/151 Controller behaves sequentially, i.e., it can do only one thing at a time.

If not done so, the reader needs to unzip the provided ZIP file containing all the example codes for this book and transfer them onto the Chromebook (see Appendix A.2). The program named "SequenceControl-1.tsk3" will be explained next.

Fig. 3.1 is a screen capture of this program which has 5 lines of "relevant" code (Lines 5-9). More formally, they are called "Statements". All TASK program begins with the "START PROGRAM" Statement (Line 3), next an opening curly bracket "{" Statement (Line 4) and ends with a closing curly bracket "}" Statement (Line 10).

All statements between the two matching curly brackets "{}" are considered "legitimate" instructions (but may not be "logically correct") for the CM-150/151 Controller to execute in sequence, top to bottom, at run time. But if they are preceded by the double forward slashes, "//", then they are considered as "Comments", i.e., information for the human reader only and not actual instructions for the Controller. For example, Line 1 just tells the reader about the goals of this program. Later, the reader will be shown that it is sometimes necessary to "nest" the curly brackets to create "compound statements" to accomplish certain logical flows needed in a wanted algorithm.

```
1   // Learning Sequence Control - Printing & Turning Motor
2   // All Rights Reserved - CNT Robotics LLC 2022
3   START PROGRAM
4   {
5       Print Screen = 123
6       Print Screen with Line = 2 Random Number
7       Speed = IR Center
8       Print Screen with Line = Speed
9       PORT[1]:Geared Motor = Speed
10  }
```

Fig. 3.1 Program "SequenceControl-1.tsk3".

Statements 5 through 9 are called LOAD statements which have this form (i.e., Syntax): "A = B". The "=" sign is not the regular "equality" sign that the reader may have seen in Algebra. In the world of Computer Programming, the "=" symbol stands for an "Assignment" operator. Each of the letters/symbols "A" and "B" is called "Operand". Thus "A=B" should be read as ("A" is assigned with the Value of "B"), or it can also be read as ("Value of B" is assigned to "A"). Essentially, "B" should be a Data Value from a Numerical Constant (e.g., "123" in Statement 5), or a Source of Data Input (e.g., a Random Number generated by the Controller internally in Statement 6, or a Value read from a Sensor such as IR Center in Statement 7).

Statement 7 also illustrates the application of a user-defined Named Variable called "**Speed**" which was first used as an Operand A. Afterwards, **Speed**, to be exact its numerical value, can be used as a type of Operand B in Statements 8 and 9.

In other words, the beginner TASK programmer should get used to "read" a LOAD statement from the Right-Hand-Side (RHS) to the Left-Hand-Side (LHS) (which may be "unusual" to English readers). As a side note, TASK V.3 supports a 2nd Assignment type of statement called COMPUTE which is frequently used, and also a 3rd Assignment type called POLYNOMIAL which is particularly "handy" for complex algebraic expressions. Practically 90% of all the Statements in a typical TASK program are of the "Assignment" type.

When the controller executes Line 5 at run time, it would just "assign" its RHS (e.g., the number "123") to the LHS which is the Screen Printer. The user can then see "123" printed out on the Program Output Monitor Screen of the Chromebook, but the display cursor will stay on the same output line afterwards, because the wanted command is a "Print Screen".

Next, when the controller executes Line 6 at run time, it will want to execute the RHS first, i.e., it will go to a specific memory location (Address 77 to be exact on the CM-150/151 Control Table, see https://emanual.robotis.com/docs/en/parts/controller/cm-151/#ram-area), and it will pull out whatever random number between 0 and 255 that it can find there. Next, it will "assign" this found value to the LHS which is the Screen Printer. The user can then see this random number printed out on the Program Output Monitor Screen of the CB (after the previous "123"), but then it will execute a "carriage return" to go to a new line afterwards, because the wanted command is now "Print Screen with (new) Line".

Next, Statement 6 is another Assignment statement where the current value of the IR Center Sensor (Address 95) is assigned to Variable **Speed**.

Next, Statements 8 and 9 both use the current Value of Variable **Speed** either to print it out (Statement 8) or to assign it as a speed value to the Port 1 Geared Motor (essentially to turn it on if **Speed** happens to be non-zero).

Please see this video https://youtu.be/Qu5OW4qm6eE for more details on how this TASK program "SequenceControl-1.tsk3" would behave at run time on a CM-150 (via TASK 2) or CM-151 (via TASK 3) on a Chromebook.

The previous video shows the user that because of all the time delays needed for Bluetooth connections to work, the run time behavior of this TASK program was far from satisfactory with the CB. At runtime, we did not see the display of "123" and the random number, and we did not hear the Port 1 Motor turn itself on during run time when **Speed** had a non-zero value. We could not see any "Print Screen" output at all. The "programming logic" used was correct as the Controller could compile and execute its machine code perfectly fine on the robot, so what was going on with our robot or setup? Basically, we could not match the timing of the execution of the binary code on the CM controller with the Output Monitor Screen on the CB. **The key understanding here is that everything in real life, and therefore in robotics also, "takes time" to get done, so we'll have to figure out a way for the controller not to "rush through" its program, so that we, slow-poke humans, can "see" what is going on.** So, one way to solve these problems is to make the robot repeats this sequence of actions "endlessly" as suggested by the Sense-Think-Act paradigm (see Figs. 1.7 and 1.8).

3.1.2 Using Main Endless Loop

The "SequenceControl-2.tsk3" program just puts the actions sequence of "SequenceControl-1.tsk3" inside an "Endless Loop" and with an additional feature of Variable **Counter** which is started at "0" outside the Endless Loop and incremented by "1", every time a loop gets done. Fig. 3.2 shows this completed program.

```
4  START PROGRAM
5  {
6      Counter = 0
7      ENDLESS LOOP
8      {
9          Print Screen = 456
10         Print Screen with Line = Random Number
11         Speed = IR Center
12         Print Screen with Line = Speed
13         PORT[1]:Geared Motor = Speed
14         Counter = Counter + 1
15         Print Screen with Line = Counter
16     }
17 }
```

Fig. 3.2 Program "SequenceControl-2.tsk3".

In "SequenceControl-2.tsk3", besides the new "Endless Loop" structure used (Statements 7, 8 and 16, and a nested set of "{}"), another new computing concept of an incremental counter is also illustrated:

- First, Variable **Counter** needs to be initiated to zero, before entering the Endless Loop. This action can be considered as part an "Init Event", previously mentioned in Section 2.5.5 and Fig. 2.28.
- Next, Variable **Counter** needs to be updated for each loop executed, by using a COMPUTE statement to make this Variable "self-add" "1", each time that the Endless Loop is executed at run time (see Statement 14). The COMPUTE statement is of the form "A = B *Op* C" which has the standard Assignment operator " = " separating the RHS expression (i.e. "B *Op* C") and LHS (i.e. "A"). The RHS now has two Operands "B" and "C" linked by a second operator *Op* which can be of 7 types:

  " + " standing for the arithmetic addition operation (used in Line 13).

  " - " standing for the arithmetic subtraction operation.

  " * " standing for the arithmetic multiplication operation.

  " / " standing for the arithmetic division operation.

  " % " standing for the remainder operation (i.e. yields the "remainder" when doing "manual" integer division.

  " & " standing for a bit-wise logical AND operation.

  " | " standing for a bit-wise logical OR operation.

- Statement 15 is used to appraise the Operator on the current value of Variable **Counter**.

Running "SequenceControl-2.tsk3" would show that thanks to the Endless Loop, the "Speed" command to the Port 1 Motor (Statement 13) was repeatedly applied to this motor so that we could now see the resulting physical effects of its turning. Essentially, Statement 9 in Program "SequenceControl-1.tsk3" (Fig. 3.1) **did** turn on Motor 1, but only for a few microseconds and then the whole robot program terminated, so we could not observe this phenomenon as it was too fast for us to see, and too short to overcome the friction inside the motor and the mechanical inertia of the robot.

However, the reader should notice that the visual feedback of the printed output was still too fast for us humans to handle, thus we need to introduce some time delays in the whole process with the use of the Hi-Resolution Timer feature of the CM-150 in the next program "SequenceControl-3.tsk3" (see Fig. 3.3). The Hi-Res Timer is set at Line 13 to 1.0 sec (or 1000 ms). It functions similarly to a stopwatch or a timer on a microwave oven. To use the Hi-Res Timer, we need first to load the Hi-Res Timer Counter with the wanted time delay (1 second in Line 13, or Address 74 from the Control Table for CM-150/151 - https://emanual.robotis.com/docs/en/parts/controller/cm-150/#ram-area). The Hi-Res Timer Counter would then automatically start to count down to zero with a 1 ms decrement. Next, we need to use a WAIT WHILE statement such as in line 14 to create user-defined time slots associated with certain events. The WAIT WHILE statement is "unusual" in the way that it keeps on executing itself, i.e., "does nothing", as long as the CONDITION specified within the two parentheses "()" remains TRUE, for example (Hi-Res Timer Counter > 0.000 sec) in Line 14. Let's now do a "thought experiment" to see how Lines 13 and 14 can work together at run-time:

```
5   {
6       ENDLESS LOOP
7       {
8           Print Screen = 789
9           Print Screen with Line = Random Number
10          Speed = IR Center
11          Print Screen with Line = Speed
12          PORT[1]:Geared Motor = Speed
13          High-resolution Timer = 1.000sec
14          WAIT WHILE ( High-resolution Timer > 0.000sec )
15          // Delay = 1.000sec
16      }
17  }
```

Fig. 3.3 Program "SequenceControl-3.tsk3".

- When Line 13 is executed, the Hi-Res Timer counter is loaded with 1 second = 1000 ms. Next Line 14 is executed, i.e., the CM-150/151 is commanded to go to the Hi-Res Timer Counter and retrieve its current value which would be a few milliseconds less than 1000 (let's say 990), as some actual milliseconds have elapsed since the execution of Line 13. Thus, the condition (Hi-Res Timer Counter > 0.000 sec) now becomes (990 > 0) and would evaluate to TRUE, therefore the CM-150/151 would be required to execute line

14 once more. So, this cycle of actions would continue for a while, but you can see that eventually this counter would be at 0. At this point, the condition (Hi-Res Timer Counter > 0) evaluates to FALSE (as "0 is not greater than 0") and the CM-150/151 can now quit Line 14 and then redo another loop restarting at Line 8.

- The Hi-Res Timer facility is available for both TASK V.2 and V.3, but the **Delay Function shown on Line 15 is available only on TASK 3.** Line 15 does the same job as Lines 13 and 14 together.

At runtime, the reader would notice that the Speed/IR Center Sensor maximum values were mostly in the 500's range and that the motor was turning at a rather slow speed. This was because the typical Geared Motor can be set to a maximum value of 1023 (i.e., 100% power) and here it was set with values in the 500's (i.e., at 50% power).

Thus the program "SequenceControl-4.tsk3" was created with a small change in Statement 12 whereas the value of **Speed** was multiplied by a factor of 4 before assigning this result as the actual speed value for Port 1 Motor – i.e. using a COMPUTE statement instead of a LOAD statement (see Fig. 3.4).

A reminder that the Hi-Res Timer is available for both TASK V.2 and V.3, but the **Delay()** Function shown on Line 15 is available only on TASK 3.

```
6    ENDLESS LOOP
7    {
8        Print Screen = 1111
9        Print Screen with Line = 2 Random Number
10       Speed = IR Center
11       Print Screen with Line = Speed
12       PORT[1]:Geared Motor = Speed * 4
13       // High-resolution Timer = 1.000sec
14       // WAIT WHILE ( High-resolution Timer > 0.000sec )
15       Delay = 1.000sec
16   }
```

Fig. 3.4 Program "SequenceControl-4.tsk3".

3.1.3 Program Modularization with Functions

Let's start with the "Maneuvers-1.tsk3" program (Fig. 3.5) which uses previous concepts already explained in Sub-Section 3.1.1. But this time, they are combined to make the robot go Forward at Speed = 512 for 0.5 second, and a COMPUTE type of statement is used to separately set the DIRECTION and SPEED for each of the Motor Ports, 2 (i.e., Right) and 1 (i.e., Left), in Lines 9 and 10 in Fig. 3.5.

```
4   START PROGRAM
5   {
6       Speed = 512
7       DelayTime = 0.500sec
8
9       PORT[2]:Geared Motor = CW:0 (0.00%) + Speed
10      PORT[1]:Geared Motor = CCW:0 (0.00%) + Speed
11      // High-resolution Timer = DelayTime
12      // WAIT WHILE ( High-resolution Timer > 0.000sec )
13      Delay = DelayTime
14  }
```

Fig. 3.5 Program "Maneuvers-1.tsk3".

```
START PROGRAM
{
    Speed = 512
    DelayTime = 0.500sec

    PORT[2]:Geared Motor = CCW:0 (0.00%) + Speed
    PORT[1]:Geared Motor = CW:0 (0.00%) + Speed
    // High-resolution Timer = DelayTime
    // WAIT WHILE ( High-resolution Timer > 0.000sec )
    Delay = DelayTime
}

FUNCTION Forward
{
    PORT[2]:Geared Motor = CW:0 (0.00%) + Speed
    PORT[1]:Geared Motor = CCW:0 (0.00%) + Speed
    RETURN
}
```

Fig. 3.6 Program "Maneuvers-2.tsk3".

A "COMPUTE" statement is used here (e.g., line 9 in Fig. 3.5) instead of a "LOAD" statement (like in Line 12 of Fig. 3.3) because **Speed** is a Variable Parameter, currently set to 512 but it can be changed to something else in other parts of the computer program (to be shown later in this Sub-Section 3.1.3). If the need was only for a fixed numerical constant that would not change throughout the robot program, then a "LOAD" statement would be more efficient to use. Part 1 of this video (https://www.youtube.com/watch?v=ylwGyHskWF0) shows how to edit COMPUTE statements inside the program "Maneuvers-1.tskx/tsk3", how this program performed at run-time and how to modify it into "Maneuvers-2.tskx/tsk3" which used a new type of computer programming structure called "FUNCTION" (see Fig. 3.6).

The later part of this video also shows how to use the MAIN PROGRAM area of "Maneuvers-2.tsk3" as a "sandbox" area where the user can try out different coding ideas, while the FUNCTION areas were used as a repository of coding ideas that had worked well. **This is a "shallow" understanding/application of the FUNCTION structure which serves as a useful but intermediary step towards a more complete understanding of the issues later.**

Once the basic Functions **Forward()**, **Backward()**, **Right()** and **Left()** are defined, the program "Maneuvers-3.tsk3" shows how to use them to create a sequence of robot maneuvers, perhaps needed to negotiate a maze, e.g., going Forward, turn Left then going Forward again (see Fig. 3.7). These tasks were done by CALLing those predefined Functions "Forward" and "Left" as needed (Lines 9, 12 and 15).

This video (https://www.youtube.com/watch?v=r3y5dyC84qw) explains the details of how the controller jumped from a CALL statement in the Main program (such as Line 9 in Fig. 3.7) into the Function definition area, executed the statements defined in this Function, then returned to the Main program to execute the next statement (e.g., Line 10).

```
4   START PROGRAM
5   {
6       Speed = 512
7       DelayTime = 0.500sec
8
9       CALL Forward
10      Delay = DelayTime
11
12      CALL Left
13      Delay = DelayTime
14
15      CALL Forward
16      Delay = DelayTime
17  }
```

Fig. 3.7 Program "Maneuvers-3.tsk3".

The reader can see that by using Functions, the logic flow in the Main program is easier to understand as it is presented in a more modular fashion, perhaps "logically coarser", while pushing the "gory" operational details into the Function definitions themselves. For example, this CM-150/151 robot has only two motors, but let's say that you buy 2 additional motors, so your robot now has four motors to control instead of two, and you still need to perform the same "mission" (Forward/Left/Forward). In this case, you would not have to modify the main program at all, but you will need to add new statements to take care of Motors 3 and 4 inside the appropriate Functions **Forward()**, **Backward()**, **Right()** and **Left()**.

The program "Maneuvers-4.tskx" shows how to use the TASK 3's built-in Function **Delay()** (Fig. 3.8) in combination with Variables **Speed** and **DelayTime** to create maneuvers with varying speeds and delay times, i.e. a more adaptive algorithm to help deal with complex mazes.

Fig. 3.8 also shows a good coding practice of setting the value for a Variable needed for a given function, just before calling that function: for example, set up **Speed** (Line 8) before calling Function **Forward()** (Line 9).

```
6   {
7       //  Going Forward for 0.5 s at Speed 512
8       Speed  =  512
9       CALL  Forward
10      DelayTime  =  0.500sec
11      Delay  =  DelayTime
12
13      //  Turn right for 0.25 s at Speed 256
14      Speed  =  256
15      CALL  Right
16      DelayTime  =  0.250sec
17      Delay  =  DelayTime
18
19      //  Go Backward for 0.35 s at Speed 400
20      Speed  =  400
21      CALL  Backward
22      DelayTime  =  0.350sec
23      Delay  =  DelayTime
24
25      //  Stop robot before ending program
26      CALL  Stop
27  }
```

Fig. 3.8 Program "Maneuvers-4.tsk3".

In this Sub-Section 3.1.3, the robot's maneuvers were "timed maneuvers", thus they may not work consistently for different/changing environments such as different ground surfaces that the robot may run on, or for different charge levels of its batteries (low or high). In later Sub-Sections, we'll use more "conditional" programming techniques so that the same robot code can work in environments different than the one that the code was originally created in.

3.1.4 Conditional Structures for Spinning Top

In Sub-Section 3.1.4, we will look at two basic ways to use Conditional Structures which allow programmers to create "Logical Branching" in the Overall Sequential Behavior of a TASK program.

The new "Spinning Top" tasks assigned to the Avoider robot are as follows:

- At start, the robot plays a musical melody to let the user know that it is "ready for action".
- Next, the behaviors of the robot depend on the position of an object relative to the IR Center Sensor, thus the Condition-Activity list to consider is as follows:
    - If the IR Center Sensor reading is less than 50, the robot stays still.
    - If the IR Center Sensor reading is between 50 and 300, the robot spins right with its speed proportional to the IR Center Sensor reading and it plays the musical scale Do# for 0.1 second.
    - If the IR Center Sensor reading is between 300 and 500, the robot spins left with its speed proportional to the IR Center Sensor reading and it plays a Re# for 0.1 second.
    - If the IR Center Sensor reading is larger than 500, the robot rolls forward at Speed = 1023 to escape from the object and it plays a Fa# for 0.5 second.

This Condition-Activity list indicates that we are dealing with a situation whereas multiple possible events may occur to the robot at runtime, but one and only one event can happen at a given point during runtime.

Using the built-in musical resources of the CM-150/151, this challenge will be solved in two ways:

- Using a "Parallel" IF structure ("**IR-based-Maneuvers-IFs.tsk3**").
- Using an "IF-ELSE-IF" structure ("**IR-based-Maneuvers-IF-ELSE-IF.tsk3**").

The program "**IR-based-Maneuvers-IFs.tsk3**" uses a direct "translation" of each item from the above Condition-Activity list into appropriate **IF compound statements** in the TASK code:

- The Initial Event is solved by playing Melody 2 out of the CM-150/151 buzzer (see Fig. 3.9). First, "Melody Time" is assigned to the "Buzzer Timer" (Line 6), next the "Buzzer Index" is assigned with the value "Melody 2" (Line 7). Then a WAIT WHILE statement is used to let the melody plays itself out (Line 8). Lastly, Variable **Speed** is initialized to zero (Line 9).

```
3  START PROGRAM
4  {
5      // Init Action
6      Buzzer Timer = Melody Time
7      Buzzer Index = Melody No.2 (2)
8      WAIT WHILE ( Buzzer Timer > 0.0sec )
9      Speed = 0
```

Fig. 3.9 Initial Event for program "IR-based-Maneuvers-IFs.tsk3".

This solution's algorithm (Figs. 3.10 and 3.11) uses an Outer Endless Loop to represent the World-Robot Interactions Cycle (previously depicted in Fig. 1.8). This loop contains a set of "Parallel" IF structures representing each separate Condition-Activity item from the complete listing as detailed out previously.

During each loop, the controller would evaluate/execute this "Parallel IFs" structure in the following manner:

- It acquires the latest IR Center Sensor reading and saves it in Variable **IR_Sensor** (Line 13 in Fig. 3.10).
- Next, it enters the "Parallel IFs" structure and checks on the first IF statement (Line 14 in Fig. 3.10), i.e., the Conditional Expression **(IR_Sensor < 50)**. If this Expression evaluates FALSE, i.e. **(IR_Sensor >= 50)**, the Controller goes on to check on the next IF statement (Line 18). But if this Conditional Expression **(IR_Sensor < 50)** is TRUE, Function **Stop()** is called (i.e., the robot stays still) and then the Controller will go on to check on the next IF statement at Line 18 of Fig. 3.12 any way (**even though logically it does not have to**). The astute reader would detect "logical inefficiencies" here already, but for now let's just go on with the rest of the algorithm.

```
10  // Main Loop
11  ENDLESS LOOP
12  {
13      IR_Sensor = IR Center
14      IF ( IR_Sensor < 50 )
15      {
16          CALL Stop
17      }
18      IF ( IR_Sensor >= 50 && IR_Sensor <= 300 )
19      {
20          Speed = IR_Sensor
21          CALL Right
22          Buzzer Timer = 0.1sec
23          Buzzer Index = Do# (4)
24          WAIT WHILE ( Buzzer Timer > 0.0sec )
25      }
```

Fig. 3.10 Part 1 of Main Algorithm used in program "IR-based-Maneuvers-IFs.tsk3".

- Line 18 checks to see if **IR_Sensor** is between 50 and 300, if this is TRUE, it makes the robot Turn Right and Play Do# for 0.1 second (Lines 20-24) – please note that the Buzzer Timer is doing "double-duty" here, it waits for the Do# Note to finish, but it also allows the Turn Right Maneuver to last 0.1 second also. Then the algorithm goes on to check on Line 26 of Fig. 3.11. If this Condition Expression (Line 18) turns out to be FALSE, the algorithm would also go to Line 26 next.

```
26  IF ( IR_Sensor >= 300 && IR_Sensor <= 500 )
27  {
28      Speed = IR_Sensor
29      CALL Left
30      Buzzer Timer = 0.1sec
31      Buzzer Index = Re# (6)
32      WAIT WHILE ( Buzzer Timer > 0.0sec )
33  }
34  IF ( IR_Sensor > 500 )
35  {
36      Speed = 1023
37      CALL Forward
38      Buzzer Timer = 0.5sec
39      Buzzer Index = Fa# (9)
40      WAIT WHILE ( Buzzer Timer > 0.0sec )
41  }
```

Fig. 3.11 Part 2 of Main Algorithm used in program "IR-based-Maneuvers-IFs.tsk3".

- Similarly, Line 26 checks to see if **IR_Sensor** is between 300 and 500, if this is TRUE, it makes the robot Turn Left and Play Re# for 0.1 second (Lines 29-32), then the algorithm goes on to check on Line 34 of Fig. 3.11. If this Conditional Expression (Line 26) turns out to be FALSE, the algorithm would also go to Line 34 next.
- Finally, at Line 34, the algorithm checks to see if **(IR_Sensor > 500)**, if this turns out to be TRUE, the robot is commanded to run Forward at Full **Speed** = 1023 (essentially to get away from the user/obstacle), and to Play Fa# for 0.5 second. Then it will repeat the Endless Loop one more time at Line 13 with a fresh **IR_Sensor** value. If this Conditional Expression (Line 34) is FALSE, then the algorithm also loops back to Line 13.

Essentially, with this solution approach, **all conditions/events** are tested at run time, even though logically **one and only one** of them can be TRUE at any one time. Overall, the reader can check that this algorithm does the job fine. The reader would also readily notice that when **IR_sensor** was between 50 and 300, we could hear the Do# note being played out, but the robot was not Turning Right as it was supposed to. This was because the **Speed** value was set equal to **IR_Sensor** in the algorithm, and these specific **IR_Sensor/Speed** values happened to be too low to overcome the friction and inertia inherent in this robot design. However, when the IR_Sensor reached into the range above 300, it could turn Left or Run Away from the user fine.

The IF-ELSE-IF structure is best suited for a situation where logically the robot can only choose **one and only one** alternative condition/action route **among several** potential condition/action routes (which is the current case).

The program "**IR-based-Maneuvers-IF-ELSE-IF.tsk3**" describes the 2nd solution approach that can be used (see Fig. 3.12):

```
11  ENDLESS LOOP
12  {
13      IR_Sensor = IR Center
14      IF ( IR_Sensor <= 50 )
15      {
16          CALL Stop
17      }
18      ELSE IF ( IR_Sensor <= 300 )
19      {
20          Speed = IR_Sensor
21          CALL Right
22          Buzzer Timer = 0.1sec
23          Buzzer Index = Do# (4)
24          WAIT WHILE ( Buzzer Timer > 0.0sec )
25      }
26      ELSE IF ( IR_Sensor <= 500 )
27      {
28          Speed = IR_Sensor
29          CALL Left
30          Buzzer Timer = 0.1sec
31          Buzzer Index = Re# (6)
32          WAIT WHILE ( Buzzer Timer > 0.0sec )
33      }
34      ELSE IF ( IR_Sensor > 500 )
35      {
36          Speed = 512
37          CALL Forward
38          Buzzer Timer = 0.5sec
39          Buzzer Index = Fa# (9)
40          WAIT WHILE ( Buzzer Timer > 0.0sec )
41      }
```

Fig. 3.12 Main Algorithm used in program "IR-based-Maneuvers-IF-ELSE-IF.tsk3".

- This solution's algorithm also uses an Outer Endless Loop to represent the World-Robot Interactions Cycle of Fig. 1.8. But this Loop now contains an IF-ELSE-IF structure representing the Condition-Activity listing as detailed out previously.
- During each loop, the controller would evaluate/execute this IF-ELSE-IF structure in the following manner:
    - It acquires the latest IR Center Sensor reading and saves it in Variable IR_Sensor (Line 13).
    - Next, it enters the IF-ELSE-IF structure and checks the first IF statement (Line 14), i.e., the Expression **(IR_Sensor <= 50)**. If this Expression evaluates TRUE, Function **Stop()** is called and the robot stays still. Next, the Controller will exit the whole IF-ELSE-IF structure at Line 41 and then it will perform another loop from Line 13. But if this condition is FALSE, i.e. **(IR_Sensor > 50)**, the controller goes on to check the next ELSE-IF statement (Line 18).
    - Please note that according to the considered Condition-Activity listing, the Controller is supposed to check if **IR_Sensor** is between 50 and 300 at Line 18, but

it now uses a Conditional Expression that involves only checking for **(IR_Sensor <= 300)**. Is that sufficient to do what is required? And the answer is that it is OK to just check for the condition **(IR_Sensor <= 300)** here, because the condition **(IR_Sensor > 50)** was implicitly TRUE when Line 18 was chosen (see previous paragraph). Thus, if **(IR_Sensor <= 300)** turns out to be TRUE (meaning that IR_Sensor is **between 50 and 300**), the Controller assigns the value of **IR_Sensor** to **Speed** and calls the Function **Right()** to make the robot Spin Right at a **Speed** proportional to **IR_Sensor**'s value. The Controller also Plays Do# for 0.1 second, then exits this IF-ELSE-IF structure and makes another loop starting at Line 13. However, if the condition **(IR_Sensor <= 300)** turns out to be FALSE, i.e. **(IR_Sensor > 300)**, then the Controller goes on to execute the next ELSE-IF statement at Line 26.

- Similarly, executing Line 26 means that the controller is checking to see if **IR_Sensor is between 300 and 500**. If this is TRUE, then the robot spins Left with a **Speed** proportional to **IR_Sensor's** value, Plays Re# for 0.1 second, exits this IF-ELSE-IF structure and continues with its looping task starting at Line 13. If the condition in Line 26 turns out to be FALSE (i.e. **(IR_Sensor > 500)**, then Line 34 is executed next.
- Let's have a pop-quiz for Line 34. Was it necessary to check for **(IR_Sensor > 500)** on line 34? Or a simple ELSE would do the job here? The author would leave it to the readers to experiment on their own here. If **(IR_Sensor > 500)** turns out to be TRUE, then the robot would run forward at **Speed**=512 and for 0.5 second while Playing Fa#. This maneuver is designed to make the robot run away from under the object/user. Once away from the object's influence, the IR_Center Sensor would then read zero, and the robot would stop and be ready to "play again".

The reader can verify that the runtime performance of this solution "IR-based-Maneuvers-IF-ELSE-IF.tsk3" was not that different from the solution "IR-based-Maneuvers-IFs.tsk3". In theory, the IF-ELSE-IF version is computationally more efficient and therefore "faster" than the Parallel IFs version. In the Example Codes package, the reader can use a "timed" version of both solutions to concretely do the comparison (the results will surprise the reader!).

3.1.5 Using IFs and Loops with Avoider/Follower

In Sub-Section 3.1.5, we'll use the same Avoider platform as before, but now it will be using its Left and Right IR Sensors to detect a single object and to accomplish tasks described in the following Condition-Activity list:

- If there is no obstacle in front of Avoider, it can go forward.
- If there is an object on the Left of Avoider, it should back away and turn to the Right to avoid the obstacle.
- If there is an object on the Right of Avoider, it should back away and turn to the Left to avoid the obstacle.
- If there is an object in front of Avoider, it should go straight Back to avoid the obstacle.

At first analysis, this seems to be a familiar programming situation of choosing one and only one alternative among several possible alternatives, thus an IF-ELSE-IF structure should do an adequate job when combined with an Outer Endless Loop.

The corresponding TASK solution is named "Avoider-1.tsk3". It starts with the playing of Melody 2 (same code as shown in Fig. 3.9). The main algorithm is described in Fig. 3.13:

- The Endless Loop starts out by reading in the latest values from the IR Left and Right Sensors and saves them respectively into Variables **IR_Left** (Line 14) and **IR_Right** (Line 15).

```
13  {
14      IR_Left  =  IR Left
15      IR_Right  =  IR Right
16      IF  ( IR_Left  <  10  &&  IR_Right  <  10 )
17      {
18          CALL  Forward
19      }
20      ELSE IF  ( IR_Left  >=  100  &&  IR_Right  <  10 )
21      {
22          CALL  RightBack
23      }
24      ELSE IF  ( IR_Right  >=  100  &&  IR_Left  <  10 )
25      {
26          CALL  LeftBack
27      }
28      ELSE IF  ( IR_Left  >=  100  &&  IR_Right  >=  100 )
29      {
30          CALL  Backward
31      }
32  }
```

Fig. 3.13 Main Algorithm used in program "Avoider-1.tsk3".

- The condition **((IR_Left >=100) AND (IR_Right < 10)),** used for the ELSE-IF statement at Line 20, represents a situation where there is an object on the Left of the robot and none of its Right side. If this condition evaluates to TRUE, the robot will Turn Right by Backing Away to avoid the obstacle on its Left (Line 22). The Controller next exits this IF-ELSE-IF structure, and loops again at Line 14. If this condition evaluates to FALSE, then Line 24 is executed next.
- Similarly, the condition **((IR_Right >=100) AND (IR_Left < 10))**, used for the ELSE-IF statement at Line 24, represents a situation where there is an object on the Right of the robot and none of its Left. If this condition evaluates to TRUE, the robot would Back Away and turn Left to avoid the obstacle on its Right (Line 26). The Controller next exits this IF-ELSE-IF structure, and loops again at Line 14. If this condition evaluates to FALSE, then Line 28 is executed next.
- The ELSE-IF statement at Line 28 uses this condition **((IR_Left >=100) AND (IR_Right >= 100))** to represent a situation when there is an object close enough to the robot's Front such that **both IR Sensors** report data values larger than 100. If this condition is TRUE, the robot would roll straight back away from the obstacle, otherwise the Controller loops back to Line 14.

During run-time, the robot handled well the situations when the object was on either side of itself. But it did not handle well the situation when a "flat" object was right in front of it, as it kept on going back and forth at the same spot, i.e., it got "stuck". **This is a side-effect of the current**

algorithm that resulted in the controller "bouncing" back and forth between the conditions represented by Line 16 and Line 28.

This problem can be easily fixed by adding a Turning maneuver to the current Backward maneuver (Line 30). The turn direction does not matter, the author happens to use a Left Turn in the revised solution "Avoider-2.tsk3" (see Fig. 3.14, Lines 30-31). Surprisingly, this solution showed much improved performance for frontal objects and corners.

```
28  ELSE IF ( IR_Left >= 100 && IR_Right >= 100 )
29  {
30    CALL Backward
31    CALL LeftBack
32  }
```

Fig. 3.14 Adding Left Turn to Backward Maneuver in program "Avoider-2.tsk3".

The next program "Avoider-3.tsk3" offers a different solution technique to these maneuvering tasks. **This new approach still relies on the same Condition/Activity listing as described at the beginning of this Sub-Section 3.1.5, but it distills different behavior patterns from this list:**

- We must identify a "default" action which in this case is "Going Forward".
- We will be using several simple IFs in parallel instead of a single big IF-ELSE-IF structure like before. **We also need to do a deeper analysis of the Condition/Activity list to retain only the "essential" Condition/Action Pairs.**

We also **need to apply a Delay Time to each action made** (this part was not needed when the IF-ELSE-IF structure was used). This Delay Time can be tuned for optimal performance.

Fig. 3.15 shows the relevant section of the algorithm used:

- Line 10 showed that a **DelayTime** of 100 ms was used (the reader should try different values such as 0, 50, 150 etc... to see the effect on the overall performance of the robot).
- An ENDLESS LOOP is still used (Line 13) and starts by updating Variables **IR_Left** and **IR_Right** with the latest data input from the IR Left and Right Sensors (Lines 15 and 16).
- The first IF statement (Lines 17-21) represents the condition when there is an object on the robot's Left (Line 17), then the robot would Back Away and Turn Right (as before) to steer away from the obstacle (Line 19). A short delay (Line 20) is necessary for the robot to stay in this "action" to make this algorithm work properly.

```
10   Speed = 256
11   DelayTime = 100
12   // Main Loop
13   ENDLESS LOOP
14   {
15      IR_Left = IR Left
16      IR_Right = IR Right
17      IF ( IR_Left >= 100 )
18      {
19         CALL RightBack
20         Delay = DelayTime
21      }
22      IF ( IR_Right >= 100 )
23      {
24         CALL LeftBack
25         Delay = DelayTime
26      }
27      CALL Forward
28   }
```

Fig. 3.15 Alternate Maneuvering Algorithm using "Parallel IFs" in program "Avoider-3.tsk3".

- The second IF statement (Lines 22-26) represents the condition when there is an object on the robot's Right (Line 22), then the robot would Back Away and Turn Left (as before) to steer away from the obstacle (Line 24). A short delay (Line 25) is also necessary for the robot to stay in this "action" to make this algorithm work properly.
- **Please note that the "default" maneuver is always programmed as the "Last Action" ("Going Forward" in this case – Line 27).**

It is important to perform a "thought experiment" here to "preview" the possible results of this Endless Loop at run time:

- **Every time that the Endless Loop is executed, Line 27 will be executed**, i.e., the robot will always go forward (a little bit), regardless of everything else that may happen.
- Regarding the two parallel IFs, during some execution cycle(s), none of them could be activated (when there is no obstacle), or one of them could be triggered (when the obstacle is on either side of robot), or both could be triggered (a frontal obstacle). The alert reader may have noticed that seemingly the case for "a frontal obstacle" was not dealt with in this algorithm, so how would the robot behave at run time when encountering a frontal obstacle? The answer is that the robot would "wiggle" left-right and backward to deal with a frontal obstacle, i.e., the Functions **RightBack()** and **LeftBack()** were alternatively called. **This effect also explains why a DelayTime=0 would not work**

for the robot, as it would be switching so fast between the Functions RightBack(), LeftBack() and Forward() (within the Endless Loop) such that the robot looks "paralyzed" from the outside.

At this point, some of the more advanced readers may have wondered whether the IF statements used in Lines 17 and 22 of Fig. 3.15 could be changed to LOOP WHILE statements? And would the robot still work OK for this challenge?

The program "Avoider-4a.tsk3" implements this idea (see Fig. 3.16), and the author found that this idea did not work! What went wrong? Was it wrong to use LOOP WHILE for this challenge?

```
13  ENDLESS LOOP
14  {
15      IR_Left  =  IR Left
16      IR_Right  =  IR Right
17      LOOP WHILE  ( IR_Left >= 100 )
18      {
19          CALL RightBack
20          Delay = DelayTime
21      }
22      LOOP WHILE  ( IR_Right >= 100 )
23      {
24          CALL LeftBack
25          Delay = DelayTime
26      }
27      CALL Forward
28  }
```

Fig. 3.16 Incorrect use of "LOOP WHILE" in program "Avoider-4a.tsk3".

As it turns out, the error was not because LOOP WHILE was used per se, but it was more because LOOP WHILE was implemented incorrectly, as we forgot to "feed" them with the latest IR sensor data. We were using the Variables **IR_Left** and **IR_Right** to control the loops and we forgot to update their values every time we went through the LOOP WHILE (as we were supposed to!). It was a simple change to get to the program "Avoider-4b.tsk3" (see Fig. 3.17) which does work properly. Please look at Lines 17 and 22 in Fig. 3.17 where the real-time data from the sensors' ports are now accessed and used to control the LOOP WHILE, not <u>copies</u> of their values, i.e., Variables **IR_Left** and **IR_Right**, as before. Please also note that Lines 15 and 16 are commented out as they are no longer needed.

```
ENDLESS LOOP
{
   // IR_Left = IR Left
   // IR_Right = IR Right
   LOOP WHILE ( IR Left >= 100 )
   {
      CALL RightBack
      Delay = DelayTime
   }
   LOOP WHILE ( IR Right >= 100 )
   {
      CALL LeftBack
      Delay = DelayTime
   }
   CALL Forward
}
```

Fig. 3.17 Correct Use of “LOOP WHILE” in program “Avoider-4b.tsk3”.

Another possible correction to the program “Avoider-4a.tsk3” is to use the CALLBACK Function available only in TASK to result in “Avoider-4a_CB.tsk3” (see Fig. 3.18):

- In the Main Endless Loop, Variables **IR_Left** and **IR_Right** continue to be used as before.
- Lines 71-75 show how the CALLBACK Function is implemented in this case:
  - CALLBACK Function is executed **independently** of the MAIN TASK program. It is triggered by a hardware timer every **8 ms**, thus it needs to be “small” as only 512 bytes are allocated to it. Any TASK statement can be used in the CALLBACK Function, but only **TWO** calls to hardware devices (i.e., sensors) are allowed. Lines 73 and 74 show that we already have used up our “sensors” limit!
  - Thus, every 8 ms, Line 73 reads the IR Left Sensor and saves its Current Value into Variable **IR_Left** and similarly for **IR_Right** with Line 74. Thus, these Variables are “always” updated “behind the scenes” even though it does not seem so in the Main Endless Loop!

```
ENDLESS LOOP
{
   LOOP WHILE ( IR_Left >= 100 )
   {
      CALL RightBack
      Delay = DelayTime
   }
   LOOP WHILE ( IR_Right >= 100 )
   {
      CALL LeftBack
      Delay = DelayTime
   }
   CALL Forward
}
```

```
CALLBACK
{
   IR_Left = IR Left
   IR_Right = IR Right
}
```

Fig. 3.18 Use of CALLBACK Function in program "Avoider-4a_CB.tsk3".

There is an interesting experiment that the reader can make with the previous different versions of the Avoider codes: create a maze using books or boxes and let the Avoider robot roam freely through it. You will be surprised with its behaviors depending on which TASK code you used, especially when your maze has a cul-de-sac.

To round out the reader's experience with using the Left and Right IR Sensors, let's apply a **"reverse logic"** on them, essentially let's make the robot follow an object when it detects something in front of it.

Let's first learn how to track a moving object, when it is detected. The corresponding Condition/Action list can be as follows:

- If there is no object detected, the robot should stay still (default behavior).
- If there is an object detected by the Right IR Sensor, the robot should Spin Right.
- If there is an object detected by the Left IR Sensor, the robot should Spin Left.

Should we also plan for the case when an object is detected and located right in front of the robot? Should it "stay still" or should it show that it already "locks in" on the object in some other way? From the preceding Avoider coding work, we should be able to let the robot bounce between its two actions of swinging Left and Right to show that it got the object in its sight, right?

The program "Follower-1.tsk3" is derived from the previous discussions using LOOP WHILEs. The algorithm used is shown in Fig. 3.19:

- Essentially, whenever any IR Sensor (Left or Right) picks up on an object (Lines 15 or 19), the robot is turned, respectively, towards the same side where that IR Sensor is located (Lines 17 or 21). Please note that a low detection threshold value of "5" is used so that the robot can detect an object as "far" as possible (about 5 inches away)!

```
13  ENDLESS LOOP
14  {
15      LOOP WHILE ( IR Left >= 5 )
16      {
17          CALL Left
18      }
19      LOOP WHILE ( IR Right >= 5 )
20      {
21          CALL Right
22      }
23      CALL Stop
24  }
```

Fig. 3.19 Object tracking algorithm in program "Follower-1.tsk3".

- The default maneuver is at Line 23 which is interestingly a "non-maneuver" as it keeps the robot still (CALL Stop)!

When the reader tries out the program "Follower-1.tsk3", please remember the broad angle of view supported by the built-in Left and Right NIR Sensors, and please note the effect of the object's size on how the robot would tract the object.

Here, the reader is challenged to come up with another algorithm using parallel IFs instead, however this new algorithm should yield the same object tracking result. Don't forget to use a small Delay Time to make this approach work!

In the next "Follower" application, a new condition/action pair is added to the previous list:

- When the object gets a bit closer to the robot (i.e., threshold value of 75 applied to both IR Sensors), the robot should get even closer to the object by going Forward. At this point during runtime, if the user then moves the object away from the robot at a proper speed, ones will see the effect of the robot "following" the object at a short distance away.

Surprisingly, there are two possible solutions: one using a combination of LOOP WHILE and an IF-ELSE-IF structure (Follower-2.tsk3), and another one using only an IF-ELSE-IF structure (Follower-3.tsk3). Who knows may be the readers can find other solutions too!

The program "Follower-2.tsk3" (see Fig. 3.20) uses a default maneuver (CALL Stop) and an outer LOOP WHILE which encompass an inner IF-ELSE-IF structure that determines the <u>one</u> appropriate action among 3 possible and mutually exclusive alternatives "Left", "Right" or "Forward".

```
13  ENDLESS LOOP
14  {
15      LOOP WHILE ( IR Left >= 5 || IR Right >= 5 )
16      {
17          IF ( IR Left >= 5 && IR Right < 2 )
18          {
19              CALL Left
20          }
21          ELSE IF ( IR Right >= 5 && IR Left < 2 )
22          {
23              CALL Right
24          }
25          ELSE IF ( IR Left >= 75 && IR Right >= 75 )
26          {
27              CALL Forward
28          }
29      }
30      CALL Stop
31  }
```

Fig. 3.20 Object tracking algorithm in program "Follower-2.tsk3".

Or in everyday language:

- When no obstacle is detected, the LOOP WHILE (Lines 15-29) would be skipped and Line 30 would be executed, i.e., the robot would stay still as its default behavior.
- When **<u>either</u>** IR Sensor detects an object (Line 15), the LOOP WHILE is entered, and the Controller checks further with the IF-ELSE-IF structure to see if the IR Left Sensor is the only one detecting the object (threshold "5" exceeded in Line 17), then the robot swings Left (Line 19). But if "Line 17" turns out to be FALSE, the Controller next checks to see if the IR Right Sensor is the only one detecting the object (threshold "5" exceeded in Line 21), then the robot spins Right (Line 23). But if "Line 21" turns out to be FALSE also, then the Controller next checks to see if the object is close to both IR Sensors (Line 25). If "Line 25" turns out to be TRUE (i.e., both IR Sensors exceed a threshold of "75"), then the robot would go Forward (Line 27). But if "Line 25" turns out to be FALSE, the Controller reaches the end of the IF-ELSE-IF structure without being able to determine a concrete action to do, so the Controller would just loop back to Line 15 to check on the condition of the LOOP WHILE **"((IR_Left >= 5) OR (IR_Right >= 5))"**. If this condition is still TRUE, it would search through the IF-ELSE-IF structure again. If this condition becomes FALSE, the controller finally quits the LOOP WHILE, next triggers Line 30 to stop the robot and then redo the outermost ENDLESS LOOP at Line 14 all over again.

Please note, once again, that when using LOOP WHILE, we need to use "current" IR Sensor Data inputs in all our conditional statements, so that most recent data from the sensors are used.

The algorithm used in the next program "Follower-3.tsk3" is described in Fig 3.21. It is more straight forward so it is easier to explain:

- It uses an outermost ENDLESS LOOP (as always) (Lines 13 to 33) and the first actions by the controller is to go out to the IR Sensors Ports and to read in the latest IR values and save them in Variables **IR_Left** and **IR_Right** (Lines 15 and 16).
- Next is the familiar IF-ELSE-IF structure which now represents 4 mutually exclusive Condition/Action Pairs for "Left", "Right", "Forward" or "Stop". Please note that the IF-ELSE-IF structure in the previous "Follower-2.tsk3" handles only 3 Condition/Action Pairs.
- If Line 17 is TRUE, i.e., only the IR Left Sensor sees the object, so the robot turns Left.
- Otherwise, Line 21 is next executed, and if Line 21 turns out to be TRUE, i.e., only the IR Right Sensor sees the object, then the robot turns Right.
- Still otherwise, Line 25 is executed next, and if Line 25 turns out to be TRUE, i.e., both IR sensors see the object up close, then the robot goes Forward to get even closer to the object.
- Lastly, when no condition turns out to be TRUE (ELSE in Line 29), the robot executes its default behavior, i.e., staying still.

```
13  ENDLESS LOOP
14  {
15      IR_Left  =  IR Left
16      IR_Right  =  IR Right
17      IF ( IR_Left >= 5 && IR_Right < 2 )
18      {
19          CALL Left
20      }
21      ELSE IF ( IR_Right >= 5 && IR_Left < 2 )
22      {
23          CALL Right
24      }
25      ELSE IF ( IR_Left >= 75 && IR_Right >= 75 )
26      {
27          CALL Forward
28      }
29      ELSE
30      {
31          CALL Stop
32      }
33  }
```

Fig. 3.21 Object tracking algorithm in program "Follower-3.tsk3".

There was no discernable difference in performance between these two solutions that the author could see. In theory, "Follower-3.tsk3" should be "faster" because it accesses the sensor ports only twice per each iteration of the overall Endless Loop, i.e. it uses less resources from the Controller as compared to "Follower-2.tsk3" which uses from 4 to 8 accesses to the Sensor Ports per each iteration of its overall Endless Loop. However, "Follower-2.tsk3" can be more "accurate" in its actions as it always uses "fresh" data inputs. In other words, there is no "free lunch" in life and in robotics.

Please note that ELSE is used in Line 29 of "Follower-3.tsk3", and its use is "critical" here. As an exercise for the reader, try to comment out Line 29 (i.e., to make "Call Stop" the default behavior) and you will find that the robot no longer works properly. But if you **then add proper Delay Times** (~100 ms) for the actions "Left", "Right" and "Forward", you will find that the robot is working properly again! (see "Follower-3a.tsk3"). So once again, the robot seems to work better when it is slowed down an appropriate amount!

Recapping Sub-Section 3.1.5, the author's goal was to show that programming "autonomous" behaviors can be fun and challenging at the same time! Starting from the (now familiar) Condition/Action analysis, **three techniques can be used to create TASK codes and after some initial actions, they all used an outermost ENDLESS LOOP for the main algorithm:**

- In Technique 1, for each iteration of the Endless Loop, latest readings from the used sensors were first recorded, then an IF-ELSE-IF structure was used to represent all possible but mutually exclusive Condition/Action Pairs that corresponded to robot behaviors required to fulfill a given challenge. Example programs using Technique 1 are "IR_based_Maneuvers-IF-ELSE-IF.tsk3", "Avoider-1.tsk3", "Avoider-2.tsk3", and "Follower-3.tsk3".
- In Technique 2, the sensors reading task is still required first, but the IF-ELSE-IF structure is replaced by "parallel" IFs. This technique requires a deeper analysis of the original Condition/Action list to identify a "default" behavior, and other Condition/Action Pairs that can work "on their own" or "in concert". This technique also requires a short time delay incorporated into each robot action. Compared to Technique 1, Technique 2 usually results in a smaller number of statements used in the TASK code, meaning that the Controller would take less time to do each iteration of the Endless Loop, and thus the robot may become more responsive to changes in the "World". Example programs using this technique are "Avoider-3.tsk3" and "Follower-1-IFs.tsk3".
- Technique 3 is a variation of Technique 2, and it uses "parallel" LOOP WHILEs instead of "parallel" IFs. It combines "Data Acquisition" tasks in its Conditional Statements, thus it uses the least number of statements, as compared to the other two Techniques. Example programs using Technique 3 are "Avoider-4b.tsk3" and "Follower-1.tsk3". "Follower-2.tsk3" is a hybrid solution combining Techniques 3 and 1.

3.1.6 Basic Remote-Control Techniques

On the Chromebook, the TASK 2 and 3 tools support a Virtual Remote Controller accessible via its Integral Keyboard or its attached Mouse/Pad (see Fig. 1.5).

A full "Remocon" Packet is based on a 16-bit user-designed message wrapped up in a 6-byte packet (http://support.robotis.com/en/product/auxdevice/communication/rc100_manual.htm). Applications using the "full" Packet will be shown in Chapter 4 via Python and TASK programming. For this Sub-Section 3.1.6, only the **first 10 bits** of the "Remocon" Packet are demonstrated in RC applications of the Avoider robot.

Both the Physical RC-100B Remote Controller and the Virtual RC-100 Remote Controller are designed to behave programmatically in the same manner:

- Each of the 10 buttons U-D-L-R-1-2-3-4-5-6 is represented by 1 bit in a 10-bit number/message, with the "U" button using the bit-0 position and the "6" button using the bit-9 position (see Fig. 3.22).

RemoCon 10-bit Message

Bit Position	*9*	*8*	*7*	*6*	*5*	*4*	*3*	*2*	*1*	*0*
RC Button	6	5	4	3	2	1	R	L	D	U
DataIn	0	0	0	1	1	0	0	1	0	0

Specific Button Info. in DataIn can be filtered out by ANDing it with Mask

Example Mask	0	0	1	0	0	1	0	0	0	1

Fig. 3.22 Structure of an RC-100 10-bit message.

- Whenever one or several buttons are pushed by the user, this 10-bit message is sent from the Physical and Virtual RC-100 to the robot. The "DataIn" row in Fig. 3.22 shows that Buttons "L", "2" and "3" are pushed as these bits are set to "1". All the "un-pushed" buttons of course have their bits set to 0.
- Lastly, **whenever the user releases all buttons, the Physical and Virtual RC-100 would send out the number "0"** (i.e., all bits set to 0) ONCE (which makes it hard to catch it sometimes). This feature is unique with the ROBOTIS RC-100 system, other manufacturers may not implement the same scheme.

The main issues in using the RC-100 are to know how many buttons are being pushed, and how to sort out which buttons are being pushed, as information about **all 10 buttons** are sent over inside each transmitted RC message.

First, the issue of "how many" is controlled by the programmer, as readers may well know that it is impossible to expect that actual users of one's programs will only push the buttons that the programmer would want them to! The programmer would need to use a Bit-level AND Operation to "filter" the 10 bits of **DataIn** with his/her chosen 10-bit **Mask**: for example, if the programmer is only interested in Buttons "U", "1" and "4", he/she would use the Mask shown in Fig. 3.22 having "1" at the appropriate Bit Positions.

The program "RC-1Button.tsk3" (Fig. 3.23) shows how to restrict one's program to deal only with 1 "expected" button ("U" or "D") and to ignore any other situation:

- After the usual Initial Action section whereas Melody 2 is played and Variables **Data** and **DelayTime** are initiated respectively to 0 and 250 (i.e., 250 ms), the main algorithm is shown in Fig. 3.23:

```
13  ENDLESS LOOP
14  {
15     WAIT WHILE ( Remocon Data Arrived == FALSE (0) )
16     Data = Remocon RXD
17     Print Screen with Line = Data
18     IF ( Data == U )
19     {
20        Print Screen = Data
21        Print Screen with Line = 11
22     }
23     ELSE IF ( Data == 0000 0000 0000 0000 0000 0000 0000 0010 )
24     {
25        Print Screen = Data
26        Print Screen with Line = 22
27     }
28     Delay = DelayTime
29  }
```

Fig. 3.23 Main algorithm in program "RC-1Button.tsk3".

- Statements 15 and 16 need to be used together and they represent the first actions to be done in a basic RC Loop. The CM-150/151 Controller's Firmware has a built-in process that keeps on listening on the RC communication channel and if a new message has been received, it sets the "**Remocon Data Received**" flag to TRUE, otherwise this flag is set to FALSE. Thus, at each iteration of the Endless Loop (Lines 14-29), the Controller would first check the WAIT WHILE's condition set in Line 15. If no button has been pushed, this condition evaluates to (FALSE == FALSE), therefore this WAIT WHILE's condition evaluates to TRUE, and the Controller would stay at Line 15. Eventually, the user pushes one or many buttons on the RC-100 device. This would make the flag "**Remocon Data Received**" become TRUE and the WAIT WHILE's condition now becomes (TRUE == FALSE) which evaluates FALSE. When this happens, Line 15 is exited, and Line 16 is executed next. It goes to the memory buffer "**Remocon RXD**" and save the value found in there into the Variable **Data**. Line 17 essentially prints out a decimal number that represents the status of all 10 buttons.
- As this program is only interested in the event whereas only ONE button has been pushed and furthermore this single button needs to be either "U" or "D" (but not both), the IF-ELSE-IF structure (Lines 18-27) implements these requirements. Line 18 checks to see if **Data** is equal to **Button "U"**, if this turns out to be TRUE, the CM-150/151 prints **Data** and then the number "**11**" as a marker (Lines 20-21). If Line 18 turns out to be FALSE, then Line 23 is executed next. Line 23 checks to see if **Data** is equal to **Button "D"**, here **Button "D"** is shown as its **binary-number equivalent** (showing that **Bit 1 is set**). If Line 23 turns out to be TRUE, the CM-150/151 prints **Data** and then the number "**22**" as a marker (lines 25-26). If Line 23 turns out to be FALSE, then Line 28 is executed next (i.e., a 250 ms delay so that the user can reasonably read the Output

Monitor Window inside the TASK tool). Then another iteration of this Endless Loop begins again at Line 15.

The next program "RC-2Buttons.tsk3" shows how to parse out two specific buttons "Button1" and "Button2" from the RC-100's 10-bit message. "Button1" needs to be a single-button input from the group "U-D-L-R", while "Button2" needs to be a single-button input from the group "1-2-3-4". The main algorithm is shown in Fig. 3.24, and it has two new features - a binary filtering procedure using the bit-level "&" operator (Lines 21 and 23) and a combo-condition using the logical OR operator "||", implemented in each of two parallel IFs structures:

```
15  ENDLESS LOOP
16  {
17    WAIT WHILE ( Remocon Data Arrived == FALSE (0) )
18    Data = Remocon RXD
19    Print Screen with Line = Data
20    // Filtering out U-D-L-R buttons for Button1
21    Button1 = Data & U+D+L+R
22    // Filtering out 1-2-3-4 buttons for Button2
23    Button2 = Data & 1+2+3+4
24    IF ( Button1 == U || Button1 == D || Button1 == L || Button1 == R )
25    {
26      Print Screen = Button1
27      Print Screen with Line = 11
28    }
29    IF ( Button2 == 1 || Button2 == 2 || Button2 == 3 || Button2 == 4 )
30    {
31      Print Screen = Button2
32      Print Screen with Line = 22
33    }
34    Delay = DelayTime
35  }
```

Fig. 3.24 Main algorithm in program "RC-2Buttons.tsk3".

- The Endless Loop starts with the usual check of the "**Remocon Data Received**" flag (Line 17), next it saves the most recent RC-100 input message into Variable **Data** (Line 18) and prints it (Line 19).
- **Button1** is extracted out of Variable **Data** by ANDing it with the Mask **U+D+L+R** (Line 21 in Fig. 3.24 and more details in Fig. 3.25). In Fig. 3.25, "Example Data" shows

that Buttons "L" and "3" were pushed, but after the "&" operation with the Mask "U+D+L+R", only the information from Button "L" remains.

- Similarly, **Button2** is extracted out of Variable Data by ANDing it with the Mask 1+2+3+4 (Line 23 in Fig. 3.24 and more details in Fig. 3.26). In Fig. 3.26, "Example Data" shows that Buttons "L" and "3" were pushed, but after the "&" operation with the Mask "1+2+3+4", only the information from Button "3" remains.
- Next, two parallel IFs are used to check on **Button1** and **Button2** separately for their actual "values", i.e., membership within their appropriate groups/masks (Lines 24 and 29). If each or both IFs turn out to be TRUE, then the values for **Button1** or/and **Button2** would be printed along their appropriate markers "**11**" or "**22**".

AND Operation with Mask U-D-L-R

Bit Position	*9*	*8*	*7*	*6*	*5*	*4*	*3*	*2*	*1*	*0*
Button	**6**	**5**	**4**	**3**	**2**	**1**	**R**	**L**	**D**	**U**
Example Data	0	0	0	1	0	0	0	1	0	0
						&				
Mask U+D+L+R	**0**	**0**	**0**	**0**	**0**	**0**	**1**	**1**	**1**	**1**
Result from & Op.	***0***	***0***	***0***	***0***	***0***	***0***	***0***	***1***	***0***	***0***

A	B	A & B
1	0	0
1	1	1
0	0	0
0	1	0

Fig. 3.25 Result from ANDing operation between "Example Data" and Mask (U+D+L+R).

AND Operation with Mask 1-2-3-4

Bit Position	*9*	*8*	*7*	*6*	*5*	*4*	*3*	*2*	*1*	*0*
Button	**6**	**5**	**4**	**3**	**2**	**1**	**R**	**L**	**D**	**U**
Example Data	0	0	0	1	0	0	0	1	0	0
						&				
Mask 1+2+3+4	**0**	**0**	**1**	**1**	**1**	**1**	**0**	**0**	**0**	**0**
Result from & Op.	***0***	***0***	***0***	***1***	***0***	***0***	***0***	***0***	***0***	***0***

A	B	A & B
1	0	0
1	1	1
0	0	0
0	1	0

Fig. 3.26 Result from ANDing operation between "Example Data" and Mask (1+2+3+4).

We are now ready to expand the concepts learned from previous paragraphs to work on a Remote-Control project for the Avoider which allows the user to use multiple Direction Buttons **simultaneously** among the group U-D-L-R and to additionally set Low-Speed or High-Speed Variables on the fly using Buttons 1 and 3 ("RC-MultiDirectionsSpeeds.tsk3).

The first part of the main algorithm is shown in Fig. 3.27:

```
21  ENDLESS LOOP
22  {
23      WAIT WHILE ( Remocon Data Arrived == FALSE (0) )
24      Data = Remocon RXD
25      Print Screen with Line = Data
26      IF ( Data == -- )
27      {
28          CALL Stop
29      }
```

Fig. 3.27 First Part of Algorithm used in "RC-MultiDirectionsSpeeds.tsk3".

- After the usual "**Remocon Data Received**" flag checking and the reading/printing of the incoming message (Lines 23-25), and if the incoming **Data** turns out to be "0" (Line

25), i.e. “No Button” is pushed, “**CALL Stop**” is used to stop all motions for the robot (Line 28).

- If Line 25 turns out to be FALSE, i.e., some Buttons are pushed, then the ELSE branch is taken (Line 30 – see Fig. 3.28) and each possible Direction and Speed Buttons are parsed out using the “&” Operator with the appropriate masks (see Fig. 3.29, Lines 33-38).

```
30  ELSE
31  {
32      // Filtering out specific U-D-L-R-1-3 buttons
33      ButtonUp = Data & U
34      ButtonDown = Data & D
35      ButtonLeft = Data & L
36      ButtonRight = Data & R
37      Button1 = Data & 1
38      Button3 = Data & 3
```

Fig. 3.28 Parsing out various Direction and Speed Buttons in “RC-MultiDirectionsSpeeds.tsk3”.

- Next, 6 parallel IFs are used to trigger the appropriate maneuvers and speed settings depending on which Buttons are determined to be pushed simultaneously at run-time (see Fig. 3.29).

During runtime of the program “RC-MultiDirectionsSpeeds.tsk3”, the reader can observe that Speed settings can be made in parallel with the Directional maneuvers of the robot. While the maneuvers themselves can be combined with each other in a functional way (e.g., “U” and “R” together) or in a completely ineffective way (e.g., “U” and “D” together)! One possible solution for this issue the use of IF-ELSEs between “U<>D”, and between “L<>R”: this is left as an exercise to the reader.

```
IF ( ButtonUp == U )
{
    CALL Forward
}
IF ( ButtonDown == D )
{
    CALL Backward
}
IF ( ButtonLeft == L )
{
    CALL Left
}
IF ( ButtonRight == R )
{
    CALL Right
}
IF ( Button1 == 1 )
{
    Speed = 400
}
IF ( Button3 == 3 )
{
    Speed = 800
}
```

Fig. 3.29 Parallel IFs used for setting Direction(s) and Speed in "RC-MultiDirectionsSpeeds.tsk3".

At this point, most of the TASK's basic programming tools and techniques have been explained and demonstrated, thus the next Section 3.2 will develop STA-derived concepts and techniques for analyzing project requirements and learning about project development techniques from robot's task analysis to complete TASK code generation.

3.2 Beginner's Approach for R+TASK Projects

In this section, the author strives to provide beginning "roboteers" with a general approach to thinking about robotics projects, as well as concrete sequential steps that ultimately result in an efficient R+TASK program that can perform the tasks/behaviors that are required of the DREAM II robot for a given project.

3.2.1 Sense-Think-Act Paradigm revisited

The Sense-Think-Act Paradigm (previously described in Sub-Section 1.2) can be used a General Thinking Tool for solving various TASK robotics projects. As applied to our current use environment of "Chromebook + TASK + DREAM II Robot", the Human STA Cycle can be "redrawn" as shown in Fig. 3.30:

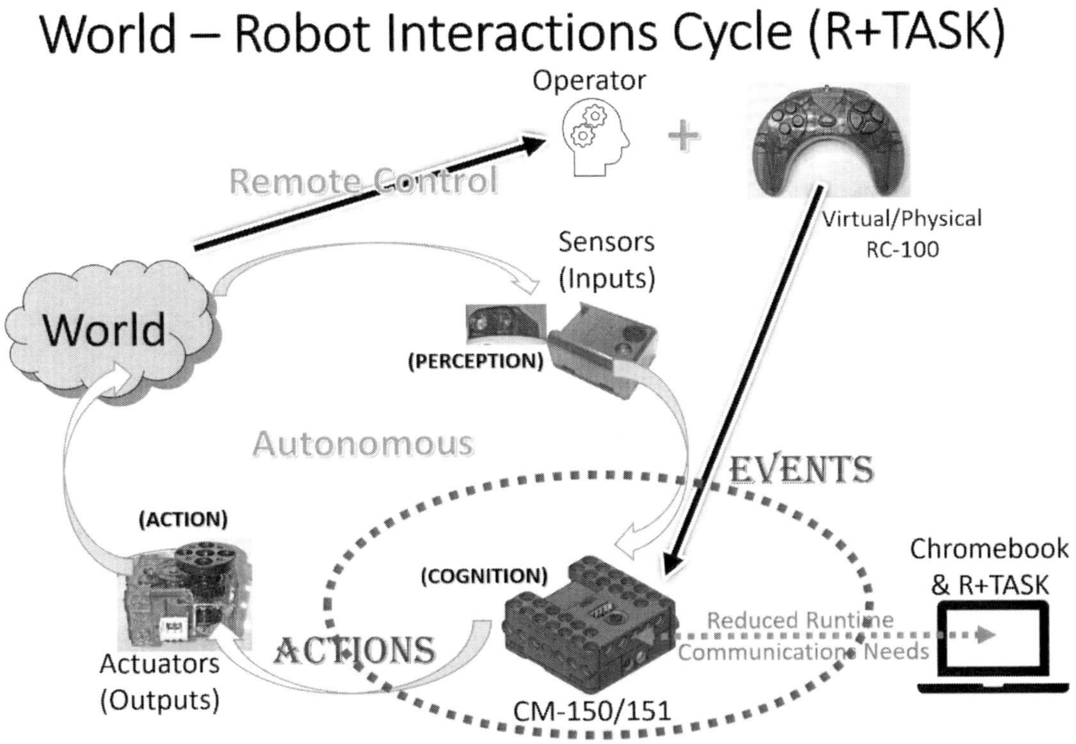

Fig. 3.30 World-Chromebook-TASK-DREAM II Robot Interactions Cycle.

- Fig. 3.30 displays the Autonomous Control Cycle, via the use of IR Sensors, that had been used previously in Section 3.1.
- R+TASK supports the use of a Remote Controller module (Virtual and Physical RC-100) which will be used to implement a Remote-Control mode with the Avoider platform.
- As the TASK binary runtime code resides and executes on the CM-150/151 (i.e., Main Controller), there is much reduced need for communications with the User/Output Interfaces on the CB at code demonstration time. The CB is now serving only as a Graphical Display at runtime.

3.2.2 Reactive-Control Approach

We will be using the "Sense-Think-Act" Paradigm to helps us keep the "big picture" of any robotics project in mind, as we start our "drilling down" journey into this paradigm to get to the actual robot-controlling code that works on the Chromebook+CM150/151 system.

Let's go through an example to see how this Sense-Think-Act paradigm can help us solve the following challenge as described in Fig. 3.31, where a CM-150/151 based Probe Car robot (*which is in Level 4*) would be able to follow an irregular black track in the shape of the number 8. Furthermore, if the Probe Car encounters a frontal obstacle (e.g., a Paper Roll) during its travel, it would need to stop in its track, pick up the Paper Roll, move it out of the way and finally return to its original direction of travel.

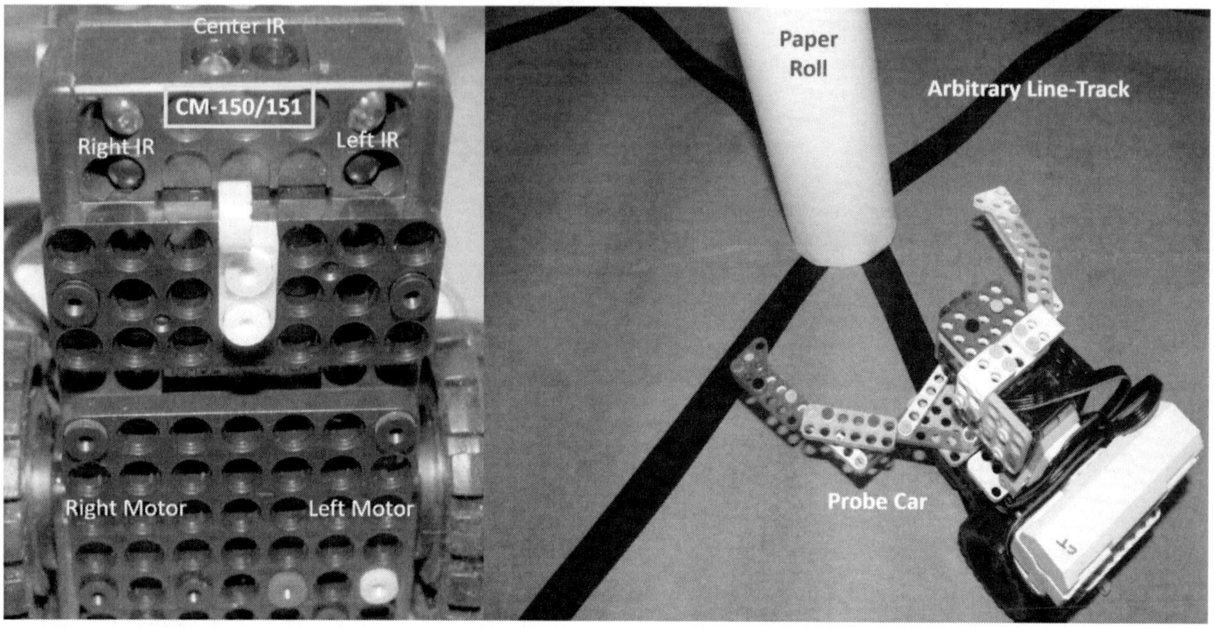

Fig. 3.31 CM-150 Probe Car following an 8-track and avoiding a frontal obstacle.

Fig. 3.31 shows that the CM-150/151 has three Integrated IR Sensors whereas the Center IR is used to detect a frontal obstacle and the Right and Left IR sensors are used to detect the Irregular Line Track. The CM-150/151 also has a Continuous-Turn Motor (GM-10) on its Left and Right sides. These motors can be programmed to provide all Car Maneuvers that would be needed to perform this challenge successfully (i.e., Forward, Turn Left and Turn Right). This System Design solution seems feasible enough, but to progress further towards the actual coding of the CM-150/151, we need to zoom in around the Robot/Computer area in Fig. 3.30 (i.e., the dashed red boundary in the lower right corner). The Sense-Think-Act paradigm still applies at this level, but some robotics practitioners would start using different labels such as "Events", "Actions" and "Reactive Control" as they are more useful in the next steps that we need to take in creating the actual computer code to control this Probe Car. Interested readers should read Matarić (2007) for a more complete description of other robotics control approaches, or Ben-Ari and Mondada (2018) for more details about "Reactive Control" and "Finite State Machines".

3.2.3 Event-Action Pairs

With the "Reactive Control" approach, ones need to match up a possible Condition/Event with an appropriate Action for the robot to do (see Fig. 3.32).

Reactive Control Approach

Given Condition (Event) >> Appropriate Robot Action

Condition/Event	Action
Car on Top of Track	Car Goes Forward
Car Left of Track	Car Turns Right
Car Right of Track	Car Turns Left
Paper Roll detected	Pick Up, Dispose of Paper Roll & Rotate Left until Original Direction

1 and Only 1 Condition (Event) can happen at any 1 time.
Or Multiple Conditions (Events) can happen at any 1 time.

Fig. 3.32 Reactive Control Approach used for the Probe Car challenge.

For example, if the Probe Car happens to be lined up on top of the track already, then it should just go forward. There are 3 other Event/Action Pairs listed in Fig. 3.32, and are these four Pairs the only ones to consider or needed for this robot challenge? My advice for beginning roboteers is just to start with a reasonable set of Event/Action Pairs and then improve on them as needed in later refinement steps.

But how would a beginner in robotics comes up with such a list as shown in Fig. 3.32 in the first place? Here, the author offers a simple "thought experiment" or "visualization exercise" approach:

- First, we can imagine to be "one" with the robot, so our Left and Right Eyes would match with the Left and Right IR Sensors, i.e., we keep our nose to the ground so that we can see the black track (to be exact, only a partial view of it). However, humans do not have a third eye right on top of our head to correspond to the Center IR, so we would really have to use our imagination and use our "mind's eye" as the third eye there!
- So now that we are the robot, how would we go about performing this 8-track challenge? Here, although not mentioned in the challenge write-up, we need to decide on where is the starting position of the robot with respect to the 8-track? At the start of the challenge, the robot could very well be positioned on a brown paper area inside or outside the track, or it could straddle the track itself. As the challenge's rules did not specify the starting event, we are going to choose the "easiest" option of having the Probe Car start its challenge already lined up with the track. Then, the reader can see why the first Event/Action Pair came about on the table list in Fig. 3.32. Next, as the 8-track would curve here and there, the robot may find itself Left or Right of the track (second and third Event/Action Pair respectively). So, with the first three Event/Action Pairs, we should feel confident, at least logic wise, that the Probe Car would be able to follow the 8-track using computer codes derived from Fig. 3.32.
- Similarly, we can come up with the 4th Event/Action Pair to take care of the situation when an obstacle comes within detection range of the Center IR. Here the alert reader would notice that the first three Actions were Simple Actions triggered when certain conditions/events occurred. But the 4th one had several Actions in sequence and its last Action was a much more complicated "Action-until-Event", i.e., Act now until some Event

happens later (e.g., Rotate Left until Probe Car returns to its original direction). In other words, there is a sort of a nested Event/Action inside an "Action-until-Event".

Before we go on to the next step, we need to have a closer look at the four Event/Action pairs (or Conditions if you will) listed in Fig. 3.32. Logically, the reader can see that the first three conditions are mutually exclusive (i.e., one and only one of them should happen at any one time at runtime), but that the 4th condition may happen in conjunction with any of the first three conditions. Or using a different viewpoint, it is only when the 4th condition is false (i.e., no obstacle in front of the Probe Car) that we would care about how to handle the first three conditions. **So, the "take-home" lesson is that not all conditions "discovered" during this step have the same logical importance or ranking.** Please keep this hierarchy of conditions in mind when ones get to the later steps of creating the actual TASK code for this type of challenge.

3.2.4 Sensor-Actuator Pairs

The next step is to build upon the results of the previous step by mirroring its Event/Action Pair into a matching Sensor/Actuator Pair as shown in Fig. 3.33, as an Input/Output table. **The goal is to specify the kind of Sensors/Actuators to be used and to prescribe how they are to be used.**

Input-Output Table

Given Input Sensor(s) >> Activate Appro. Actuator(s)

Input Sensors	Output Actuators
Right IR & Left IR see "Black"	Right Motor CW Left Motor CCW
Right IR sees "Black" & Left IR sees "Brown"	Right Motor Stops Left Motor CCW
Right IR sees "Brown" & Left IR sees "Black"	Right Motor CW Left Motor Stops
Center IR sees "Obstacle"	Right Motor CW & Left Motor CW until Left IR sees "Black"?

Fig. 3.33 Input/Output Table mirroring list of Event/Action Pairs as shown in Fig. 3.32.

For example, let's look at the second Sensor/Actuator Pair in Fig. 3.33 which corresponds to the second Event/Action Pair in Fig. 3.32 which specifies that if the Probe Car is found on the left of the black track, then it should turn right so that it can get back onto the track. Thus, for the Probe Car to be on the Left of the track, its Left IR should be seeing "Brown" and its Right IR should be seeing "Black" (see 2nd row of the Input Sensors column in Fig. 3.33). The corresponding Action for the Probe Car is then to Turn Right, i.e., the Right motor should stop, and the Left motor should go Forward, i.e., keep on turning CCW (see 2nd row of the Output Actuators column in Fig. 3.33).

Let us next have a closer look at the specifications for the Output Actuators for the 4th Event when a frontal obstacle comes within the detection range of the Center IR Sensor. This 4th Event would trigger a sequence of complex actions as described in Fig. 3.32, but for lack of space in Fig. 3.33, we will just consider the very last Action "Rotate Left until Back on Track": i.e., the Probe Car is supposed to spin Left until it straddles the black track again and in the original travel direction, so the specifications were to run the Right motor Forward (CW) and to run the Left motor going Backward (CW also) so that the Probe Car would rotate Left, until the Left IR Sensor sees Black. Or would the reader think that it should really be "until the **Right** IR Sensor sees **Black**" instead, to make sure that the Probe Car is completely straddling the black track?

Please note that the previous procedures do not involve the use of any computer nor learning about how to use a specific software tool, and the author highly recommends that beginner roboteers to always start with these procedures instead of jumping directly into coding or try to adapt someone else codes, as these procedures would help the roboteer refine in his/her mind the tasks to be done by the robot, before committing them into hard codes, thus saving the programmer lots of time in the long run.

In the "Computational Thinking" practice, the previous two Pairing steps (Event/Action and Sensor/Actuator) would correspond to the "Decomposition" phase (Beecher, 2017; Denning and Tedre, 2019).

3.2.5 Event-Programming Approach

Another benefit of using the Event/Action and Input/Output Tables is that they will help with the actual writing of the TASK code (as shown later in Chapter 3) as the TASK syntax used would support practically a one-to-one correspondence of all Event/Action Pairs from these tables into similar "physical" locations within the final program, except for two aspects that had not been discussed yet (see Fig. 3.34):

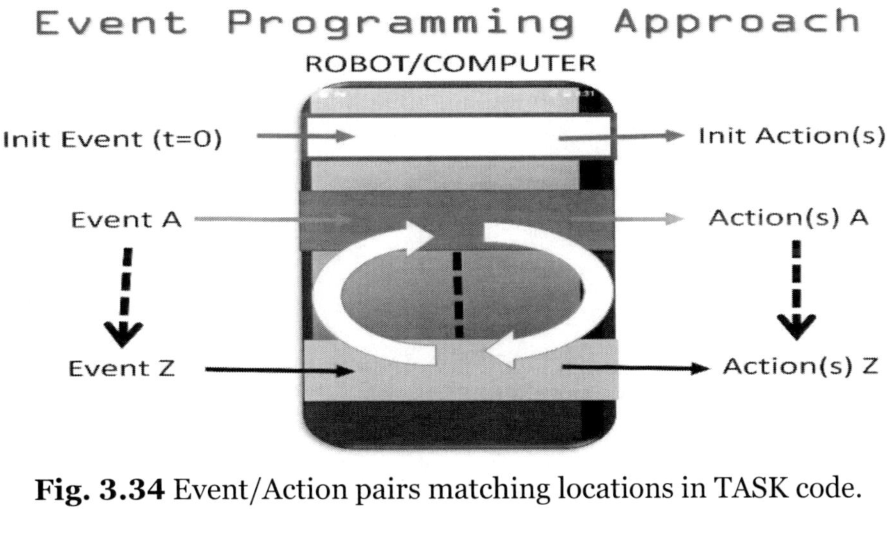

Fig. 3.34 Event/Action pairs matching locations in TASK code.

- A main code looping structure that allows the controller to go through the list of possible Events A to Z that may occur during run time, as it listens to what the Sensors tell it about the current state of the World. This is the overall loop (i.e., Interactions Cycle) shown in Fig. 3.30.

- The Init Event which occurs once at time=0, i.e., when power is applied to the robot or when the TASK program starts running. The Init Event would usually result in a known state for the robot which could be some audio output to let the user knows that it is ready for "action", or it could be a display of a menu of tasks to let the user choose which task for the robot to perform when using R+m.PLAY700 App (see Thai (2018)).

In "Computational Thinking" practice, these steps would correspond to the "Pattern Recognition" phase.

3.2.6 Application to Line Tracker with Speed Control

For this project, the Line Tracker robot is constructed by rotating the Avoider robot 90 degrees "down" and by attaching a small wheel to its front (see Fig. 3.35), to keep the four IR Sensors within a good working range for surveying the ground surface, although only the Internal IR Left and Right Sensors will be used for following the Black Track.

Fig. 3.35 Line Tracker robot.

This project's requirements are as follows:

- At start, the robot is to be lined up with the Black Track before its power is enabled.
- At Power On, the robot plays a Buzzer Melody to notify the Operator that it is ready, but it **stays still**, until it receives an RC packet from the Operator tapping on the **"U" Button** of the RC-100, then it can go Forward at a Default Speed, while staying On-Track.
- When the robot is in Motion, the Operator can tap on **Button "D"** at any time to stop the robot. Then, at will, the Operator can restart the robot at Default Speed by tapping the **"U" Button**.
- When the robot is traversing the Track, the Operator can tap **Button "1"** to increase the robot's Speed by 5 units for each Button "1" Tap, or **Button "3"** to decrease the robot's Speed by 5 units for each Button "3" Tap. The goal of this feature is to determine the maximum Speed that the robot can use to keep On Track for all parts of a Given Black Track.

For this "Line Tracker With Speed Control" (LTWSC) Project, we have two types of Events: one from the IR Sensors to appraise the robot about its On-Track status ("Line Tracking" algorithm), and the other is from the potential RC Packets that may be sent over from the Operator ("Speed Control" algorithm).

For the "Line Tracking" algorithm, the first three Event/Action Pairs listed in Fig. 3.32 apply for this Project. Noting that these Event/Action Pairs are mutually exclusive, they can be transformed to Sensor/Actuator Pairs:

- If the Left IR Sensor sees the Black Track and the Right IR Sensor sees the White Background, then the robot should Turn Left in an attempt to re-align itself with the Black Track.
- If the Right IR Sensor sees the Black Track and the Left IR Sensor sees the White Background, then the robot should Turn Right in an attempt to re-align itself with the Black Track.
- Otherwise, the robot should continue to go Forward.

The reader can use the provided sample program "Test_IR_Sensors.tsk3" to determine the appropriate IR Threshold values to use for the detection of the Black Track.

For the "Speed Control" algorithm, we can apply the "1-Button" RC technique so as to allow only 1-Button inputs from the Operator, i.e., only 1 Button allowed from the list of "U", "D", "1" and "3" Buttons (see Fig. 3.23).

In "Computational Thinking" practice, these steps would correspond to the "Abstraction" phase.

The solution for the LTWSC Project is the program "Line_Tracker_SC.tsk3", illustrated via Figs. 3.36 through 3.39.

```
7   // Init Action
8   Buzzer Timer = Melody Time
9   Buzzer Index = Melody No.2 (2)
10  WAIT WHILE ( Buzzer Timer > 0.0sec )
11  Speed = 0
12  Threshold = 250
13  Data = 0
```

Fig. 3.36 Init Event for "Line_Tracker_SC.tsk3".

Fig. 3.36 describes the Init Event:

- First, Melody 2 is played (Lines 8-10).
- Line 11 initializes Variable **Speed** to 0 which is used as the robot's maneuvering speed.
- Line 12 initializes Variable **Threshold** to 250 which is used to detect the Black Track.
- Line 13 initializes Variable **Data** to 0 which is used to receive the RC packet sent from the Operator.

Fig. 3.37 displays Part 1 of the Main Endless Loop which handles the "Line Tracking" algorithm:

- The First IF construct matches with Sensor-Actuator Pair #1, i.e., when IR Left Sensor detects the Black Track and IR Right Sensor detects the Background, then the robot should Turn Left (Lines 17-21).
- The Second IF construct (ELSE-IF) matches with Sensor-Actuator Pair #2, i.e., when IR Left Sensor detects the Background and IR Right Sensor detects the Black Track, then the robot should Turn Right (Lines 22-26).
- The Final ELSE construct matches with Sensor-Actuator Pair #3, i.e., when the robot should keep Going Forward (Lines 27-31).

```
15  ENDLESS LOOP
16  {
17      IF ( IR Left < Threshold && IR Right > Threshold )
18      {
19          // Turns left when ONLY the left sensor detects black.
20          CALL LeftTurn
21      }
22      ELSE IF ( IR Left > Threshold && IR Right < Threshold )
23      {
24          // Turns right when ONLY the right sensor detects black
25          CALL RightTurn
26      }
27      ELSE
28      {
29          // Moves forward for all other cases.
30          CALL Forward
31      }
```

Fig. 3.37 "Line Tracking" algorithm for "Line_Tracker_SC.tsk3".

Figs. 3.38 and 3.39 displays Part 2 of the Main Endless Loop which handles the "Speed Control" algorithm. Fig. 3.38 shows how to pick up and handle Buttons "1" and "3":

- Line 32 checks on the possible arrival of a new RC packet via the "**Remocon Data Arrived**" flag. If this condition evaluates FALSE, the Controller returns to the beginning of the Main Endless Loop at Line 17 in Fig. 3.37. If this condition evaluates TRUE, Line 34 is activated next, and the value of the received RC packet is saved into Variable **Data**.
- Next, an IF-ELSE-IF structure is used to handle Buttons "1-3-U-D" (one at a time only).
- Lines 35-42 handles the case when Button "1" is received via **Data**:
    - **Speed** is incremented by 5 units (Line 37).
    - Lines 38-41 are used to make sure that **Speed** does not exceed the value of **1023** which corresponds to 100% for the Ports 1 and 2 Gear Motors.

```
32      IF ( Remocon Data Arrived == TRUE (1) )
33      {
34        Data = Remocon RXD
35        IF ( Data == 1 )
36        {
37          Speed = Speed + 5
38          IF ( Speed >= 1023 )
39          {
40            Speed = 1023
41          }
42        }
43        ELSE IF ( Data == 3 )
44        {
45          Speed = Speed - 5
46          IF ( Speed <= 0 )
47          {
48            Speed = 0
49          }
50        }
```

Fig. 3.38 Handling Buttons “1” and “3” in “Speed Control” algorithm for “Line_Tracker_SC.tsk3”.

- Lines 43-50 handles the case when Button “3” is received via **Data**:
  - **Speed** is decremented by 5 units (Line 45).
  - Lines 46-49 are used to make sure that **Speed** does not go negative, as negative values will reverse the Turning Direction of the Ports 1 and 2 Geared Motors.

Fig. 3.39 shows how to pick up and handle Buttons “U” and “D”:

- Lines 51-54 handles the case when “U” is received via **Data:**
  - **Speed** is then set to 350 by Line 53 to restart motion on the robot.
- Lines 55-58 handles the case when “D” is received via **Data:**
  - **Speed** is then set to 0 by Line 57 to stop the robot.
- Line 59 informs the Operator on the current value for **Speed**.
- Line 60 clears out Variable **Data** for another iteration of the Main Endless Loop.

```
    ELSE IF ( Data == U )
    {
        Speed = 350
    }
    ELSE IF ( Data == D )
    {
        Speed = 0
    }
    Print Screen with Line = Speed
    Data = 0
}
```

Fig. 3.39 Handling Buttons "U" and "D" in "Speed Control" algorithm for "Line_Tracker_SC.tsk3".

In "Computational Thinking" practice, the previous steps would correspond to the "Algorithm" phase.

Using the "Smooth Curve" Track that comes with the DREAM School Set, the author's LTWSC robot could get to a **Speed** setting of around 900-1000.

3.2.7 Application to Smart Avoider

For this "Smart Avoider" Project, the Avoider robot is reset to its original Up-Right configuration as shown in Fig. 1.1. The Smart Avoider will use Ports 3 and 4 IR Sensors to have a longer detection range.

The "Smart Avoider" Project shows how to combine Autonomous-Behavior and Remote-Control techniques so that the Avoider robot can "save itself" when, on purpose, the user drives the robot straight into an obstacle. Essentially, we'll try to program a little bit of "intelligence" into the robot!

The overall Control Algorithm can be expressed as a pseudo-code shown below:

Main Endless Loop {

If (Port 3 & 4 NIR Sensors detect no obstacle) then ***{*** *// Take "RC" Path*

When new RC packet comes in, save it in ***Data***

Depending on the Actual value of ***Data****, activate the appropriate Robot Maneuver (Stop, Forward, Backward etc...)*

} *// End of RC Path*

Else ***{*** *// Obstacle within NIR detection ranges so take "Escape" Path*

Stop the robot

Figure out whether Obstacle is Frontal, on Left or on Right of robot and activate appropriate robot's avoidance moves <u>*until obstacle out of NIR detection ranges*</u>

Use Musical Melody to signal robot's successful obstacle avoidance

} // End of Escape Maneuver

} (End of Main Endless Loop)

If the reader checks this pseudo code against the one used for the "Clap-based Smart Avoider" Project in Sub-Section 2.6.3, the reader will notice that no Delay Time is needed for each maneuver to activate the motors anymore, because the RC Packet flow is a much faster process, and it has the special "0" packet feature when the Operator releases all RC Buttons (see Sub-Section 3.1.6).

In "Computational Thinking" practice, this step would correspond to the "Abstraction" phase.

This Project adapts the previous program "RC-MultiDirectionsSpeeds.tsk3" into "Smart_Avoider.tsk3" while implementing the pseudo code described earlier (see Figs. 3.40 through 3.43).

```
5    // Init Action
6    Buzzer Timer = Melody Time
7    Buzzer Index = Melody No.3 (3)
8    WAIT WHILE ( Buzzer Timer > 0.0sec )
9    Data = 0
10   Direction = 0
11   DelayTime = 0.500sec
12   Speed = 512
13   Threshold_1 = 400
14   Threshold_2 = 350
15   Threshold_3 = 300
```

Fig. 3.40 Initialization Steps in "Smart_Avoider.tsk3".

Fig. 3.40 shows the Init Event for "Smart_Avoider.tsk3":

- Lines 6-8 play Melody 3.
- Lines 9-10 zero out Variables **Data** and **Direction** which are used in the processing of RC packets.
- Line 11 initializes Variable **DelayTime** to 0.5 second which is used in implementing the "Escape" Maneuver for the robot.
- Line 12 shows that we'll be using a constant **Speed** setting for this project.
- Lines 13-15 initialize the 3 **Threshold** values used for Ports 3 and 4 IR Sensors when implementing the "Escape" Maneuver.

```
17  ENDLESS LOOP
18  {
19      IF ( PORT[3]:IR Sensor < Threshold_1 && PORT[4]:IR Sensor < Threshold_1 )
20      {
21          // No obstacle in front so OK to abide by user's RC inputs
22          IF ( Remocon Data Arrived == TRUE (1) )
23          {
24              Data = Remocon RXD
25              Print Screen with Line = Data
26              IF ( Data == -- )
27              {
28                  CALL Stop
29              }
30              ELSE
31              {
32                  // Filtering out specific Direction buttons or 1-3 buttons
33                  Direction = Data & U+D+L+R
34                  IF ( Direction == U )
35                  {
36                      CALL Forward
37                  }
```

Fig. 3.41 Processing user's RC Inputs **only when** no obstacle located in front of robot ("Smart_Avoider.tsk3").

Fig. 3.41 shows Part 1 of the Main Endless Loop which handles the "RC" Path when "no obstacle" is detected by Ports 3 and 4 IR Sensors (Line 19):

- In this case, Lines 22-25 handles the received RC Packet by saving its content in Variable **Data** and printing it out for the Operator's monitoring.
- Lines 26-19 handle the situation when the Operator releases ALL RC Buttons which is to Stop the robot (Line 28).
- Otherwise, a big ELSE structure (Lines 30-50) is activated to handle the "mutually exclusive" Events corresponding the potential "U", "D", "L" or "R" Button Pushes performed by the Operator at runtime. But for lack of space, only the case for "U" is shown in Fig. 3.41.
- Line 33 filters out **Data** and only keeps potential UDLR inputs in Variable **Direction**.
- The case for "U" Button is handled by Lines 34-37 which is to command the robot to Go Forward. For the other cases, the reader is referred to the provided sample code "Smart_Avoider.tsk3".

Fig. 3.42 describes the steps programmed to make the robot steer clear of the detected Obstacle:

- First, the robot stops with Line 57.

```
    // End of RC procedure
  ELSE
    // Obstacle is getting close - Robot stops first then tries to avoid obstacle on its own
  {
    CALL Stop
    LOOP WHILE ( PORT[3]:IR Sensor > Threshold_2 || PORT[4]:IR Sensor > Threshold_2 )
    {
      NIR_Diff = PORT[3]:IR Sensor - PORT[4]:IR Sensor
      IF ( NIR_Diff >= -10 && NIR_Diff <= 10 )
      {
        CALL Backward
        CALL Delay
      }
      LOOP WHILE ( PORT[3]:IR Sensor > Threshold_3 && PORT[3]:IR Sensor > PORT[4]:IR Sensor )
      {
        CALL Right
      }
      LOOP WHILE ( PORT[4]:IR Sensor > Threshold_3 && PORT[4]:IR Sensor > PORT[3]:IR Sensor )
      {
        CALL Left
      }
    }
```

Fig. 3.42 Obstacle Avoidance Procedure in "Smart_Avoider.tsk3".

- Essentially, the "Escape" algorithm needs to figure out which one among three possible Event/Action Pairs that may have happened at run time:
    - If the obstacle happens to be "Frontal", Ports 3 and 4 IR Sensors would provide Values that are "about the same". Then the robot should roll Backward a bit, next it should keep on Turning Right or Left until the obstacle is no longer detected.
    - If the obstacle is on the Left of the robot, the Port 3 IR Sensor would yield a Higher Value than the Port 4 IR Sensor. Then the robot should Turn Right until the obstacle is no longer detected.
    - If the obstacle is on the Right of the robot, the Port 4 IR Sensor would give a Higher Value than the Port 3 IR Sensor. Then the robot should Turn Left until the obstacle is no longer detected.

- In Fig. 3.42, the "Escape" algorithm is implemented as an Outer WHILE Loop that contains one IF structure to deal with the case of a "Frontal Obstacle", and two Inner WHILE Loops to deal with the cases of "Left Obstacle" and "Right Obstacle" respectively.
- However, to understand the design of the Outer WHILE Loop, we need to refer back to the Conditional Expression used in the IF structure found in the Main Endless Loop (see Line 19 of Fig. 3.41). When an obstacle is detected, this Conditional Expression becomes **(Port 3 IR Sensor ≥ Threshold_1) OR (Port 4 IR Sensor ≥ Threshold_1)** where **Threshold_1=400**. Thus, the Conditional Expression used for the Outer WHILE Loop follows the **same structure**, except now **Threshold_2=350** is used (see Fig. 3.42 – Line 58), which is "stronger" logic-wise to "guarantee" that the Outer WHILE Loop will be activated when an obstacle is detected (from a previous section of code in the Main Endless Loop – see Fig. 3.42).
- Once this Outer WHILE Loop is entered, Variable **NIR_Diff** is computed (Line 60) and next tested to see if it is **between -10 and 10** (Line 61). If this turns out to be TRUE (i.e., "frontal" obstacle), then the robot rolls Backward for 0.5 second - otherwise no action is done, and the First Inner WHILE Loop is activated next.
- The First Inner WHILE Loop (Lines 66-69) is designed to check if the obstacle is found "only" on the Left of the robot, meaning that **(Port 3 IR Sensor > Port 4 IR Sensor)**. The "extra" condition **(Port 3 IR Sensor > Threshold_3)** where **Threshold_3=300** is used to make sure that the robot would rotate itself "well away" from the obstacle. If this situation turns out to be true at run time, then the robot is instructed to keep on Rotating Right until the obstacle is no longer detectable (i.e., **Port 3 IR Sensor ≤ Threshold_3**). Please note that the 3 Threshold values used **(400, 350, 300)** suited the author's setup only, the reader will need to determine the appropriate values for her/his situation.
- Similarly, the Second WHILE Loop (Lines 70-73) is designed to check if the obstacle is found "only" on the Right of the robot, meaning that **(Port 4 IR Sensor > Port 3 IR Sensor)**, and the extra condition **(Port 4 IR Sensor > Threshold_3)** is used for the same reason as explained above. If this turns out to be the case at run time, then the robot is instructed to keep on Rotating Left until the obstacle is no longer detectable (i.e., **Port 4 IR Sensor ≤ Threshold_3**).
- Please note that for a single obstacle, once the First Inner WHILE Loop is activated, the Second Inner WHILE Loop would not apply at all (i.e., no chance for the robot to rotate back into the obstacle).
- Please note that for a "Single Frontal Obstacle", the robot would roll backwards first, then would likely rotate right to evade further from the obstacle, because of the current ordering of the two Inner WHILE Loops.
- Once the obstacle is evaded, i.e., break-out from the Outer WHILE Loop, the robot exits this Loop at Line 75.

When the algorithm gets out of the outer WHILE Loop at Line 75 of Fig. 3.43, "back in the physical world" everything is clear in front of the robot, then the robot stops (Line 76) and just waits there for the user to "gather his/her wits" after such a close call! At that time, assuming the Operator was still pressing on some button(s) on the Remote Controller, the last WHILE loop (Lines 77-81) is designed to wait the user to release all buttons on the RC-100, i.e., it keeps on looping as long as (**Data** NOT EQUAL to "**No Button**") – conditional statement used on Line 77 (and the author apologizes for using a "double negative" here, but it was necessary!).

```
75          // Here, no more obstacle in front - Robot stops and wait for user to release buttons on RC-100
76          CALL Stop
77          LOOP WHILE ( Data != -- )
78          {
79             WAIT WHILE ( Remocon Data Arrived == FALSE (0) )
80             Data = Remocon RXD
81          }
82       }
83       // End of ELSE for Escape Procedure
84    }
85    // End of Main Endless Loop
```

Fig. 3.43 End of Obstacle Avoidance Procedure in "Smart_Avoider.tsk3".

When the Operator finally releases all RC buttons, the algorithm goes back to the original ENDLESS LOOP (Line 19), and essentially allows the user to use the RC-100 once again to move the robot around.

In "Computational Thinking" practice, the previous steps would correspond to the "Algorithm" phase.

This "Escape" algorithm was designed to respond to a single obstacle such as a flat wall or fence, but the reader can try to send this robot into a wall/fence corner to see how the robot would maneuver itself then. The author had found that most of time this algorithm would handle a corner well, but there are instances where it would get "stuck" bouncing back and forth between the two walls/fences.

The reader is also encouraged to rearrange the IRSS-10 sensors into a "horizontal" configuration and note any changes in the Obstacle Detection performance.

As a comparison, the reader may check out the equivalent "Smart Avoider" solution using MIT SCRATCH-2 running on a Desktop PC (https://www.youtube.com/watch?v=o8w_YVocCpA)

3.3 More Line Tracking Applications

Two more Line Tracking applications are described in Section 3.3. The first one is a stripped-down version of the Probe Car, because the School Set comes with only 1 SM-10 Servo, while the Probe Car needs 2 SM-10s. The second one is an Enhanced Line Tracker using 4 IR Sensors (2 Internal and 2 IRSS-10s out of Ports 3 and 4) which are needed to work on a "Yin-Yang" Track.

3.3.1 Line Tracker & Obstacle Avoider

For this Sub-Section 3.3.1, we will be using a CarBot configuration as described in Fig. 3.44. Construction wise, this CarBot derives from the Avoider (Fig. 2.2), but the whole robot gets rotated down 90 degrees so that the Internal Left & Right IR Sensors are now aimed towards the ground and with additional "sprocket" wheels for front end support. The goal is to program this CarBot to go forward and follow an arbitrary track created using standard black electrical tape laid out on brown Craft paper. This CarBot would have to use its Left and Right NIR sensors to keep it "on-track". During its progress along this track, if the CarBot finds a frontal obstacle, it should be able to swing around, re-discover the black track and follow it again but in the opposite direction.

Fig. 3.44 CarBot used as Autonomous Line Tracker & Obstacle Avoider.

Let's first analyze how the three built-in NIR Sensors behave when the CarBot is laid on the black track under the four main conditions that would occur in this challenge:

- When the track is under the bot's Left IR Sensor.
- When the bot is centered on the track
- When the track is under the bot's Right IR Sensor
- When the bot is gone completely off-track (see Fig. 3.45).

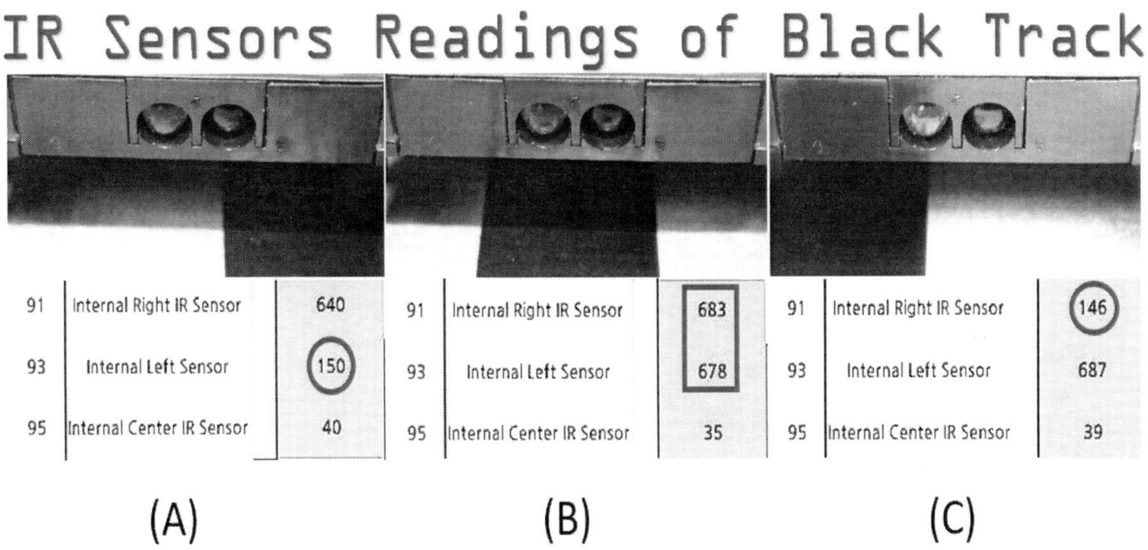

Fig. 3.45 Values for L-R-C IR Sensors for 3 configurations between CarBot and Black Track.

In Fig. 3.45 and going from Column (A) to Column (C) to check for the readings from the Right IR Sensor (Address 91): ones can see that its value is around 640-683 whenever it sees the Background Craft paper, and that its value is reduced to around 146 when this sensor is "seeing" the Black Track. Similarly, for the Left IR Sensor (Address 93), it reads around 150 when it "sees" the Black Track and between 678 and 687 when it "sees" the Craft Paper surface. Most importantly, Column (B) shows that the spacing between the Left and Right IR Sensors is larger than the width of the black electrical tape as both sensors read "high" here (e.g., 683 and 678) - in other words, both sensors are only seeing the brown paper in this situation. **Thus, it is physically impossible for this CarBot to figure out via its Left and Right NIR sensors whether it is centered on the black track (Case B) or that it is completely off-track (Case D) – not shown in Fig. 3.45!**

Also, in Fig. 3.45, an analysis of the IR Center Sensor data shows that this CarBot was aiming a bit low as it still can see a bit of the "brown paper" resulting in values between 35 and 40.

The previous experimental analysis shows that there are only two Conditions (mutually exclusive) that can be monitored unequivocally by this CarBot hardware configuration:

- If the Left IR Sensor can see the Black Track, then it should Turn Left in its attempt to re-align the bot with the Black Track.
- If the Right IR Sensor can see the Black Track, then it should Turn Right in its attempt to re-align the bot with the Black Track.

Otherwise, this robot really cannot be "sure" of anything else, and so it should continue its default maneuver (i.e., Going Forward). From previous discussions, we are then expecting to see an IF-ELSE-IF structure for three mutually exclusive alternatives in the programming solution for this challenge.

Fig. 3.46 shows the Initial Actions for the solution "LineTracker-ObstacleAvoider.tsk3", whereas the CM-150/151 plays Melody 3 (Lines 5-8) then sets initial values for 4 Variables - **Speed, TimeDelay, LineDistance** and **ObjectDistance** (Lines 9-12). The "200" value assigned to Variable **LineDistance** is just a number between the values "680" and "140" that were reported for both IR Left and Right Sensors (see Fig. 3.45 and blue circles in columns (A) and (C)). **ObjectDistance** is set to "100", a number sufficiently larger than the typical 35-40 values found in Fig. 3.45.

```
// Init Action
Buzzer Timer = Melody Time
Buzzer Index = Melody No.3 (3)
WAIT WHILE ( Buzzer Timer > 0.0sec )
Speed = 350
TimeDelay = 125
LineDistance = 200
ObjectDistance = 100
```

Fig. 3.46 Initial Actions for program "LineTracker-ObstacleAvoider.tsk3".

Fig. 3.47 shows the EndLess Loop where the Main Algorithm is found:

- If a frontal object is found (Line 17), the CarBot should stop then swing around to the reverse direction (Lines 19-20).
- Otherwise, the programming logic flows into an IF-ELSE-IF structure whereas:
    - If only the Left IR Sensor "sees" black (Line 25) then Function **LeftTurn()** is called (Line 28).
    - Otherwise, the Right IR Sensor is checked next (Line 30), and if only this Right Sensor "sees" black then Function **RightTurn()** is called (Line 33).
- For all other situations (i.e. the ELSE on Line 35), Function **Forward()** is executed (Line 38).

```
ENDLESS LOOP
{
    // If there is an obstacle in front (using IR Center)
    IF ( IR Center > ObjectDistance )
    {
        CALL Stop
        CALL SwingAround
    }
    ELSE
        // No obstacle in front of CarBot, so OK to follow black track
    {
        IF ( IR Left < LineDistance && IR Right >= LineDistance )
        {
            // Turns left when ONLY the left sensor detects black.
            CALL LeftTurn
        }
        ELSE IF ( IR Left >= LineDistance && IR Right < LineDistance )
        {
            // Turns right when ONLY the right sensor detects black
            CALL RightTurn
        }
        ELSE
        {
            // Moves forward for all other cases.
            CALL Forward
        }
```

Fig. 3.47 Main Algorithm for program "LineTracker-ObstacleAvoider.tsk3".

Function **SwingAround()** (Fig. 3.48) is a combination of "Timed" and "Conditional" Right Turns to accommodate for the fact that the "obstacle" can be found at any point along the Black Track:

- First, a **0.125s** Timed Right-Turn is used (Lines 83-86) because at this point in time, the CarBot is mostly lined up with the Black Track and it just needs to be nudged a bit to the right to make sure that Right IR Sensor is now "seeing" the "whiter" Craft paper so that the Conditional Expression **(IR_Right > LineDistance)** evaluates TRUE.

```
80 FUNCTION SwingAround
81 {
82     // Turn right for 0.125 second to clear IR-Right sensor off the track
83     PORT [2]:Geared Motor = CCW:0 (0.00%) + 250
84     PORT [1]:Geared Motor = CCW:0 (0.00%) + 250
85     High-resolution Timer = 0.125sec
86     WAIT WHILE ( High-resolution Timer > 0.000sec )
87 
88     // Keep on turning right as long as IR-Right sees white
89     WAIT WHILE ( IR Right > LineDistance )
90 
91     // IR-Right now sees Black, but OK to keep on turning right until IR-Right sees White again
92     WAIT WHILE ( IR Right <= LineDistance )
93     // Bot should be lining up with the track but in the reverse direction at this point
94 }
```

Fig. 3.48 Function **SwingAround()** for program "LineTracker_ObstacleAvoider.tsk3".

- Next, the reader needs to recall that the previous commands for the CarBot to turn right (Lines 83-84) are still in effect. The 2 WAIT-WHILE Loops (Lines 89 and 92) are "conditional" loops so that the CarBot can adjust to "any" section of the arbitrary track where the "SwingAround" maneuver may be needed. These 2 loops are used to ensure that an orderly process is followed to detect whether the Right IR Sensor is seeing "black" or "white" (e.g., brown Craft paper):
    - The first loop (Line 89) is active whenever Right IR Sensor sees "white", i.e., whenever the CarBot has **not** yet crossed back into the Black Track. When this situation happens, IR Right Sensor sees "black", the First Loop's Conditional Expression becomes FALSE and this loop is exited, and the Second Loop is entered (Line 92).
    - The purpose for the Second Loop (Line 92) is to keep the CarBot Turning Right some more until IR Right Sensor sees "white" (i.e., it has just passed over the other edge of the Black Track). At this point in time, the CarBot should be mostly lined up with the Black Track (but in the reverse direction), thus Function **SwingAround()** is exited and another iteration of the Main Endless Loop is executed, resulting in the Carbot's Default Maneuver of Going Forward.

The reader can try this algorithm on different shapes of a closed-circuit Black Track. Ones can see that this algorithm cannot handle "sharp" turns, as it would get off-track at these spots.

3.3.2 Yin-Yang Line Tracker with Speed Control

The Yin-Yang Line Track is described in Fig. 2.56. For the "Yin-Yang Line Tracker with Speed Control" (YYLTSC) Project, we will be using all 4 IR Sensors mounted on the Line-Tracker robot (see Fig. 2.56): the two Internal IR Sensors, and the two IRSS-10s mounted to Ports 3 (Left) and 4 (Right).

In the YYLTSC Project, the Event/Action Pairs to monitor are based on the ones used for the "Simple Line Tracker" Project (Sub-Section 3.26) as the robot is still required to follow a "track", but this track can now flip from Black to White. Additionally, we need to use the IRSS-10s to figure out which type of Background is the robot running on at any point in time and we'll add a way to vary the robot's speed via RC Buttons "U" and "D".

Again, we are going to assume that the Operator has already lined up the robot on the Yin-Yang Track as shown in Fig. 2.56 before activating the robot "Yin-Yang Line Tracking" algorithm. This assumption reduces the Event/Action Pairs to monitor to five Pairs that are mutually exclusive:

- If the Left Internal IR Sensor sees the Black Track, and both IRSS-10s report seeing a White Background, then the robot should Turn Left in an attempt to re-align the robot with the Black Track.
- If the Left Internal IR Sensor sees the White Track, and both IRSS-10s see a Black Background, then the robot should Turn Left to re-align itself with the White Track.
- If the Right IR Sensor sees the Black Track, and both IRSS-10s see a White Background, then the robot should Turn Right to re-align itself with the Black Track.
- If the Right IR Sensor sees the White Track, and both IRSS-10s see a Black Background, then the robot should Turn Right to re-align itself with the White Track.
- Otherwise, the robot should continue to go Forward.

```
ENDLESS LOOP
{
    Button = 0
    Print Screen with Line = Speed
    IF ( Remocon Data Arrived == TRUE (1) )
    {
        Button = Remocon RXD
        IF ( Button == U )
        {
            CALL Increase_Speed
        }
        ELSE IF ( Button == D )
        {
            CALL Decrease_Speed
        }
    }
```

Fig. 3.49 Setting Robot Speed Functions in "Yin_Yang_Line_Tracker.tsk3".

The TASK solution is named "Yin_Yang_Line_Tracker.tskx" (described in Figs. 3.49 through 3.51). This TASK solution also offers the feature where the user can push on RC Button "U" to increase robot speed and conversely uses RC Button "D" to decrease robot speed, while the robot is traversing the Yin-Yang Track.

The algorithm uses a standard Endless Loop where the First Process (see Fig. 3.49) is to clear out Variable **Button** (Line 13) and check for any new RC input from the user, i.e., check the flag "**Remocon Data Received**" (Line 15). If there is new RC input, then save the data received into **Button** (Line 17). If **Button** is found to be "**U**", then call Function **Increase_Speed()** (Lines 18-21). Otherwise, if **Button** is "**D**", then call Function **Decrease_Speed()** (Lines 22-25).

```
86  FUNCTION Increase_Speed
87  {
88      IF ( Speed < 900 )
89      {
90          Speed = Speed + 10
91          Buzzer Timer = 0.1sec
92          Buzzer Index = La# (1)
93          WAIT WHILE ( Buzzer Timer > 0.0sec )
94      }
95  }
96
97  FUNCTION Decrease_Speed
98  {
99      IF ( Speed > 120 )
100     {
101         Speed = Speed - 10
102         Buzzer Timer = 0.1sec
103         Buzzer Index = Do# (4)
104         WAIT WHILE ( Buzzer Timer > 0.0sec )
105     }
106 }
```

Fig. 3.50 Functions **Increase_Speed()** and **Decrease_Speed()** in "Yin_Yang_Line_Tracker.tsk3".

Fig. 3.50 describes Functions **Increase_Speed()** and **Decrease_Speed()**:

- Each time that Function **Increase_Speed()** is called, it increments **Speed** by 10 units (Line 90), but only if **Speed** is less than 900 (Line 88). So, its Maximum Value is around 890 at runtime. It also sounds off a La# Note for 0.1 s if the Speed Increase process was successful (Lines 92-93). In other words, if the Operator pushes on Button "U" at runtime, and he/she hears no audio feedback, then the robot has reached its maximum speed.
- Similarly, each time that Function **Decrease_Speed()** is called, it decrements **Speed** by 10 units (Line 101), but only if **Speed** is more than 120 (Line 99). So, its Minimum Value is around 120 at runtime. It also sounds off a Do# Note for 0.1 s if the Speed Decrease operation was successful (Lines 102-104). In other words, if the Operator pushes on Button "D" at runtime, and he/she hears no audio feedback, then the robot has reached its minimum speed.

Fig. 3.51 illustrates the Yin-Yang Line Tracking algorithm which uses the familiar IF-ELSE-IF structure to implement the 5 mutually exclusive Event/Actions Pairs described previously (the robot needs to start its trek straddling on the Black/White Line):

```
27  // Algorithm to traverse Yin-Yang Track
28  IF ( IR Left <= 100 && PORT [3] :IR Sensor >= 600 && PORT [4] :IR Sensor >= 400 )
29  {
30      // Turns left when ONLY the left sensor detects black track & NIR_3+NIR_4 detect white background
31      CALL LeftTurn
32  }
33  ELSE IF ( IR Left >= 400 && PORT [3] :IR Sensor <= 300 && PORT [4] :IR Sensor <= 150 )
34  {
35      // Turns left when ONLY the left sensor detects white track & NIR_3+NIR_4 detect black background
36      CALL LeftTurn
37  }
38  ELSE IF ( IR Right <= 150 && PORT [3] :IR Sensor >= 600 && PORT [4] :IR Sensor >= 400 )
39  {
40      // Turns right when ONLY the right sensor detects black track & NIR_3+NIR_4 detects white background
41      CALL RightTurn
42  }
43  ELSE IF ( IR Right >= 400 && PORT [3] :IR Sensor <= 300 && PORT [4] :IR Sensor <= 150 )
44  {
45      // Turns right when ONLY the right sensor detects white track & NIR_3+NIR_4 detects black background
46      CALL RightTurn
47  }
48  ELSE
49  {
50      // Moves forward for all other cases.
51      CALL Forward
52  }
```

Fig. 3.51 Line Tracking algorithm used in "Yin_Yang_Line_Tracker.tsk3".

- If the **Left IR Sensor** sees Black Track and **Ports 3 & 4 IR Sensors** see White Background, then call Function **LeftTurn()** (Lines 28-32).
- Else If the **Left IR Sensor** sees White Track and **Ports 3 & 4 IR Sensors** see Black Background, then call Function LeftTurn() (Lines 33-37).

- Else If the **Right IR Sensor** sees Black Track and **Ports 3 & 4 IR Sensors** see White Back-ground, then call Function RightTurn() (Lines 38-42).
- Else If the **Right IR Sensor** sees White Track and **Ports 3 & 4 IR Sensors** see Black Background, then call Function RightTurn() (Lines 43-47).
- Else the robot needs to go Forward – Lines 48-52.

At runtime, the author had found that it is harder for the robot to stay on the Yin-Yang Track than for the Simple Line Track of Sub-Section 3.2.6.

3.4 Servo Motor Applications

The DREAM II School Set comes with 1 SM-10 Servo, so we'll strive to make good use of in three Projects: two involving a TriCycle Platform and one involving a (wood) Dowel Scanner.

The reader is referred to Section 2.7 for a brief description of the working of a Servo Motor such as the SM-10, and to Sub-Section 2.7.1 or Appendix B.2 for the construction of the TriCycle robot.

3.4.1 TriCycle Steering Control with Servo Motor

Fig. 3.52 looks at the front of a Servo Motor that has an IRSS-10 Sensor attached to it. **In TASK codes**, the Servo Motor's position can be controlled by sending it a positive integer between 0 and 1023 whereas the Position Number "512" corresponds the Default "0 degree" Position, while "819" corresponds to +90 degrees and "205" corresponds to -90 degrees.

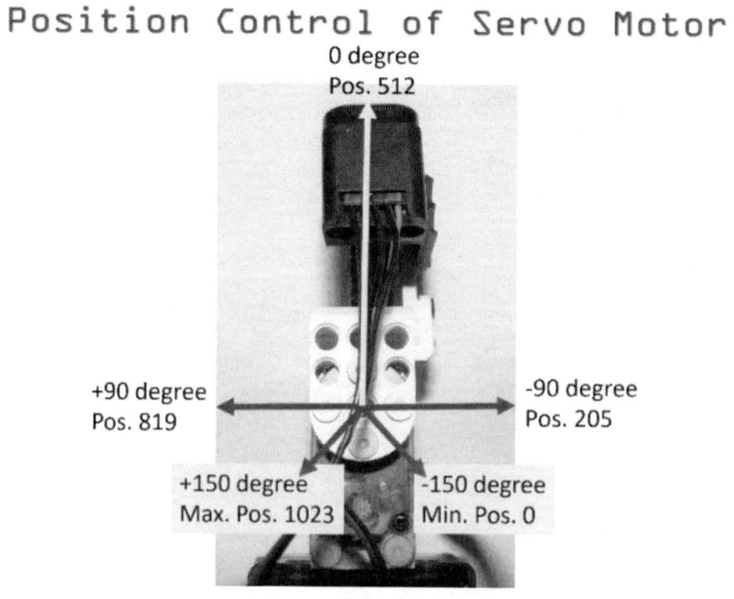

Fig. 3.52 TASK's Position Control of Servo Motor with attached NIR Sensor.

The projects in Sub-Sections 3.4.1 and 3.4.2 use an SM-10 Servo Motor via Port 3 and an IRSS-10 Sensor via Port 4. The TriCycle's powered back end is the same as the previous Avoider, but on its front end is mounted a Servo Motor/Front Wheel (Fig. 3.53). The Front Wheel is passive, and it is attached via a U-shaped frame to the horn of the Servo Motor. The NIR Sensor is then attached to the Front Wheel assembly. Thus, the Servo Motor (Port 3) controls the direction of the Steering Wheel and the View Angle of the NIR Sensor (itself connected to Port 4).

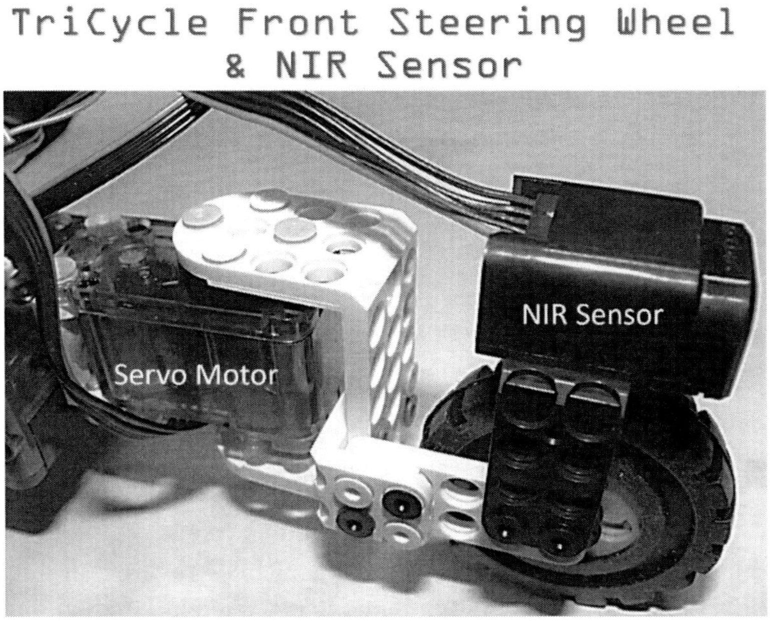

Fig. 3.53 Details of TriCycle's Front End.

Although the Smart Avoider (Section 3.2.7) and the TriCycle share the same triangular shape for their ground contact points (see Fig. 3.54), the Smart Avoider has a more compact triangular shape allowing it to be nimble during turn maneuvers. The TriCycle has an elongated triangular shape, i.e., a longer moment arm between the rear drive wheels and front steering wheel (so-called "Delta" configuration). Furthermore, its rubber front wheel can generate more sideways friction forces with the ground surface than the little plastic support used by the Smart Avoider, so mechanically the TriCycle won't be as nimble as the Smart Avoider. Thus, we will need a different way of activating the Left and Right drive motors, more gradually in conjunction with the Front Wheel's turn angle, to assist the TriCycle during its Left and Right turns.

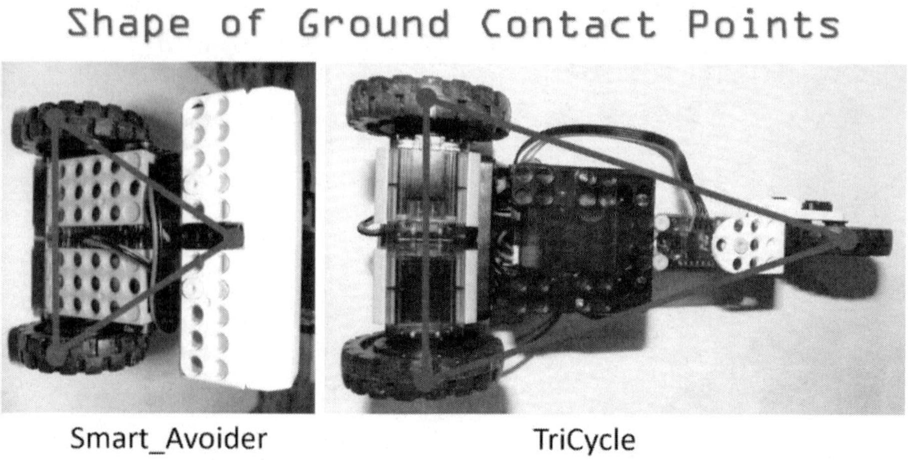

Fig. 3.54 Triangular Shapes of Ground Contact Points for Smart_Avoider and TriCycle.

The first TriCycle Project is about its Remote-Control operations but with a new control approach as implemented in the program “RC-TriCycle.tsk3”, which itself is based on the previous program “RC-MultiDirectionsSpeeds.tsk3”.

The INIT section (see Fig. 3.55) shows how to set up and use the Servo Motor out of Port 3 programming. Line 11 sets the Servo’s Drive Mode to TRUE (1), i.e., puts it in Position Control mode *(Note: set this mode to FALSE (0) to use this Servo Motor like a regular Geared Motor).*

Next, Line 12 sets the Servo’s Speed to 1023 out of an allowable range of [0-1023]. This Speed controls how fast the Servo can move between Goal Positions which are also in the same [0-1023] range (see their corresponding physical positions as illustrated in Fig 3.52). This Speed value also is a measure of how strong a torque this servo can hold at a Goal Position, under the weight/inertia of whatever structure/device that is attached to the same servo. The user could use a mid-value for Speed to achieve less sudden Servo Positions and thus to reduce structure vibrations, but the user needs to use Speed=1023 to hold a steady Servo Position once the servo gets where it is supposed to be.

Once the **Mode** and **Speed** Variables are set (Lines 11-12), the user can put the Servo Motor to good use by setting its next **Goal Position** value (Line 13) which should be an integer between 0 and 1023 (see Fig. 3.52). There is one important point that beginner users need to be cognizant of when programming servo positional changes: as it will take some actual time periods for the servo to get from one physical location to the next, the user needs to program in some WAIT time for the current servo operation to finish, before issuing a new command to move the same servo to another location.

The INIT section also contains 3 new Variables: **SpeedL, SpeedR** and **SpeedDif** (see Fig. 3.55). **SpeedL** is used exclusively by the Left Geared Motor (Port 1), while **SpeedR** is reserved for the Right Geared Motor (Port 2). Please see Fig. 3.56 for the use of **SpeedL** and **SpeedR** in the new Functions **Forward()** and **Backward()** as examples. **SpeedDif** will be set proportionally to the Turn Angle of the Steering Wheel’s Servo Motor (controlled via Port 3) when Left or Right Turns are required.

```
    PORT [3]:Servo Motor Drive Mode = TRUE (1)
    PORT [3]:Servo Joint Speed = 1023
    PORT [3]:Servo Joint Position = 512
    Speed = 256
    SpeedL = 0
    SpeedR = 0
    SpeedDif = 0
    Data = 0
```

Fig. 3.55 Init Event with Variables **SpeedL, SpeedR** & **SpeedDif** in "RC-TriCycle.tsk3"

```
FUNCTION Forward
{
    PORT[2]:Geared Motor = CW:0 (0.00%) + SpeedR
    PORT[1]:Geared Motor = CCW:0 (0.00%) + SpeedL
    RETURN
}

FUNCTION Backward
{
    PORT[2]:Geared Motor = CCW:0 (0.00%) + SpeedR
    PORT[1]:Geared Motor = CW:0 (0.00%) + SpeedL
    RETURN
}
```

Fig. 3.56 Functions **Forward()** and **Backward()** in "RC-TriCycle.tsk3"

Next the procedure to parse the Remocon Packet "Data" into separate Variables **Direction, Button1** and **Button3** remains the same as in Sub-Section 3.1.6, but the Actions that correspond to the cases when **(Direction == "U")** or when **(Direction == "D")** are now quite different (see Fig. 3.57).

Let's take the case when **(Direction == "U")** is TRUE in Line 40, then **Port 3 Servo Position** is set to "512" (Line 42 - i.e. Front Steering Wheel set to go straight) and **SpeedDif** is set to "0" (Line 43), thus **SpeedL** and **SpeedR** would have the same numerical value as Variable **Speed** (which can be either "400" or "800") – see Lines 44 and 45. Finally Function **Forward()** is called to make the TriCycle go Forward (Line 46 of Fig.3.57).

A similar sequence of actions would be executed when **(Direction == "D")** is TRUE in Line 48, whereas the TriCycle would go Backward in a "straight" line (Lines 50 to 54 in Fig. 3.57). If for some reasons, there are some mechanical inaccuracies in the robot assembly, **SpeedDif** may have to be a non-zero value and need to be manually adjusted to make this TriCycle go really "straight" in its Forward or Backward motions (the reader would need to experiment with his/her own TriCycle robot to obtain the optimal value for **SpeedDif**).

```
IF ( Direction == U )
{
    PORT[3]:Servo Joint Position = 512
    SpeedDif = 0
    SpeedL = Speed - SpeedDif
    SpeedR = Speed + SpeedDif
    CALL Forward
}
ELSE IF ( Direction == D )
{
    PORT[3]:Servo Joint Position = 512
    SpeedDif = 0
    SpeedL = Speed - SpeedDif
    SpeedR = Speed + SpeedDif
    CALL Backward
}
```

Fig. 3.57 Actions executed when "Direction" == "U" or "D" in "RC-TriCycle.tsk3"

When **(Direction == "L" or "R"),** i.e., when Left or Right Turns are required, the appropriate sequences of required Actions are quite a bit different than for Forward or Backward (see Fig. 3.58). For example, Line 58 shows that **each time** that **(Direction == "L")** is TRUE, the **Current Servo Position is increased by "50" units** which make the **Steering Wheel turn Left about an "extra" 15-degrees angle each time**. In other words, if the user keeps on pressing the "L" button when wanting to make a Left Turn, like in all previous applications, this TriCycle will make tighter and tighter turns "on its own", so the user needs to adjust his or her "driving" behavior for this "new design" (key-tapping only).

Furthermore, **SpeedDif** is also increased by "5" units each time that the condition **(Direction == "L")** is TRUE (see Line 59 of Fig. 3.58). Next, Lines 60 and 61 show that **SpeedL** is decreased by **SpeedDif** (Left Motor slows down) while **SpeedR** is increased by **SpeedDif** (Right Motor speeds up), i.e., making the robot turn **Left**. In other words, the Left and Right Motors are also differentially commanded to assist in the Left Turn motion initiated by the Steering Wheel, because the TriCycle has a longer moment arm to work with as compared to the Avoider.

```
56 ELSE IF ( Direction == L )
57 {
58     PORT[3]:Servo Joint Position = PORT[3]:Servo Joint Position + 50
59     SpeedDif = SpeedDif + 5
60     SpeedL = Speed - SpeedDif
61     SpeedR = Speed + SpeedDif
62     CALL CheckSpeeds
63     CALL Left
64 }
65 ELSE IF ( Direction == R )
66 {
67     PORT[3]:Servo Joint Position = PORT[3]:Servo Joint Position - 50
68     SpeedDif = SpeedDif + 5
69     SpeedL = Speed + SpeedDif
70     SpeedR = Speed - SpeedDif
71     CALL CheckSpeeds
72     CALL Right
73 }
```

Fig. 3.58 Actions executed when “Direction” == “L” or “R” in “RC-TriCycle.tsk3”

As the substracting and adding operations in Lines 60-61 may make the Variables **SpeedL** and **SpeedR** go out of their “allowed” range [0-1023], the Function **CheckSpeeds()** (Line 62 in Fig. 3.58) is always executed (see code details in Fig. 3.59) before the Function **Left()** is called to perform the actual Left Turn maneuver in Line 63 of Fig 3.58.

A similar procedure is used when the condition **(Direction** == **“R”)** is TRUE to make the Tri-Cycle performs an increasing Right Turn.

```
FUNCTION CheckSpeeds
{
    IF ( SpeedL < 0 )
    {
        SpeedL = 0
    }
    ELSE IF ( SpeedL > 1023 )
    {
        SpeedL = 1023
    }
    IF ( SpeedR < 0 )
    {
        SpeedR = 0
    }
    ELSE IF ( SpeedR > 1023 )
    {
        SpeedR = 1023
    }
}
```

Fig. 3.59 Function **CheckSpeeds()** in “RC-TriCycle.tsk3”.

Last but not least, the maneuvers for the “in-between” positions like U+R, U+L, D+R and D+L are also implemented very differently for the TriCycle from what they were in previous RC applications (see Figs. 3.60 and 3.61). As an example, for the Chromebook’s Virtual RC-100, the user can push on the Up and Right Arrows at the same time to obtain the “U+R” RC command.

For example, in Fig. 3.60 (Lines 76-79), when “**U+R**” is TRUE, the current Servo Position is increased by 25 only as compared to the previous cases for “R” and “L” (“milder” change in Steering Wheel angle). **SpeedDif** also increases only by 3: i.e., “milder” changes in Motor **Speed** are used with “milder” **Steering** changes. Function **CheckSpeeds()** is still used to clamp down **SpeedL** and **SpeedR** to their “legal” limits as needed (Line 80). Please note that only “CALL Forward” is used with a **0.25 second delay** to create “smooth” and “mild” maneuvers.

A similar procedure is used for “U+L” in Lines 86-92 of Fig. 3.60 to create “smooth and mild” Left turns.

Similar maneuvers are also implemented for the other cases “D+R” and “D+L”, except now the TriCycle will use the Function **Backward()** instead (see Fig. 3.61).

```
ELSE IF ( Direction == U+R )
{
    PORT[3]:Servo Joint Position = PORT[3]:Servo Joint Position - 25
    SpeedDif = SpeedDif + 3
    SpeedL = Speed + SpeedDif
    SpeedR = Speed - SpeedDif
    CALL CheckSpeeds
    CALL Forward
    CALL Delay
}
ELSE IF ( Direction == U+L )
{
    PORT[3]:Servo Joint Position = PORT[3]:Servo Joint Position + 25
    SpeedDif = SpeedDif + 3
    SpeedL = Speed - SpeedDif
    SpeedR = Speed + SpeedDif
    CALL CheckSpeeds
    CALL Forward
    CALL Delay
}
```

Fig. 3.60 Maneuvers for "U+R" and "U+L" inputs in "RC-TriCycle.tsk3".

```
ELSE IF ( Direction == D+R )
{
    PORT[3]:Servo Joint Position = PORT[3]:Servo Joint Position - 25
    SpeedDif = SpeedDif + 3
    SpeedL = Speed + SpeedDif
    SpeedR = Speed - SpeedDif
    CALL CheckSpeeds
    CALL Backward
    CALL Delay
}
ELSE IF ( Direction == D+L )
{
    PORT[3]:Servo Joint Position = PORT[3]:Servo Joint Position + 25
    SpeedDif = SpeedDif + 3
    SpeedL = Speed - SpeedDif
    SpeedR = Speed + SpeedDif
    CALL CheckSpeeds
    CALL Backward
    CALL Delay
}
```

Fig. 3.61 Maneuvers for "D+R" and "D+L" inputs in "RC-TriCycle.tsk3".

The reader is also encouraged to try his or her own control scheme whereas each turn corresponds to a "fixed" angular change for the Steering Wheel and a "fixed" numerical change in the Variable Speed Differential.

3.4.2 Smart TriCycle

The Smart TriCycle Project has the same operational features as the Smart Avoider, that is it can evade an obstacle on its own, when the Operator drives it into an obstacle (Sub-Section 3.2.7).

The author developed the "Smart_Tricycle.tsk3" solution to illustrate the adaptations/solutions in Mechanical Engineering and Robotics Programming issues that differ between the two robots:

- The Avoider robot used two IR Sensors to determine the Presence and Location of an Obstacle in Front, on the Left or on the Right sides of the robot. The TriCycle robot will need to provide the same operational capabilities, but with a Servo Motor and only 1 IR Sensor, thus its "Obstacle Avoidance" algorithm will need to be quite different.
- As previously discussed, the Avoider robot is more maneuverable than the TriCycle which has to deal with larger Ground Frictional forces due to the Rubber Tire's location with respect the Powering Motors. Thus, at runtime, the Operator will have the option of Increasing or Decreasing the robot's **Speed** as needed via **Button 1** and **Button 3**. But RC Directional changes to the robot are simplified by removing the combo options "U+R", "U+L", "D+R" and "D+L" as compared to Sub-Section 3.4.1.

Fig. 3.62 shows the relevant Variable Initializations needed in the solution "Smart_Tricycle.tsk3":

- Only Variables **Threshold_1** and **Threshold_3** are needed in this Project (Lines 20 and 22).
- Boolean Variable **Obstacle** is also needed (Line 25).

```
20    Threshold_1 = 300
21    // Threshold_2 = 200
22    Threshold_3 = 10
23    Button1 = 0
24    Button3 = 0
25    Obstacle = FALSE (0)
```

Fig. 3.62 Selected Initializations in "Smart_TriCycle.tsk3".

Fig. 3.63 displays the beginning of the Main Endless Loop:

- Lines 29 and 30 print out current values for **Port 4 IR Sensor** and **Speed** for the Operator's reference at runtime.
- Line 31 checks for a "close" detection of an Obstacle via **Port 4 IR Sensor** and **Threshold_1**.
- If there is no obstacle detected - **(Port 4 IR Sensor < Threshold_1)**, the RC algorithm begins and it is unchanged from its descriptions for "U", "D", "L", "R", "1" and "3" as discussed in the previous Sub-Section 3.4.1.

```
27  ENDLESS LOOP
28  {
29      Print Screen = PORT[4]:IR Sensor (PortNickName)
30      Print Screen with Line = Speed
31      IF ( PORT[4]:IR Sensor (PortNickName) < Threshold_1 )
32      {
33          // No obstacle in front so OK to abide by user's RC inputs
34          IF ( Remocon Data Arrived == TRUE (1) )
35          {
36              Data = Remocon RXD
37              Print Screen with Line = Data
38              IF ( Data == -- )
39              {
40                  CALL Stop
41              }
42              ELSE
43              {
44                  // Filtering out specific Direction buttons or 1-3 buttons
45                  Direction = Data & U+D+L+R
46                  Button1 = Data & 1
```

Fig. 3.63 Beginning of Main Endless Loop in "Smart_TriCycle.tsk3".

Let us recall that the "Obstacle Avoidance" algorithm is activated when Port 4 IR Sensor reports a Value that is **greater or equal** to **Threshold_1** (=300), i.e., when an obstacle is detected (see Fig. 3.63 Line 31). However, because of the use of a Single IR Sensor to do the Obstacle Detection work, much is still "**unknown**" about the "**geometrical**" relationship between the robot and the obstacle at this point in time, such as:

- Is the Steering Wheel angled to the Right or to the Left of the robot's Longitudinal Axis?
- Or is the Steering Wheel set Straight Forward? In this case, is the robot Longitudinal Axis "perpendicular" to the Obstacle Plane? Or is this Axis at an angle to this plane, and then which way? To the Left or to the Right?
- All these possible Event/Action Pairs will of course be implemented with a big IF-ELSE-IF structure.

The "Obstacle Avoidance" algorithm is described in Figs. 3.64, 3.65 and 3.66. Fig. 3.64 illustrates the steps taken when the robot encounters the obstacle, and the Steering Wheel is already steered to the Left or to the Right:

- First, Variable **Obstacle** is set to TRUE (Line 106) and the robot stops (Line 107).
- Line 108 saves the Current Value of **Port 3 Servo Position** into Variable **Servo3_Pos**.
- Next, an IF-ELSE-IF structure (Lines 109-118) is used to determine whether the Steering Wheel is already turned to the Left or to the Right of the Longitudinal Axis of the TriCycle robot:

- If **(Servo3_Pos $\leq$ 470)** meaning that the Front Wheel is already steered to the Right of the robot, then it might as well as keeps on Turning Right by calling Function **Escape_Right()** (see Fig. 3.67 for more details).
- If **(Servo3_Pos $\geq$ 555)** meaning that the Front Wheel is already steered to the Left of the robot, then it might as well as keeps on Turning Left by calling Function **Escape_Left()** (see Fig. 3.67 for more details).

```
ELSE
    // Obstacle is getting close - Robot stops first then tries to avoid obstacle
{
    Obstacle = TRUE (1)
    CALL Stop
    Servo3_Pos = PORT [3]:Servo Joint Position (PortNickName)
    IF ( Servo3_Pos <= 470 )
        // Already steered to the Right
    {
        CALL Escape_Right
    }
    ELSE IF ( Servo3_Pos >= 555 )
        // Already steered to the Left
    {
        CALL Escape_Left
    }
```

Fig. 3.64 Part 1 of "Obstacle Avoidance" Algorithm in "Smart_TriCycle.tsk3".

If the previous two Events do not apply, meaning that the Steering Wheel happens to be set Straight Forward, we then need to figure out the relative position between the robot's Longitudinal Axis and the Obstacle's Plane. There are 3 possible cases A, B and C:

- The robot's Axis is at an acute angle to the Obstacle Plane with respect to its Left side (Case A).
- This Axis is at an acute angle to the Obstacle Plane with respect to its Right side (Case B).
- This Axis is mostly perpendicular to the Obstacle Plane (Case C).

These three "mutually exclusive" Event/Action Pairs will thus be implemented as a Nested IF-ELSE-IF structure within the Outer IF-ELSE-IF structure already described in Fig. 3.64.

```
119  ELSE
120      // Steering wheel straight forward but Tricycle's body may be at an angle to the obstacle
121      {
122          Servo3_Pos_L = Servo3_Pos + 30
123          Servo3_Pos_R = Servo3_Pos - 30
124          PORT[3]:Servo Joint Position (PortNickName) = Servo3_Pos_L
125          Delay = 0.250sec
126          ObstacleDistance_L = PORT[4]:IR Sensor (PortNickName)
127          PORT[3]:Servo Joint Position (PortNickName) = Servo3_Pos_R
128          Delay = 0.250sec
129          ObstacleDistance_R = PORT[4]:IR Sensor (PortNickName)
130          IF ( ObstacleDistance_L > ObstacleDistance_R )
131          {
132              CALL Escape_Right
133          }
134          ELSE IF ( ObstacleDistance_L < ObstacleDistance_R )
135          {
136              CALL Escape_Left
137          }
```

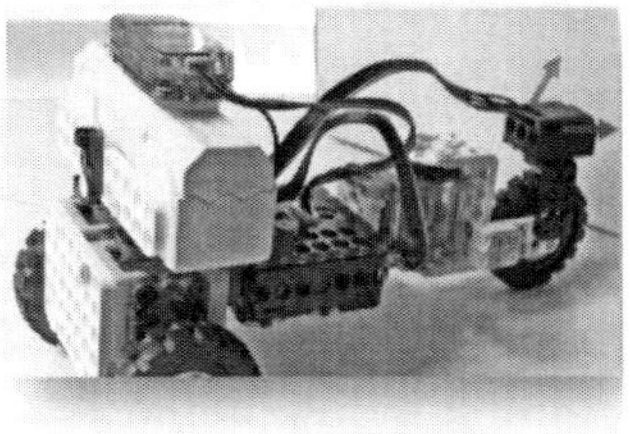

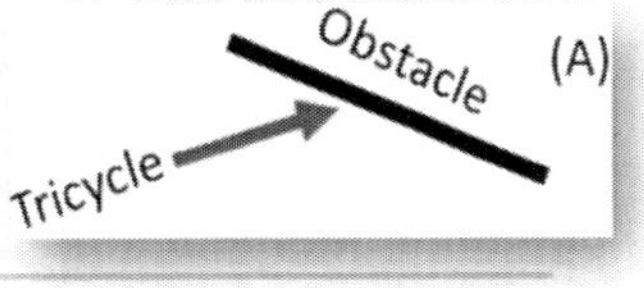

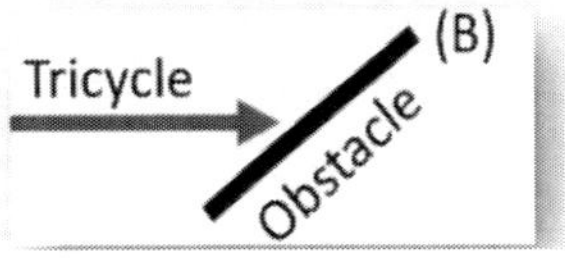

Fig. 3.65 Cases A and B when Steering Wheel is along Robot's Axis in "Smart_TriCycle.tsk3".

Fig. 3.65 describes the steps used for Case A and Case B:

- Lines 122-129 show the procedure for taking "sample" **Port 4 IR Sensor** data to figure out which Case A or B is in play at runtime:
  - Essentially, Servo 3 is moved slightly to the Left of its Current Servo Position **Servo3_Pos** (Lines 122, 124 and 125), then the Current Value of **Port 4 IR Sensor** is saved into **ObjectDistance_L** (Line 126).
  - Next, Servo 3 is moved slightly back to the Right of its Current Servo Position **Servo3_Pos** (Lines 123, 127 and 128), then the Current Value of **Port 4 IR Sensor** is saved into **ObjectDistance_R** (Line 129).
- Now, **ObjectDistance_L** and **ObjectDistance_R** can be used to figure out Case A or Case B:
  - Line 130 shows that IF **(ObjectDistance_L > ObjectDistance_R)**, the Tri-Cycle's Left Side is closer to the Obstacle Plane, so Function **Escape_Right()** is called (Line 132).
  - Otherwise, IF **(ObjectDistance_L < ObjectDistance_R)**, the TriCycle's Right Side is closer to the Obstacle Plane, so Function **Escape_Left()** is called (Line 136).

Fig. 3.66 lists the steps used for Case C (i.e., robot's Axis perpendicular to Obstacle Plane):

- The robot goes Backward for 0.1 seconds, then Function **Escape_Right()** is called (Line 142). Here, the reader can substitute Function **Escape_Left()** if so desired.

- Once the robot successfully steers clear of the Obstacle (at Line 144), it then Stops (Line 146), and the WHILE Loop (Lines 147-151) is used to wait for the Operator to release all RC Buttons, if not done so by then (review code explanations in Sub-Section 3.2.7 if needed).
- Then another iteration of the Main Endless Loop is started, and the Operator can use the RC Buttons again.

```
138             ELSE
139             {
140                 CALL  Backward
141                 Delay  =  0.100sec
142                 CALL  Escape_Right
143             }
144         }
145         //  Here, no more obstacle in front - Robot stops and wait for user to release buttons
146         CALL  Stop
147         LOOP WHILE  (  Data  !=  --  )
148         {
149             WAIT WHILE  (  Remocon Data Arrived  ==  FALSE (0)  )
150             Data  =  Remocon RXD
151         }
152     }
153     //  End of BIG Else for Escape Procedure
```

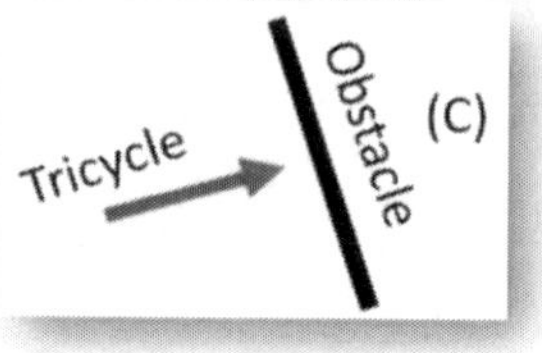

Fig. 3.66 Case C when Steering Wheel is along Robot's Axis and End of "Obstacle Avoidance" Algorithm in "Smart_TriCycle.tsk3".

Fig. 3.67 lists the programming steps used for Functions **Escape_Right()** and **Escape_Left()**. As an example, let us go through Function **Escape_Right()**:

- Recall that **Obstacle** was already set to TRUE back in Line 106 of Fig. 3.64, because the Obstacle was just detected. So the WHILE Loop (Lines 161-169) is immediately activated upon entering Function **Escape_Right()**.
- Servo 3 is then set to Goal Position 305 (Line 163), i.e., Steering Wheel hard rotated to the Left, and a Delay of 0.25 seconds is used to wait for Servo 3 to get there (Line 164).
- The robot then goes Forward for 0.3 seconds, then stops (Lines 165-167).
- Function **Check_NIR()** is then called (Line 168) to check if the Obstacle is still potentially obstructing the robot's motion. **Check_NIR()** is described in Fig. 3.68:
    - Lines 187-188 are used to straighten the Steering Wheel forward.

```
159 FUNCTION Escape_Right
160 {
161     LOOP WHILE ( Obstacle == TRUE (1) )
162     {
163         PORT [3] :Servo Joint Position (PortNickName) = 305
164         Delay = 0.250sec
165         CALL Forward
166         Delay = 0.300sec
167         CALL Stop
168         CALL Check_NIR
169     }
170 }
171
172 FUNCTION Escape_Left
173 {
174     LOOP WHILE ( Obstacle == TRUE (1) )
175     {
176         PORT [3] :Servo Joint Position (PortNickName) = 718
177         Delay = 0.250sec
178         CALL Forward
179         Delay = 0.300sec
180         CALL Stop
181         CALL Check_NIR
182     }
183 }
```

Fig. 3.67 Functions **Escape_Right()** and **Escape_Left()** in "Smart_TriCycle.tsk3".

```
185 FUNCTION Check_NIR
186 {
187     PORT [3] :Servo Joint Position (PortNickName) = 512
188     Delay = 0.250sec
189     ObstacleDistance = PORT [4] :IR Sensor (PortNickName)
190     IF ( ObstacleDistance > Threshold_3 )
191     {
192         Obstacle = TRUE (1)
193     }
194     ELSE
195     {
196         Obstacle = FALSE (0)
197     }
198 }
```

Fig. 3.68 Function **Check_NIR()** in "Smart_TriCycle.tsk3".

- Line 189 is used to acquire the latest **Port 4 IR Sensor** reading and save it in Variable **ObstacleDistance**.

- Then the IF-ELSE structure of Lines 190-197 is used to determine if the Obstacle is still "there". If in Line 190, the Expression **(ObstacleDistance > Threshold_3)** evaluates TRUE, **Obstacle** is set to TRUE, and Function **Check_NIR()** returns to either **Escape_Right()** or **Escape_Left()** which will help the robot steer further away from the Obstacle.
- Eventually, the robot steers itself far away enough from the Obstacle that **Obstacle_Distance** becomes less than **Threshold_3**, then **Obstacle** is set to FALSE, and **Check_NIR()** returns (for the last time) to either Function **Escape_Right()** or **Escape_Left()** which also terminates, because the Conditional Expression controlling their WHILE Loops no longer evaluates TRUE (Lines 161 and 174 in Fig. 3.67).

Fig. 3.69 lists a sequence of pictures recording an actual Obstacle Avoidance Event.

Fig. 3.69 Runtime Performance of "Smart_TriCycle.tsk3".

3.4.3 Dowel Scanner

For this project, we would need a nimble robot, so it's back to the Avoider's platform, but this time with a Servo Motor (SM-10) "Head" that is slewing a forward-looking IRSS-10 Sensor (see Fig. 3.70), yielding a "Dowel Scanner" robot. This robot's tasks are to identify two Wood Dowels placed at random in front of it (see Fig. 3.71), specifically to identify which Dowel is closest to it and approach it within a certain distance, then it would need to switch its attention to find the second Dowel and likewise to approach it within a given distance.

As TASK codes run directly on the CM-150/151, we do not have to deal with the **I/O Systemic Time Delays** issue like for R+BLOCK (see Sub-Section 2.7.2), and this robot can now "scan on the fly". Thus, the reader will see that the TASK's Dowel Scanning approach is different from the one used in R+BLOCK.

Fig. 3.70 “Dowel Scanner” Robot.

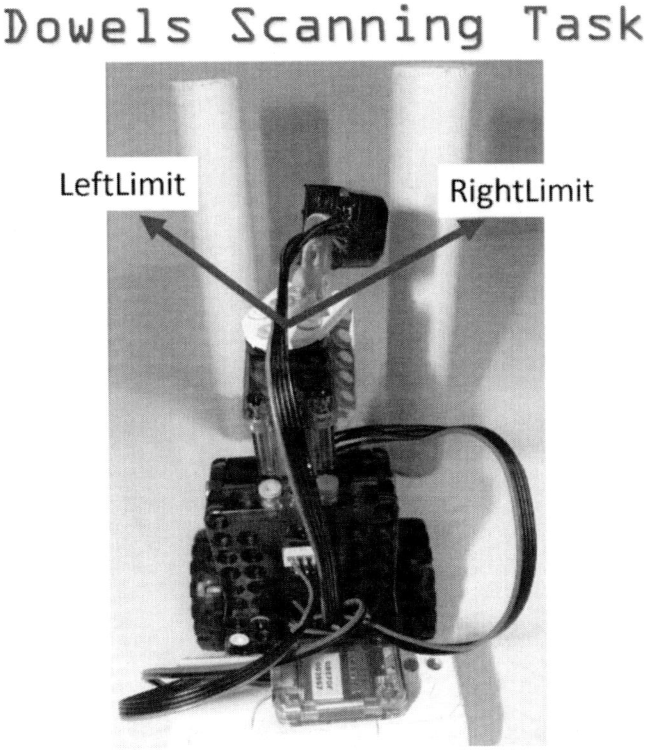

Fig. 3.71 Dowel Scanning Task Layout.

In this project, the IRSS-10 Sensor can be slewed in two ways:

- By the Servo Motor via Port 3 (while the robot is standing still): this technique will be used to identify which Dowel (Left or Right) is closest to the robot at the start of the tasks.

- With the Servo Motor locked into Position 512, thus with the IRSS-10 Sensor facing straight forward, the whole robot platform is rotated left-right as needed by the two Geared Motors via Ports 1 and 2 to perform the Dowel Scanning task using this IR Sensor, this mode is used in the final identification of each Dowel and in the final approach to each Dowel.

Figs. 3.70 and 3.71 also show that the IR Sensor is mounted on its "side" to take full advantage of the fact that the Dowels are "vertical" objects. Fig. 3.72 is an output list of the Servo's Position data (first column) and the IR Sensor's values (second column) at the same positions, as the IR Sensor is slewed from Left to Right (i.e., Position values from ~700 to ~380), using a simple TASK program.

Program Output Monitor

703	3
705	6
705	3
697	3
685	5
674	4
662	5
650	5
635	62
621	154
609	294
595	289
581	125
567	30
552	6
539	4
523	3
508	5
494	4
479	27
463	37
448	53
433	28
419	7
406	4
392	5
379	3

Fig. 3.72 Data Output for Servo's Positions and corresponding IR Values as the IR Sensor is slewed across two Wood Dowels from Left to Right.

The reader can readily see that the Background reads as "3 to 7" and that the Dowel's Presence is easily determined by an increase followed by a decrease in IR values. The First Dowel can be found around Position 609 (i.e., on Left of robot) corresponding to a Local Maximum IR value of 294, while the second Dowel can be found around Position 448 (i.e., on Right of robot) corresponding to a lower Local Maximum of 53. Thus, we can ascertain that the "Left" Dowel (max IR value = 294) is closer to the robot than the one on its "Right" (max IR value = 53). This is the strategy used in developing a **search algorithm in TASK code to find the two Dowels**, to be described in later parts of this Sub-Section 3.4.3.

The complete program for this project is named "Dowel_Scanner.tsk3" and its main parts are described below:

- Part 1 of The INIT event is described in Fig. 3.73, whereas many variables are initialized. First, "**Re#**" is played for 1 second (Lines 8-10), then Port 3 Servo Motor is defined for Position Control mode (Line 11), with its Speed set to 512 (50% power – Line 12), and its initial Position set to 512 (Line 13). Variables **SpeedL** and **SpeedR** are used for Ports 1 and 2 Geared Motors as before (Lines 14-15). The **LeftLimit** and the **Right-Limit** Variables are used to define the slewing range for the Port 3 Servo Motor (Lines 16-17). The other Variables **ProximityLimit, PreviousNIR, CurrentNIR, DowelDistance1, DowelDistance2, Position1, Position2** and **FoundDowel** will be described, and their usage explained in later paragraphs.

```
7    // Init Action
8    Buzzer Timer = 1.0sec
9    Buzzer Index = Re# (6)
10   WAIT WHILE ( Buzzer Timer > 0.0sec )
11   PORT [3] :Servo Motor Drive Mode = TRUE (1)
12   PORT [3] :Servo Joint Speed = 512
13   PORT [3] :Servo Joint Position = 512
14   SpeedL = 200
15   SpeedR = 200
16   LeftLimit = 720
17   RightLimit = 350
18   ProximityLimit = 0
19   PreviousNIR = 0
20   CurrentNIR = 0
21   DowelDistance1 = 0
22   DowelDistance2 = 0
23   Position1 = 0
24   Position2 = 0
25   FoundDowel = 0
```

Fig. 3.73 Part 1 of the INIT Event in "DowelScanner.tsk3".

- Next, the robot waits for the user to make some noise (clapping hands or loud whistling) (Line 26 of Fig. 3.74). Once the **Detected Clap Count** becomes non-zero, **Back-groundNIR** is initialized to zero (Line 27). Next, a FOR loop takes in 20 consecutive Data Values from Port 4 IR Sensor (Lines 28-31) and adds them up into **Background-NIR**, next this cumulative sum is averaged out and saved back into the same Variable **BackgroundNIR** (Line 32). Variable **DowelThreshold** is defined as having 3 units higher than **BackgroundNIR**. The final Init Action is to play Melody 0 (Lines 34-36).

- The Main Actions are to CALL Functions **ServoScan()** and **VerifyDowels()** (Lines 38-39). **ServoScan()** does the first scan using the "Port 3 Servo + IR Sensor" combo to figure out which Dowel is closest to the robot. **VerifyDowels()** accomplishes the other tasks for Final Identification and Robot Approach using Ports 1 and 2 Geared Motors in conjunction with the IR Sensor.

```
26  WAIT WHILE ( Detected Clap Count == 0 )
27  BackgroundNIR = 0
28  LOOP FOR ( I = 1 ~ 20 )
29  {
30      BackgroundNIR = BackgroundNIR + PORT[4]:IR Sensor
31  }
32  BackgroundNIR = BackgroundNIR / 20
33  DowelThreshold = BackgroundNIR + 3
34  Buzzer Timer = Melody Time
35  Buzzer Index = Melody No.0 (0)
36  WAIT WHILE ( Buzzer Timer > 0.0sec )
37
38  CALL ServoScan
39  CALL VerifyDowels
```

Fig. 3.74 Part 2 of the INIT Event and Main Actions in "DowelScanner.tsk3".

- Next, let us look further into Function **ServoScan()** with Fig. 3.75. First, Port 3 Servo is sent to its **LeftLimit** position (= 720) via Line 79. Next, **LeftLimit** is decreased by 20 units (= 700) by Line 80 and this new value is used in a WAIT WHILE Loop on Line 81 (essentially making the CM-150/151 "do nothing" while waiting for the Servo Motor to reach its assigned position of 720). Please note that this Servo Move is done at Speed = 512 (see Line 12 of Fig 3.73). Once Servo 3 passes over Position 700, Variable **FoundDowel** is initialized to 0 (Line 82), **Servo 3 Speed** is slowed down to 200 (Line 83), and Servo 3 is once again sent away to Position **RightLimit** (= 350 in Line 17 of Fig. 3.73). This "trip" would take a longer time period to finish, and it is during this "trip" that the two Dowels would be searched for by the IR Sensor and their closeness to the robot determined. This task is performed by Function **FindDowel()** on Line 87 for the First Dowel (i.e., the Left one), and its (max) IR Distance to the robot is saved into Variable **DowelDistance1**. The WAIT WHILE in Line 91 is used for the Controller to spend some time, while the IR Sensor scans over the **remainder** of the Left Dowel and **starts getting into** the empty space after the Left Dowel. In this transitional zone, the IR Sensor would read in values that are greater than **BackgroundNIR.** Once the IR Sensor "sees" the empty space between the two Dowels, the Controller would quit this WAIT WHILE Loop, and execute Line 94 next to find the Second Dowel on the Right and save its (max) IR distance from the robot into Variable **DowelDistance2**. Line 96 stops Servo 3 and Line 97 "freezes" its Current Position (basically sets its Present Position as its Goal Position). Lines 98 and 99 display the Dowel Distances found for the user to help keeping track of the robot's progress.

```
77  FUNCTION ServoScan
78  {
79      PORT[3]:Servo Joint Position = LeftLimit
80      LeftLimit = LeftLimit - 20
81      WAIT WHILE ( PORT[3]:Servo Joint Position < LeftLimit )
82      FoundDowel = 0
83      PORT[3]:Servo Joint Speed = 200
84      PORT[3]:Servo Joint Position = RightLimit
85
86      // Finding Dowel 1
87      CALL FindDowel
88      DowelDistance1 = PreviousNIR
89
90      // Let IR Sensor scan by the rest of Dowel 1 and until IR Sensor sees Background
91      WAIT WHILE ( PORT[4]:IR Sensor > BackgroundNIR )
92
93      // Finding Dowel 2
94      CALL FindDowel
95      DowelDistance2 = PreviousNIR
96      PORT[3]:Servo Joint Speed = 0
97      PORT[3]:Servo Joint Position = PORT[3]:Servo Joint Position
98      Print Screen = DowelDistance1
99      Print Screen with Line = DowelDistance2
100 }
```

Fig. 3.75 Function **ServoScan()** in "DowelScanner.tsk3".

- Function **FindDowel()** (see Fig. 3.76) implements the search strategy previously mentioned. The reader would need to remember that the IR Sensor started to scan from Servo Position **LeftLimit** just before Function **FindDowel()** is called. This means that the IR Sensor would read in values mostly around the current value of Variable **BackgroundNIR** during this time. As **DowelThreshold** was set to be "3" units above **BackgroundNIR,** it was used in the WAIT WHILE on Line 104 to let the Controller "do nothing" until the IR Sensor starts to detect the Left Edge of the Left Dowel. At that time, the IR Sensor would read in values greater than **DowelThreshold**, thus the Controller would quit this WAIT WHILE loop, and begin the actual search algorithm alluded to earlier with Line 105. As shown back in Fig. 3.72, the goal is to search for and keep the Local Maximum IR value found (to be saved as **DowelDistance1** later). The algorithm used is rather straightforward:

- We'll keep up with two "consecutive" and "running" IR Sensor readings named **PreviousNIR** and **CurrentNIR**.
- If **CurrentNIR** is greater than **PreviousNIR**, we have not found the Local Maximum value yet, i.e., the "physical" centerline of the dowel is not yet found – well a good estimate of it any how!
- When **CurrentNIR** is less than **PreviousNIR** for the first time, we **just pass over** the Local Maximum, so we are going to save the value of **PreviousNIR** as **DowelDistance1** (or **DowelDistance2** depending on **when** one uses Function **FindDowel()**).

```
102 FUNCTION FindDowel
103 {
104     WAIT WHILE ( PORT[4]:IR Sensor <= DowelThreshold )
105     PreviousNIR = BackgroundNIR
106     CurrentNIR = PORT[4]:IR Sensor
107     NIRDifference = CurrentNIR - PreviousNIR
108     LOOP WHILE ( NIRDifference >= 0 )
109     {
110         PreviousNIR = CurrentNIR
111         Print Screen = PreviousNIR
112         CurrentNIR = PORT[4]:IR Sensor
113         Print Screen with Line = CurrentNIR
114         NIRDifference = CurrentNIR - PreviousNIR
115     }
116     Print Screen = PORT[3]:Servo Joint Position
117     Print Screen = PreviousNIR
118     FoundDowel = FoundDowel + 1
119     Print Screen with Line = FoundDowel
120 }
```

Fig. 3.76 Function **FindDowel()** in "DowelScanner.tsk3".

Let's go through Fig. 3.76 to explain this algorithm line by line:

- First, we need to "seed" the Variables **PreviousNIR** and **CurrentNIR** with "logical" values. For **PreviousNIR**, **BackgroundNIR** is definitely a logical one to use (Line 105), because at this time, we know that the IR Sensor starts to "see" the edge of the dowel so its values would be for sure greater than **BackgroundNIR**. For **CurrentNIR**, it is of course logical to assign the latest reading from **Port 4 IR Sensor** to it, so Line 106 is used. **NIRDifference** is computed next (Line 107), and we **know for sure** that the WHILE LOOP on Line 108 would be entered next because of our previous choices for **PreviousNIR** and **CurrentNIR**.
- The actions defined between Lines 110 and 114 are designed to swap the **CurrentNIR** value into **PreviousNIR** (Line 110) and print it (Line 111). Next, **CurrentNIR** is updated with the latest reading from **Port 4 IR Sensor** and it is also printed out. Finally, **NIRDifference** is updated, and the loop's condition **(NIRDifference > 0)** is checked, and another iteration would be started if needed.
- Eventually, **NIRDifference** becomes negative when the Dowel's Centerline gets passed over, and the loop exits into Line 116 to print out the current Servo Position (for the user's benefit only), and then **PreviousNIR** gets printed. Finally, **FoundDowel** gets

incremented by "1" and printed. When Function **FindDowel()** gets called for the second time by Function **ServoScan()** on Line 94 in Fig. 3.75, the previous steps are executed and **FoundDowel** gets incremented to "2" and printed.

- There is a **subtle point regarding the Function FindDowel()** that the reader needs to be aware of. The PRINT commands in Lines 111 and 113 of Fig. 3.76 seem to be used there just to help the user monitor the progress of the algorithm, and for "full speed" performance the user may want to "comment" them out. However, if so done, the user will find out that the **Port 4 NIR Sensor** readings are no longer correct and tend to be smaller in numerical values than they really are. Essentially, the process for sensor data acquisition may be a little slower than the speed that this algorithm is capable of, thus creating some synchronization problems. There is another more robust approach using Function CALLBACK which will be described later in this Sub-Section.
- Once the previous Function ScanServo() is done (Line 38 of Fig. 3.74), Variables **DowelDistance1** and **DowelDistance2** are assigned with their actual values, then Function **VerifyDowels()** is started (Line 39 of Fig. 3.74). Fig. 3.77 describes the details of Function **VerifyDowels()** which is essentially an appropriate sequence of Function Calls with Functions **CenterSensor(), Right(), Left(),** and **FindApproachDowel()**. For example, if **DowelDistance2** is larger than **DowelDistance1** (Line 124), this means that the Right Dowel is closer to the robot than the Left Dowel, thus the robot first straightens up the IR Sensor Forward (Line 126) and then Turns Right towards the Right Dowel. Once the robot finds the Right Dowel with the IR Sensor, it approaches this dowel and when it is close enough, it will play Melody 3 signaling the user to remove the Right Dowel (Line 129). Next, it swings to the Left looking for the Left Dowel (Line 131) and performs similar tasks of finding and approaching the Left Dowel (Line 132). If the other situation happens instead (Line 134), i.e., the Left Dowel is closer to the robot than the Right Dowel, then the robot would execute similar find-approach tasks for the Left Dowel first (Lines 136-139) then Turns Right to take care of the Right Dowel (Lines 141-142).

```
122 FUNCTION VerifyDowels
123 {
124     IF ( DowelDistance2 > DowelDistance1 )
125     {
126         CALL CenterSensor
127         // Approach closer dowel on right
128         CALL Right
129         CALL FindApproachDowel
130         // Approach farther dowel on left
131         CALL Left
132         CALL FindApproachDowel
133     }
134     ELSE
135     {
136         CALL CenterSensor
137         // Approach closer dowel on left
138         CALL Left
139         CALL FindApproachDowel
140         // Approach farther dowel on right
141         CALL Right
142         CALL FindApproachDowel
143     }
144 }
```

Fig. 3.77 Function VerifyDowels() in "DowelScanner.tsk3".

The detailed codes for Functions **CenterSensor()** and **FindApproachDowel()** are listed in Figs. 3.78 and 3.79:

- Function **CenterSensor()** is quite simple (see Fig. 3.78). It just sets Servo 3 Speed to 512 and Servo 3 Position to Position 512, and then waits for 0.5 second for the Servo to get there.

```
FUNCTION CenterSensor
{
    PORT[3]:Servo Joint Speed = 512
    PORT[3]:Servo Joint Position = 512
    High-resolution Timer = 0.500sec
    WAIT WHILE ( High-resolution Timer > 0.000sec )
}
```

Fig. 3.78 Function **CenterSensor()** in "DowelScanner.tsk3".

```
FUNCTION FindApproachDowel
{
    WAIT WHILE ( PORT[4]:IR Sensor <= DowelThreshold )
    PreviousNIR = BackgroundNIR
    CurrentNIR = PORT[4]:IR Sensor
    NIRDifference = CurrentNIR - PreviousNIR
    LOOP WHILE ( NIRDifference >= 0 )
    {
        PreviousNIR = CurrentNIR
        Print Screen = PreviousNIR
        CurrentNIR = PORT[4]:IR Sensor
        Print Screen with Line = CurrentNIR
        NIRDifference = CurrentNIR - PreviousNIR
    }
    High-resolution Timer = 0.010sec
    WAIT WHILE ( High-resolution Timer > 0.000sec )
    CALL Stop
    ProximityLimit = PORT[4]:IR Sensor
    ProximityLimit = ProximityLimit + 150
    CALL Forward
    WAIT WHILE ( PORT[4]:IR Sensor < ProximityLimit )
    CALL Stop
    Buzzer Timer = Melody Time
    Buzzer Index = Melody No.3 (3)
    WAIT WHILE ( Buzzer Timer > 0.0sec )
```

Fig. 3.79 Function **FindApproachDowel()** in "DowelScanner.tsk3".

- Function **FindApproachDowel()** (see Fig. 3.79) uses the same search strategy employed in Function **FindDowel()** to find the needed Dowel, but this time by rotating the whole robot Left or Right as needed using the Geared Motors on Ports 1 and 2 (i.e. the various Functions **Left(), Right(), Forward()** and **Stop()** using Variables **SpeedL** and **SpeedR** previously set in the INIT code section – see Fig. 3.73). When **FindApproachDowel()** was called from inside **VerifyDowels()**, the robot was already turning Left or Right appropriately to get to its targeted Dowel (Lines 128, or 131, or 138, or 141 in Fig. 3.77). Thus, the first statement inside **FindApproach-Dowel()** is a WAIT WHILE (Line 148), active as long as the IR Sensor reads in a value that is less than **DowelThreshold**, i.e., the Controller code is letting the robot scan "empty space" until the IR Sensor starts "seeing" the Dowel's edge, resulting in readings larger than **DowelThreshold**.
- At that point, the controller quits the WAIT WHILE Loop and starts on the search for the Local Maximum value of the IR Sensor (Lines 149-159). This search strategy is already explained in a previous paragraph. Once this search is finished (Line 159), the robot should be stopped as it should be now facing the Targeted Dowel "more or less" squarely. But, as it is less precise to stop the Geared Motors with the inertia of the whole robot to deal with, than to stop the much lighter IR Sensor device as before. A small Time Delay (=0.01 s, but the reader can use a different value as needed) is used (Lines 160-161) before issuing the STOP command in Line 162. Next, the current value of the IR Sensor is read in and its value saved in **ProximityLimit** (Line 163). Line 164 increments **ProximityLimit** by 150 units and this new **ProximityLimit** value is used later in Line 166 to tell the robot when to stop its approach towards the targeted dowel (Lines 165-167). The robot then plays Melody 3 to convey "Mission Accomplished"!

The advanced Function named **CALLBACK,** which was already described in Sub-Section 3.1.5 and Fig. 3.18, is used to develop a revised version of this algorithm in the program named "Dowel_Scanner_CB.tsk3".

```
CALLBACK
{
    CB_NIR = PORT[4]:IR Sensor
    CB_POSITION = PORT[3]:Servo Joint Position
}
```

Fig. 3.80 CALLBACK Function used in "DowelScanner_CB.tsk3".

Fig. 3.80 shows that for every 8 ms, CALLBACK collects the current values of the Port 4 NIR Sensor and the Port 3 Servo Motor Position and saves them respectively in Variables **CB_NIR** and **CB_POSITION**. Consequently, the codes from inside the MAIN and REGULAR Functions need to be changed accordingly: i.e., each instance of "**Port 4 NIR Sensor**" is replaced by "**CB_NIR**", and each instance of "**Port 3 Servo Motor Position**" is replaced by "**CB_POSITION**", but ONLY when it is used as a **READ** Variable (i.e., when it is used on the RHS of an Assignment Statement or inside a Conditional Statement). Please compare Fig. 3.76 to Fig. 3.81 to see the appropriate changes made to Function **FindDowel()**. The side benefit from using the CALLBACK function is that the PRINT statements are no longer needed in the LOOP WHILE code and that the IR Sensor readings will be in the correct range of values.

```
FUNCTION FindDowel
{
    WAIT WHILE ( CB_NIR <= DowelThreshold )
    PreviousNIR = BackgroundNIR
    CurrentNIR = CB_NIR
    NIRDifference = CurrentNIR - PreviousNIR
    LOOP WHILE ( NIRDifference >= 0 )
    {
        PreviousNIR = CurrentNIR
        // Print Screen = PreviousNIR
        CurrentNIR = CB_NIR
        // Print Screen with Line = CurrentNIR
        NIRDifference = CurrentNIR - PreviousNIR
    }
    Print Screen = CB_POSITION
    Print Screen = PreviousNIR
    FoundDowel = FoundDowel + 1
    Print Screen with Line = FoundDowel
}
```

Fig. 3.81 New Function **FindDowel()** using CALLBACK Variables **CB_NIR** and **CB_POSITION** in "DowelScanner_CB.tsk3".

In summary, for the Projects shown in Chapter 3, the author recommends that the reader should make time to compare TASK techniques and results as compared to the ones used for R+BLOCK in Chapter 2. So, it is worth repeating the old adage "When there is a will, there's a way", or perhaps an adapted version of another old adage "The old robot can be taught new tricks!".

3.5 References

Beecher K (2017) Computational Thinking: A Beginner's Guide to Problem-Solving and Programming. BCS Learning & Development Ltd, Swindon, UK.

Ben-Ari M, Mondada F (2018) Elements of Robotics. Springer Publishers International, Cham, Switzerland.

Bergin J (2012) Pedagogical patterns: advice for educators. CreateSpace Independent Publishing Platform, p. 115.

Denning PJ, Tedre M (2019) Computational Thinking. The MIT Press, Cambridge.

Matarić MJ (2007) The robotics primer. The MIT Press, Cambridge.

Thai CN (2017) Exploring Robotics with ROBOTIS Systems (2nd Edition). Springer Publishers International, New York.

Thai CN (2018) Advanced Application Programming with ROBOTIS DREAM System. Amazon Kindle Direct Publishing.

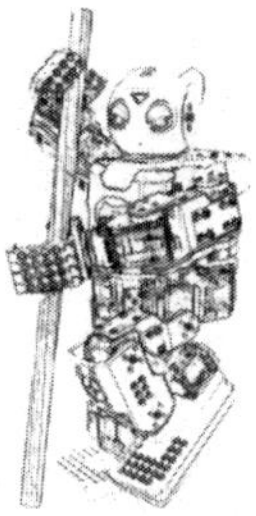

Chapter 4: Using Python (and TASK)

The author's motivation for writing this Chapter was to develop an approach for readers of Chapters 2 (R+BLOCK) and 3 (TASK) to carry the concepts and techniques learned there, and to apply them to Python on Chromebooks.

The author's approach was to "cater" to R+BLOCK users and used "Analogy" as an instructional tool. Thus, as much as possible, each BLOCK interacting with the CM-150/151 will match with ONE Python Function using the same numbers and types of inputs to the matching BLOCK in R+BLOCK. As an example, for the CM-150/151 Buzzer, Fig. 4.1 shows on the left an R+BLOCK code that plays a Music Scale for 2 seconds, next plays a Musical Melody before exiting, and on the right an equivalent Python program:

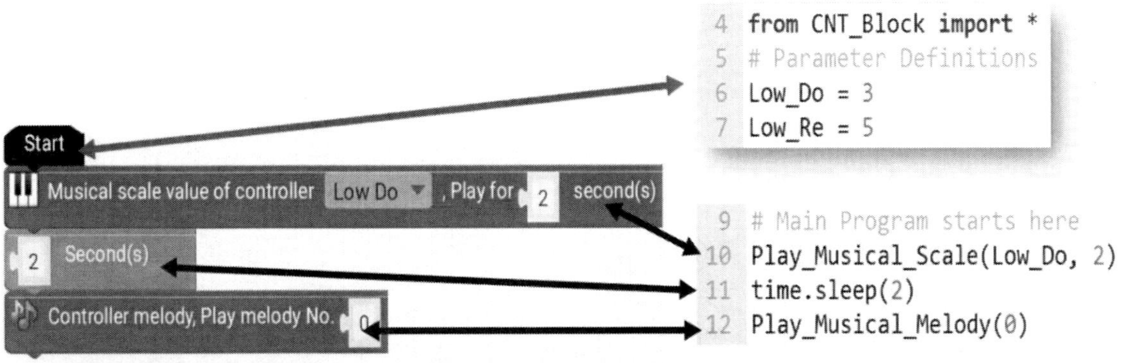

	Do	Re	Mi	Fa	So	La	Ti
Low Scale	3	5	7	8	10	12	14
Basic Scale	15	17	19	20	22	24	36
High Scale	27	29	31	32	34	36	38

Fig. 4.1 Using Buzzer in R+BLOCK and Python.

- The reader can see that the Main Function/Program in Python uses the same structure and almost the same syntax as the R+BLOCK equivalent.
- **Play_Musical_Scale()** uses similar arguments "**Low_Do**" for the musical note, and "**2**" for the **2** second duration as the R+BLOCK's Equivalent Block. The table at the bottom of Fig. 4.1 also shows the Constants to use for the other Musical Scales in Python.
- **Time.sleep(2)** is equivalent to the **2** seconds Delay Block.
- **Play_Musical_Melody()** uses the same argument "**0**" to play **Melody 0** in R+BLOCK.
- This "ANALOGICAL" approach is enabled by using a special Python module/library named "**CNT_Block.py**" which can be used as a "black-box" for beginner Python users, to ease their transitions from R+BLOCK to Python. In a way, the reader can think of the START Block as enabling/activating a similar module/library "behind the scenes" for R+BLOCK.

Later in Section 4.3, we will get into the "nuts and bolts" of this module. There is a companion TASK program "**CNT_Block.tskx/tsk3**" running on the CM-150/151 upon powering up, which the Python Programmer does not need to change in any way. This TASK program will also be explained further in Section 4.3. The reader can view this TASK program as an extension to the ROBOTIS Firmware on the CM-150/151, so that Python codes on the Chromebook can work more efficiently with the CM-150/151 Firmware than the combination of R+BLOCK/Chrome OS can currently provide.

In this Chapter 4, we will be using Standard Python 3 for the robotics projects. Python 3 will allow the user to tap into the vast resources of other Python packages available at the Python Software Foundation (https://pypi.org/ and https://docs.python.org/3/contents.html). The web site Python Central is also a great resource for learning and practicing Python programming (https://www.pythoncentral.io/).

The author also assumes that the reader already has some computer programming experiences in Python, if not the author would like to suggest the following references, some are for younger beginners, while others are good references to keep in the reader's library (by the way, plenty of "free" web resources abound for Python):

- "Python for Tweens and Teens: Learn Computational And Algorithmic Thinking" (Bouras and Ainarozidou, 2017).
- "Introducing Python", 2nd Edition (Lubanovic, 2020).
- "Robot Adventures in Python and C" (Bräunl, 2020).
- "Programming in Python 3", 2nd Edition (Summerfield, 2010).
- "The Python 3 Standard Library by Example", 1st Edition (Hellmann, 2017).

4.1 Setting up Chromebook for Python

The reader is referred to Appendix A for proper procedures to install the Linux Virtual Environment on his/her Chromebook, along with the Thonny IDE which is used by the author to develop the Python projects in this book. The reader can of course use any other Python editor of his/her choice.

The typical hardware configuration for using Python with Chromebooks and the DREAM II system is described in Fig. 4.2:

- The reader can see the Thonny IDE and Linux Terminal Apps on the Chromebook's Display.
- A USB hub is definitely needed so that the USB BT-410 Dongle can be used. Multiple dongles can be used – the first one maps into the Linux device named "**ttyUSB0**", while the next one maps into "**ttyUSB1**", etc... Please note that currently this Dongle works properly only with Linux, and it is not useable at the Chrome OS level. The reader can recall that R+BLOCK and TASK 2/3 do not need a USB BT-410 Dongle to connect to the BT-410 Receiver on the CM-150/151 side.

Fig. 4.2 Typical Hardware Configuration used for Python Programming on Chromebooks.

- On the CM-150's side, a BT-410 UART Receiver is used to communicate with Linux on the Chromebook. If the reader uses a CM-151, then the BT-410 USB Dongle would connect readily to the Embedded BT-410 BLE module on the CM-151. As a matter of fact, the user will find that the CM-151 BLE Receiver has a better wireless performance than the BT-410 UART Receiver. A new ROBOTIS USB Dongle is scheduled to be released by the end of 2022 to provide up to 1 Mbps to the CM-151 BLE Receiver, so Python's runtime performance will be even better with this forthcoming USB Dongle for the CM-151 (due to replace the CM-150 by late 2022).

Software-wise, the reader only needs to download either the file "CNT_Block.tskx" (if using TASK 2) or the file "CNT_Block.tsk3" (if using TASK 3) to the CM-150/151, and restart Power on the CM-150/151. The reader should not have to modify any code in these two TASK programs in order to use the Utility Functions of the "CNT_Block.py" Module as shown later in this Chapter 4.

On the Chromebook, the reader needs to make sure to follow the instructions in Appendix A, Section A-2 for proper locations of the various R+BLOCK, TASK, and Python files.

The "Sense-Think-Act" Paradigm as applied to the above Hardware Configuration can be "redrawn" as shown in Fig. 4.3, which is a combination of the ones for R+BLOCK and R+TASK:

- The user's Python programs are developed and run (via Thonny) on the Chromebook in a similar manner to R+BLOCK codes.
- At run time, the user's Python program interacts with the combination of "**CNT_Block.py**" (running on Thonny/Linux) and "**CNT_Block.tskx/tsk3**" (running on CM-150/151 via BT communications), to provide access to the Actuators and Sensors connected to the CM Controllers. The "**CNT_Block.py"** module contains Block-like Utility Functions that send Commands via BT to the CM Controllers and also receive Data back from them if needed.
- The Python "**CNT_Block.py**" module implements the ROBOTIS Remocon Packet protocol (Virtual RC-100) mentioned earlier in Sub-Section 3.1.6, but this time the full 16-bit packet will be used (more in Section 4.3).

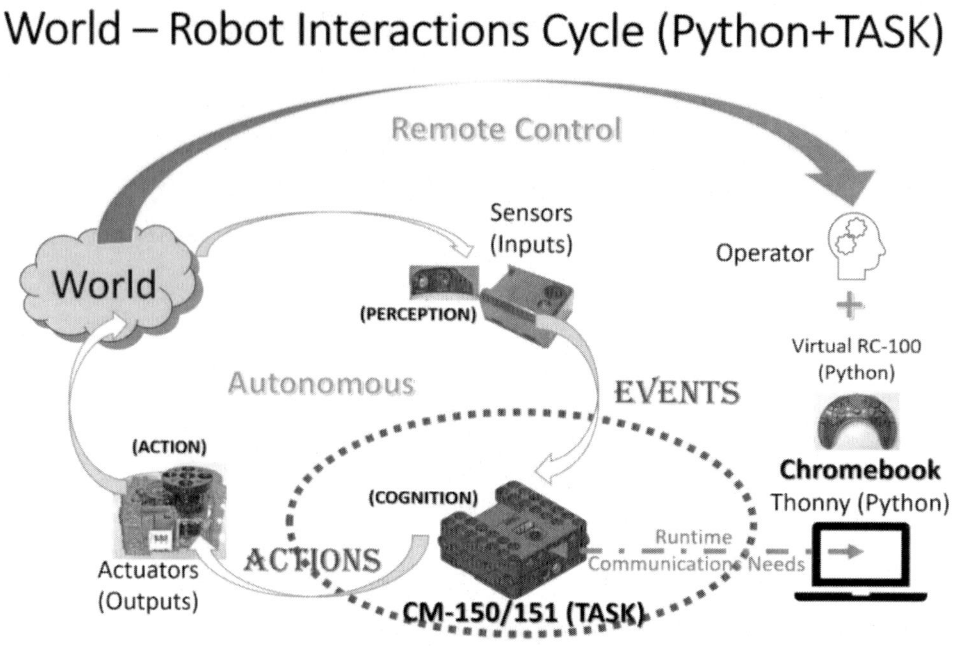

Fig. 4.3 STA Paradigm applied to Python Usage on Chromebooks.

4.2 Using "CNT_Block.py" Module

In this Section, the reader is guided to use the Python Module "CNT_Block.py" as a "black-boxy" library of Block-like Utility Functions, along with some handy Python structures and techniques for getting the best runtime performance out of DREAM II robots. The API documents for "CNT_Block.py" and "CNT_Block.tskx/tsk3" are available in Appendix C.

4.2.1 Integral IR Sensors

The program "IR_Throughput.py" shows how to access the Internal/Integral IR Sensors, Left, Middle and Right, with the Utility Function **Read_Integrated_IR(Location)**, defined within the "CNT_Block.py" Module, where the Argument **Location** is defined as follows:

- Location = 0 for Left Integral IR Sensor.
- Location = 1 for Middle Integral IR Sensor.
- Location = 2 for Right Integral IR Sensor.

Fig. 4.4 illustrates that ones can program in Python using the same style as in R+BLOCK (i.e., "IR_Throughput.rpb" – see Sub-Section 2.4.4):

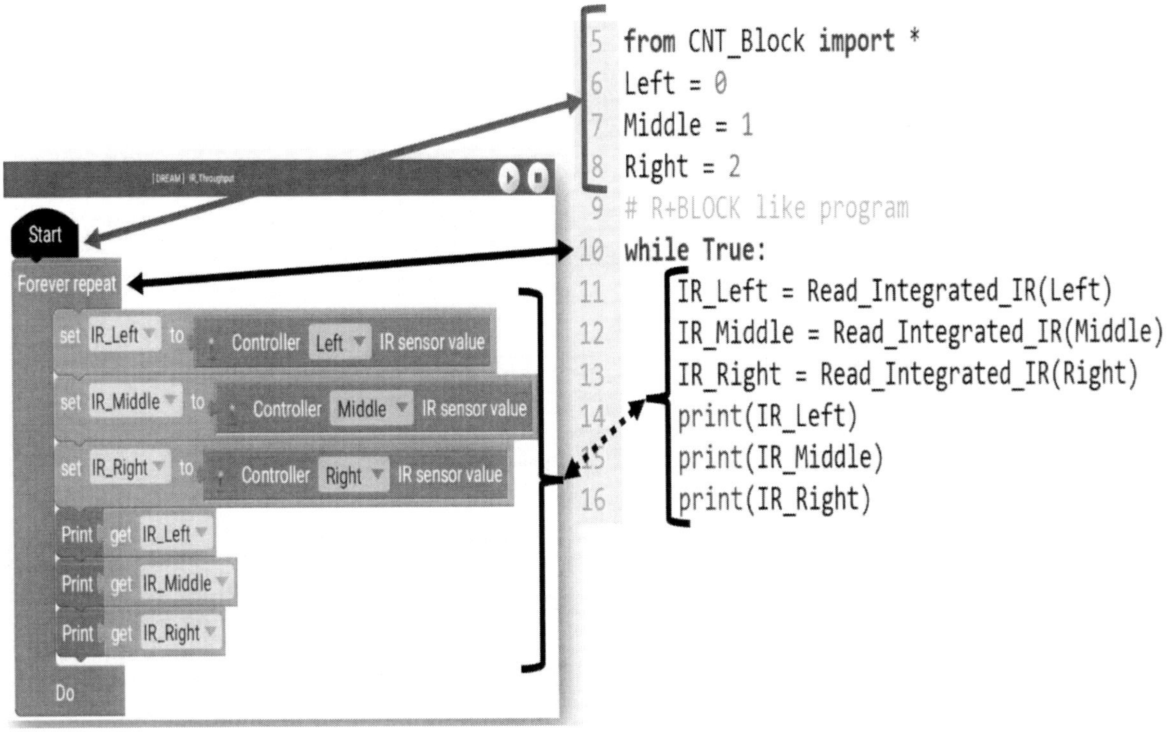

Fig. 4.4 "IR_Throughput.py" Solution using same style as R+BLOCK.

- Line 5 is an "IMPORT" statement, needed to provide access to all Utility Functions packed inside the **CNT_Block** module.
- Lines 6, 7 and 8 provide "word" equivalents to the Constant Values 0, 1 and 2.
- A "FOREVER REPEAT" Block can be translated as a "WHILE TRUE" statement in Python (Line 10).
- Lines 11 to 16 are used to provide a one-to-one match between Python statements with the Blocks used on the R+BLOCK side.

However, using Python, in the approach shown above, defeats the whole purpose of "learning a new programming language". Thus, a more Pythonic version is shown in Fig. 4.5:

- Line 19 defines a List with 3 elements having Constant String values: 'Left', 'Middle' and 'Right'. Thus Location[0] is set to 'Left', Location[1] is set to 'Middle' and Location[2] is set to 'Right'.

```
18 # A more Pythonic version with "error checking"
19 Location = ['Left', 'Middle', 'Right']
20 while True:
21     for i in range(3):  # i = 0, 1, 2
22         IR_data = Read_Integrated_IR(i)
23         #time.sleep(0.001)
24         if (IR_data != 1000):    # if no data reception error
25             print("IR Data = ", IR_data, "\t at location = ", Location[i])
```

Fig. 4.5 A more Pythonic version of "IR_Throughput.py".

- Line 20 shows the start of the familiar Endless Loop (WHILE TRUE Loop in Python).
- Line 21 show the beginning of an Inner FOR Loop (or Counting Loop) with **Index i** taking on values 0, 1 and 2 as this Loop iterates:
  - In Line 22, Function **Read_Integrated_IR(i)** brings in the latest Value from each of the three Integrated IR Sensors on the CM-150/151 (depending on the value of the current **Index i**) and saves it in Variable IR_data.
  - Function **Read_Integrated_IR(i)** has built-in error checking features (more details in Section 4.3), thus if it returns the value "**1000**", this means that it has encountered some kind of errors in the Remocon Packet returned from the CM Controller, therefore this Received Value should not be used. In this case, the IF statement in Line 24 evaluates FALSE and the PRINT statement on Line 25 is skipped. Please note that Function **Read_Integrated_IR()** has its own Error Messaging scheme, so at this point in time the Operator would see various Error Messages displayed in Thonny's Run Shell. Next, the FOR Loop re-iterates with its next Index value.
  - If Line 24 evaluates TRUE at runtime, then **IR_data** is good to use, and the PRINT statement in Line 25 will be activated to print the Current Value of the Integrated IR Sensor at the Current Location (Left, Middle or Right).

A typical output on Thonny's Run Shell is shown below in italics:

IR Data = 259 at location = Left

IR Data = 280 at location = Middle

IR Data = 416 at location = Right

The reader will see that the data throughput of "IR_Thoughput.py" is much higher than "IR_Throughput.rpb" on the same Chromebook and at the same baud rate used for BT communications, because Remocon Packets are simpler messages than R+BLOCK messages (whatever they are as their structures are proprietary to ROBOTIS).

4.2.2 IR Sensors used on Ports 3 and 4

In R+BLOCK, Ports 3 and 4 on the CM-150/151 can be used to connect to an IRSS-10 IR Sensor or a Touch Sensor. Thus, Utility Function **Read_Port_IR_Touch (PortNo, Option)**, defined within the "CNT_Block.py" Module, was created with this feature whereas:

- **PortNo=0** corresponds to **Port 3**, while **PortNo=1** corresponds to **Port 4**.
- **Option=0** corresponds to **IRSS-10 Sensor**, while **Option=1** corresponds to **Touch Sensor**.

```
5  from CNT_Block import *
6  IR_sensor = 0  # 0 = IRSS-10 sensors; 1 = Touch sensors
7  Touch_sensor = 1
8  while True:
9      for i in range(2):  # i = 0 for Port 3, 1 for Port 4
10         IR_data = Read_Port_IR_Touch(i, IR_sensor) ⟵
11         #time.sleep(0.001)
12         if (IR_data != 1000):
13             print("Port IR Data = ", IR_data, "\t at Port = ", i+3)
```

Fig. 4.6 Program "Port_IR_Throughput.py".

- While the usual Endless Loop is started at Line 8, this time, the Inner FOR Loop cycles only from 0 to 1 for **Index i** (Line 9).
- Line 10 reads in and saves the latest Value of an IRSS-10 IR Sensor out of Port 3 (**i=0**) or Port 4 (**i=1**) into Variable **IR_data**.
- Function **Read_Port_IR_Touch ()** has a similar Error Catching and Messaging system to the one used for Function **Read_Integrated_IR()** (see Sub-Section 4.2.1 and Lines 12-13).

A typical output on Thonny's Run Shell is shown below in italics:

Port IR Data = 852 at Port = 4

Port IR Data = 474 at Port = 3

The reader will also see a high data throughput for "Port_IR_Thoughput.py" as for "IR_Thoughput.py".

4.2.3 Geared Motors used on Ports 1 and 2

The program "SequenceManeuvers.py" illustrates the typical usage of Utility Function **Turn_Gear_Motor(PortNo, Direction, Power)** which is defined within the "CNT_Block.py" Module, whereas:

- **PortNo** can be set to **0** (for Port 1) or **1** (for Port 2).
- **Direction** can be set to **0** (for CCW) or **1** (for CW).
- **Power** is a value between 0 and 100.
- There is no error checking in Utility Function **Turn_Gear_Motor(PortNo, Direction, Power)** thus the programmer needs to develop his/her own range-checking procedures.

- Please note that Utility Function **Turn_Gear_Motor(PortNo, Direction, Power)** is designed to match with its equivalent Block in R+BLOCK (see Fig. 4.7).

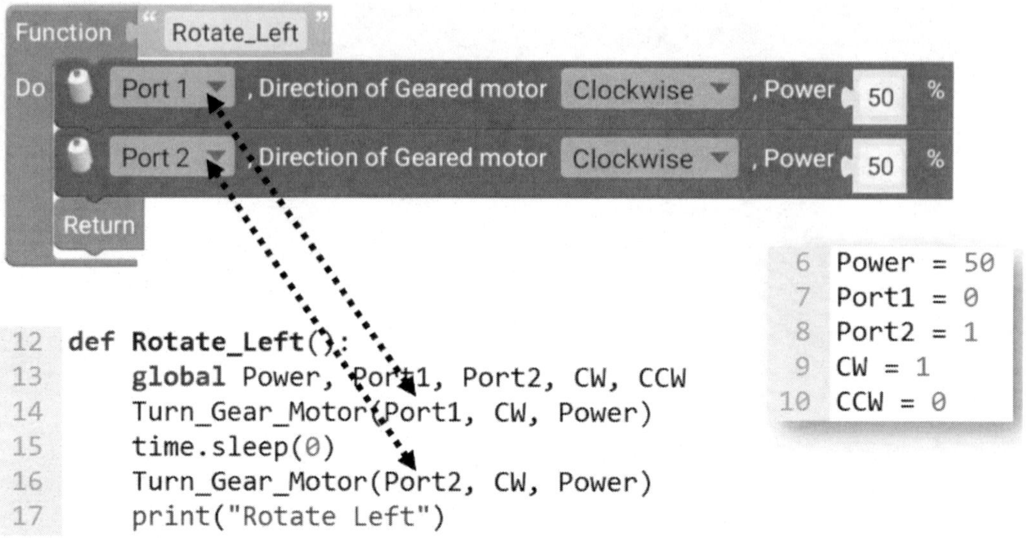

Fig. 4.7 Function **Rotate_Left()** in program "SequencedManeuvers.py".

Fig. 4.7 also shows how Utility Function **Turn_Gear_Motor()** is used in a user-created Function **Rotate_Left()**:

- First, the reader may recall that ALL Variables in R+BLOCK and TASK have a Global Scope. Thus, the author will be keeping the same practice as much as possible. Lines 6-10 declare and initialize the Global Variables Power, Port1, Port2, CW and CCW (https://www.w3schools.com/python/python_variables_global.asp), and also see Line 13 of Function **Rotate_Left().**
- Lines 12-17 define Function **Rotate_Left():**
    - Line 13 "reminds" the user that any change to these Variables (inside Function **Rotate_Left()**) will affect the Values of the Same Variables in the Main Function, and other user-defined Functions that happen to use those Same Variables.
    - Lines 14 and 16 apply Utility Function **Turn_Gear_Motor()** to make the robot Rotate Left (in the same way as in R+BLOCK). Line 15 is included in case the user wants to "slow" the robot behavior down to have the same I/O STD effects of R+BLOCK discussed back in Section 2.6.
    - Line 17 use a print-out to supplement what the Operator sees out in the physical behaviors of the robot.

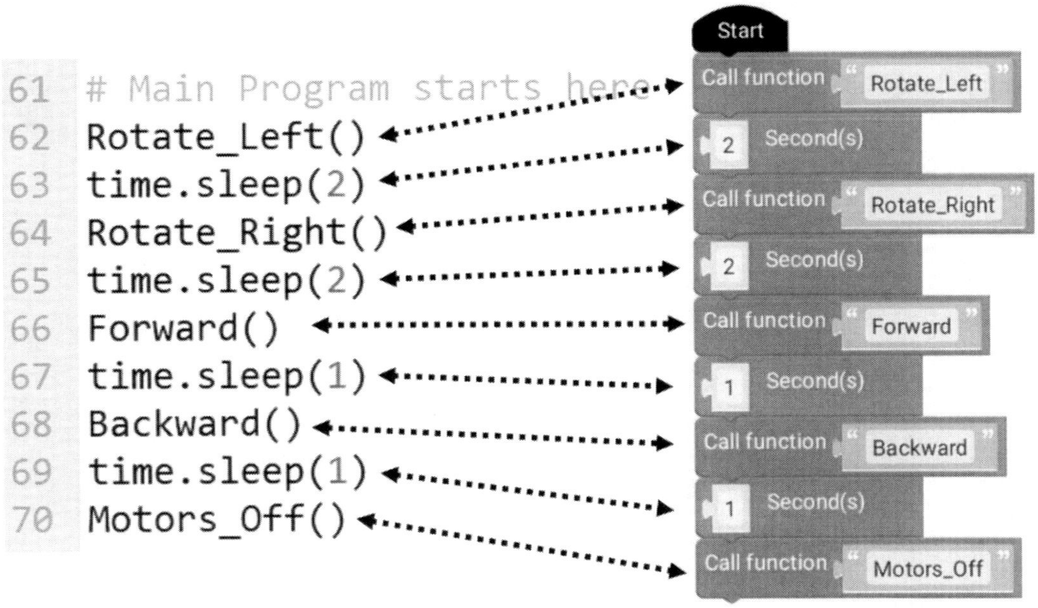

Fig. 4.8 Main Function in program "SequencedManeuvers.py".

Fig. 4.8 shows that the structure of the Main Function in Python and R+BLOCK is exactly the same. The reader can also compare "SequenceManeuvers.py" to "Maneuvers-4.tskx/tsk3" in Sub-Section 3.1.3 to see similarities and differences.

4.2.4 Escaping Spinning Top

Next, let us compare "Spinning_Top_2.py" to "Spinning_Top_2.rpb" (see Fig. 2.21 and Sub-Section 2.4.4).

Fig. 4.9 shows the complete listing for program "Spinning_Top_2.py" (except for Lines 6-10 which are displayed in Fig. 4.7 already):

- Lines 31 and 33 show the sameness in structure for the Main Function in both Python and R+BLOCK codes.
- User Function **Set_Power()** has the same goal for both versions, which is to collect the first reading on IR Sensor Middle and convert it into a Power setting between 0 and 100:
    - Lines 14 and 15 have the same "meaning" in both versions.
    - Line 17 is unique to Python because Utility Function **Read_Integrated_IR()** has an Error Checking feature previously mentioned in Sub-Section 4.2.1.
    - The constant "275" used in Line 18 represents a way to set a zone about **1" and above** the CM-150/151 Middle IR Sensor to serve as the **sensing region** used for the setting of Variable **Power**. The reader also should remember that this Python solution requires a "marker/obstacle" to be above the Middle IR Sensor before its execution, for this project to run properly.

```
12 def Set_Power():
13     global Power
14     Power = 0
15     IR_data = Read_Integrated_IR(1)    # Index 1 is used for Middle IR
16 #     time.sleep(0.001)
17     if ((IR_data != 1000) and (IR_data != 0)):    # IR data is good to use
18         Power = (int) ((IR_data / 275) * 100)
19         if (Power > 100):
20             Power = 100
21     print("Power = ", Power)
22
23 def Rotate_Left():
24     global Power
25     Turn_Gear_Motor(Port1, CW, Power)
26 #     time.sleep(0.001)
27     Turn_Gear_Motor(Port2, CW, Power)
28
29 # Main Function
30 while True:
31     Set_Power()
32 #    time.sleep(0.001)
33     Rotate_Left()
34 #    time.sleep(0.001)
```

Fig. 4.9 Program "Spinning_Top_2.py".

Next, let us see how Parallel IFs are used in Python. Fig. 4.10 shows the complete listing for program "Spinning_Top_3_Par_IFs.py" which can be compared to "Spinning_Top_3_Par_IFs.rpb" (Fig. 2.29 in Sub-Section 2.5.6):

- Comparing Fig. 2.29 to Fig. 4.10 shows that the Main Functions are practically the same in terms of structure and syntax, except for Line 86 which calls for a Time Delay of 0.8 s. Without this Time Delay, the author had found that, randomly, Utility Function **Read_Integrated_IR(1)** (inside User Function **Set_Power()**) would not work properly with the CM-150, but NOT with the CM-151: in other words, this issue seemed to be related to BT communication throughput capabilities/differences between the CM-150 and CM-151.
- Line 18 inside Function **Set_Power()** is now using the value **600** to scale **IR_Data**, meaning that the sensing region used for the setting of Variable **Power** is now a zone about **0.25" and above** the CM-150/151 Middle IR Sensor.

```
# Main Function
Play_Musical_Melody(3)
time.sleep(3)
while True:
    Set_Power()
#    time.sleep(0.001)
    if (Power <= 5):
        Motors_Off()
    if ((Power > 5) and (Power <= 50)):
        Rotate_Right()
    if ((Power > 50) and (Power <= 95)):
        Rotate_Left()
    if (Power > 95):
        Play_Musical_Scale(32,1)    # four
        time.sleep(0.8)
        Forward()
```

```
def Set_Power():
    global Power
    Power = 0
    IR_data = Read_Integrated_IR(1)    # Index
#    time.sleep(0.001)
    if ((IR_data != 1000) and (IR_data != 0)):
        Power = (int) ((IR_data / 600) * 100)
        if (Power > 100):
            Power = 100
    print("Power = ", Power)
```

Fig. 4.10 Program "Spinning_Top_3_Par_IFs.py".

Figs. 4.11 and 4.12 can be compared to Figs. 2.30 and 2.31 to illustrate the use of IF-ELSE-IF structures in Python and R+BLOCK which are similar in structure, but the syntax unavoidably had to be different. But the reader can agree that so far programming techniques in Chapter 2 (and 3) can be transferred to Python rather straightforwardly.

```
# Main Function
Play_Musical_Melody(3)
time.sleep(3)
while True:
    Set_Power()
#    time.sleep(0.001)
    if (Power <= 5):
        Motors_Off()
    elif ((Power > 5) and (Power <= 50)):
        Rotate_Right()
    elif ((Power > 50) and (Power <= 95)):
        Rotate_Left()
    elif (Power > 95):
        Play_Musical_Scale(32,1)    # found
        time.sleep(0.8)
        Forward()
```

Fig. 4.11 Program "Spinning_Top_3_IF_ELSE_IF_1.py".

```
# Main Function
Play_Musical_Melody(3)
time.sleep(3)
while True:
    Set_Power()
#    time.sleep(0.001)
    if (Power <= 5):
        Motors_Off()
    elif (Power <= 50):
        Rotate_Right()
    elif (Power <= 95):
        Rotate_Left()
    else:
        Play_Musical_Scale(32,1)
        time.sleep(0.8)
        Forward()
```

Fig. 4.12 Program "Spinning_Top_3_IF_ELSE_IF_2.py".

4.2.5 Avoider and Follower

Let us next compare "Avoider_3.py" (Figs. 4.13, 4.14 and 4.15) to "Avoider_3.rpb" (Figs. 2.34, 2.38 and 2.42).

In Fig. 2.34 (Sub-Section 2.5.3), the Block **WAIT UNTIL (MIRS != 0)** was used to wait for the Operator to insert a blocking object within the LOS of the Middle IR Sensor, to begin the process of acquiring the latest MIRS value and using it for setting Variable **Speed**. But, in Python, there is no WAIT UNTIL Loop, so its WHILE Loop can be used as a WAIT WHILE Loop, thus the Conditional Expression is changed to the reverse of the one used in R+BLOCK: i.e., **WAIT WHILE (MIRS == 0)** – see Line 25 of Fig. 4.13. The 0.01 s Time Delay in Line 26 is actually needed because Python does not like the body for a WHILE Loop to be "empty", and also to give time for the Middle IR Sensor to get ready for another Read operation.

Two other Python differences in Fig. 4.13 are:

- Line 29 for error checking which is implemented by Utility Function **Read_Integrated_IR()** – see Sub-Section 4.2.1.
- Line 30 for scaling the **IR_data** reading from Line 27 to **600 as 100% Power**.

```python
def Set_Power():
    global Power
    Power = 0
    Play_Musical_Scale(3,1)
    time.sleep(1)
    while (Read_Integrated_IR(1) == 0):
        time.sleep(0.01)
    IR_data = Read_Integrated_IR(1)    # Index 1
#     time.sleep(0.001)
    if ((IR_data != 1000) and (IR_data != 0)):
        Power = (int) ((IR_data / 600) * 100)
        if (Power > 100):
            Power = 100
    print("Power = ", Power)
    Play_Musical_Scale(25,1)
    time.sleep(1)
```

Fig. 4.13 Function **Set_Power()** in program "Avoider_1.py".

Fig. 4.14 shows that User Function **Read_Port_IR_Sensors()** makes use of Utility Function **Read_Port_IR_Touch()**, previously described in Sub-Section 4.2.2, to create an Error Checking procedure via Lines 71-75:

- Lines 71-72 pick up the latest data from the IRSS-10 IR Sensors mounted on Ports 3 and 4 and assign them to Variables **Port3_IR** and **Port4_IR** respectively, via the Utility Function **Read_Port_IR_Touch()** which is designed to return a Special Value "**1000",** when communication-related errors happen during the BT packet transfer from the CM Controllers to the Chromebook's Python program (more in Section 4.3).
- Lines 73-75 form a WHILE Loop that is activated when **either** Variable **Port3_IR** or **Port4_IR** has the Special Value of "**1000**", meaning that a "corrupted" BT packet had been received, then Lines 74-75 show that a new Read-Sensors operation is re-activated. As a consequence of using this Error-Checking Loop, when this User Function **Read_Port_IR_Sensors()** exits and returns to the Main Function of "Avoider_1.py", Global Variables **Port3_IR** and **Port4_IR always contain valid** sensor data. **This feature was not "accessible/affordable" in R+BLOCK** (see Fig. 2.38).
- The author had noticed that the CM-150 with a BT-410 UART Receiver would encounter "corrupted" packets occasionally at runtime, but so far, the author had no issue with the BLE on the CM-151.

```
def Read_Port_IR_Sensors():  # Read Sensors with Error Checking
    global Port3, Port4, Port3_IR, Port4_IR, Option_IR
    Port3_IR = Read_Port_IR_Touch(Port3, Option_IR)
    Port4_IR = Read_Port_IR_Touch(Port4, Option_IR)
    while ((Port3_IR == 1000) or (Port4_IR == 1000)):
        Port3_IR = Read_Port_IR_Touch(Port3, Option_IR)
        Port4_IR = Read_Port_IR_Touch(Port4, Option_IR)
    print("Port3_IR = ", Port3_IR)
    print("Port4_IR = ", Port4_IR)
```

Fig. 4.14 Function **Read_Port_IR_Sensors()** in program "Avoider_3.py".

Fig. 4.15 shows that the Main Function is structured the same way on Python and on R+BLOCK (Fig. 2.42):

- **Port3_Init** and **Port4_Init** were set to 0.
- Because the data throughput is now high, the Time Delay of 0.1 s. (Line 127) is needed to let a "commanded" maneuver to stay active for a little while before getting switched by the logic of the IF-ELSE-IF structure.

```
112 # Main Program starts here
113 Stop()
114 Set_Power()
115 time.sleep(0.5)
116 Play_Musical_Melody(2)
117 time.sleep(2)
118
119 while True:
120     Read_Port_IR_Sensors()  # Smart Sensors
121     if ( (Port3_IR - Port3_Init) >= 50 ):
122         Back_Right()          → = 0
123     elif ( (Port4_IR - Port4_Init) >= 50 ):
124         Back_Left()           → = 0
125     else:
126         Forward()
127     time.sleep(0.1)
```

Fig. 4.15 Main Function in program "Avoider_3.py".

```
# Main Program starts here
Stop()
Set_Power()
time.sleep(0.5)
Ports_Init()
time.sleep(1)
Play_Musical_Melody(2)
time.sleep(2)

while True:
    Read_Port_IR_Sensors()  # Smart Sensors
    if ( (Port3_IR - Port3_Init) >= 50 ):
        Back_Right()
    elif ( (Port4_IR - Port4_Init) >= 50 ):
        Back_Left()
    else:
        Forward()
    time.sleep(0.1)
```

Fig. 4.16 Main Function in program “Avoider_4i.py”.

Fig. 4.16 illustrates the only difference between “Avoider_4i.py” and “Avoider_3.py”, which is in the use of User Function **Ports_Init()**, described in Fig. 4.17:

```
57 def Ports_Init():
58     global Port3, Port4, Port3_Init, Port4_Init, Option_IR
59     Port3_Init = 0
60     Port4_Init = 0
61     for i in range(5):  # read 5 data input and take average
62         Port3_Init = Port3_Init + Read_Port_IR_Touch(Port3, Option_IR)
63         Port4_Init = Port4_Init + Read_Port_IR_Touch(Port4, Option_IR)
64     Port3_Init = (int) (Port3_Init / 5.)
65     Port4_Init = (int) (Port4_Init / 5.)
66     print("Port3_Init = ", Port3_Init)
67     print("Port4_Init = ", Port4_Init)
```

Fig. 4.17 Function **Ports_Init()** in program “Avoider_4i.py”.

- Lines 61-65 list a FOR Loop which is used to collect 5 consecutive Ports 3 and 4 IR Sensors readings, take their averages, and save them into Variables **Port3_Init** and **Port4_Init**. This current structure does not have an Error Checking feature, the reader is challenged to create one based on the technique shown in Fig. 4.14.

If the reader is interested in using the Integral Left, Middle and Right IR Sensors for the Avoider robot, he/she can check out the solution named "Avoider_4i_Int_IR.py" which is described in Figs. 4.18, 4.19 and 4.20.

Fig. 4.18 shows 2 user-friendly features:

- Lines 11-13 use Named Constants **Port_L, Port_R** and **Port_M** instead of "raw numbers" such as 0, 1 and 2.

```
11 Port_L = 0
12 Port_R = 2
13 Port_M = 1
14 Port_L_Init = 0
15 Port_R_Init = 0
16 Left_IR = 0
17 Right_IR = 0
18 Option_IR = 0
19
20 def Set_Power():
21     global Power
22     Power = 0
23     Play_Musical_Scale(3, 1)
24     time.sleep(1)
25     while ((Read_Integrated_IR(Port_M) == 0) or (Read_Integrated_IR(Port_M) == 1000)):
26         time.sleep(0.1)
27     IR_data = Read_Integrated_IR(Port_M)
28     if (IR_data < 1000):    # IR data is good to use
29         Power = (int) ((IR_data / 275) * 100)
30         if (Power > 100):
31             Power = 100
32         print("Power = ", Power)
33     else:
34         print("Middle IR malfunctioning - Please restart robot")
35         Stop()
36         sys.exit()
37     Play_Musical_Scale(16, 1)
38     time.sleep(1)
```

Fig. 4.18 Function **Set_Power()** in program "Avoider_4i_Int_IR.py".

- As Variable **Power** is crucial to the proper execution of the rest of the Python program, Lines 33-36 use an ELSE structure to be enabled in case of malfunction with the Middle IR Sensor:
    - Print the error message at Line 34.
    - Stop the robot for safety with Line 35.
    - And finally exit the program at Line 36.

Fig. 4.19 shows the Main Function for “Avoider_4i_Int_IR.py”:

- Function **Ports_Init()** is used in Line 96, but it is completely rewritten to use the Integral IR Sensors instead of the Ports 3 and 4 IR Sensors (Fig. 4.20).
- Function **Read_Port_IR_Sensors()** is used in Line 102, but it is also completely rewritten to use the Integral IR Sensors instead of the Ports 3 and 4 IR Sensors (Fig. 4.20).
- Lines 105 and 108 show an alternate use of the Time Delays to be applied for the **Back_Right()** and **Back_Left()** maneuvers only.

```
92  # Main Program starts here
93  Stop()
94  Set_Power()
95  time.sleep(0.5)
96  Ports_Init()
97  time.sleep(1)
98  Play_Musical_Melody(2)
99  time.sleep(2)
100
101 while True:
102     Read_Port_IR_Sensors()   # Smart Sensors Re
103     if ( (Left_IR - Port_L_Init) >= 100 ):
104         Back_Right()
105         time.sleep(0.01)
106     elif ( (Right_IR - Port_R_Init) >= 100 ):
107         Back_Left()
108         time.sleep(0.01)
109     else:
110         Forward()
```

Fig. 4.19 Main Function in program “Avoider_4i_Int_IR.py”.

```
def Ports_Init():
    global Port_L, Port_R, Port_L_Init, Port_R_Init
    Port_L_Init = Read_Integrated_IR(Port_L)
    Port_R_Init = Read_Integrated_IR(Port_R)
    print("Port_L_Init = ", Port_L_Init)
    print("Port_R_Init = ", Port_R_Init)

def Read_Port_IR_Sensors():   # Read Sensors with Error
    global Port_L, Port_R, Left_IR, Right_IR
    Left_IR = Read_Integrated_IR(Port_L)
    Right_IR = Read_Integrated_IR(Port_R)
    while ((Left_IR == 1000) or (Right_IR == 1000)):
        Left_IR = Read_Integrated_IR(Port_L)
        Right_IR = Read_Integrated_IR(Port_R)
    print("Left_IR = ", Left_IR)
    print("Right_IR = ", Right_IR)
```

Fig. 4.20 Functions **Ports_Init()** and **Read_Port_IR_Sensors()** in program “Avoider_4i_Int_IR.py”.

Fig. 4.21 shows an IF_ELSE_IF solution and a Nested WHILE Loop solution for the Avoider, using Ports 3 and 4 IR Sensors with sensor initialization, used in program "Avoider_4i_Ports_IR.py":

- Please note that the reader needs to comment out the solution NOT USED for the other solution to work properly.
- Please also note that Function **Read_Port_IR_Sensors()** needs to be used within the Nested WHILE Loops (Line 120 and Line 124) so that Variables **Port3_IR** and **Port4_IR** get "refreshed" at the end of each iteration of the INNER WHILE Loops.

```
106 while True:
107     Read_Port_IR_Sensors()  # Smart Sensors R
108 # IF-ELSE-IF Solution
109     if ( (Port3_IR - Port3_Init) >= 50 ):
110         Back_Right()
111     elif ( (Port4_IR - Port4_Init) >= 50 ):
112         Back_Left()
113     else:
114         Forward()
115     time.sleep(0.0)
116 
117 # WHILE Solution
118     while ( (Port3_IR - Port3_Init) >= 100 ):
119         Back_Right()
120         Read_Port_IR_Sensors()
121 
122     while ( (Port4_IR - Port4_Init) >= 100 ):
123         Back_Left()
124         Read_Port_IR_Sensors()
125 
126     Forward()
```

Fig. 4.21 Main Function in program "Avoider_4i_Ports_IR.py".

Figs. 4.22 and 4.33 show that the previous concepts and techniques can be applied to the Follower robot ("Follower_3.py" vs. "Follower_3.rpb" in Fig. 2.29 of Sub-Section 2.6.2):

- Fig. 4.22 illustrates the use of **either** Integral IR **or** Ports 3-4 IR Sensors for the Follower algorithm. Please make sure to comment out the solution NOT USED for the other solution to work properly.
- Fig. 4.23 shows that Error Checking is implemented for both IR Sensor types.

```
while True:
    Read_IR_Sensors()  # Smart Sensors Read Procedure when using Internal IR Sensors
    if ( ((Left_IR - Left_IR_Init) > 250) or ((Right_IR - Right_IR_Init) > 250) ):
        Forward()
    elif ( (Left_IR - Left_IR_Init) > 10 ):
        Rotate_Left()
    elif ( (Right_IR - Right_IR_Init) > 10 ):
        Rotate_Right()
    else:
        Stop()

    Read_Port_IR_Sensors()  # Smart Sensors Read Procedure when using Ports 3/4 IRSS-
    if ( ((Port3_IR - Port3_Init) > 250) or ((Port4_IR - Port4_Init) > 250) ):
        Forward()
    elif ( (Port3_IR - Port3_Init) > 10 ):
        Rotate_Left()
    elif ( (Port4_IR - Port4_Init) > 10 ):
        Rotate_Right()
    else:
        Stop()
```

Fig. 4.22 Main Endless Loop in program "Follower_3.py".

```
def Read_IR_Sensors():  # Read Sensors with Error Checking
    global Left_IR, Right_IR
    Left_IR = Read_Integrated_IR(0)  # Option 0 for Left IR
    Right_IR = Read_Integrated_IR(2)  # Option 2 for Right IR
    while ((Left_IR == 1000) or (Right_IR == 1000)):
        Left_IR = Read_Integrated_IR(0)
        Right_IR = Read_Integrated_IR(2)
    print("Left_IR = ", Left_IR)
    print("Right_IR = ", Right_IR)
```

```
def Read_Port_IR_Sensors():  # Read Sensors with Error Checking
    global Port3, Port4, Port3_IR, Port4_IR, Option_IR
    Port3_IR = Read_Port_IR_Touch(Port3, Option_IR)
    Port4_IR = Read_Port_IR_Touch(Port4, Option_IR)
    while ((Port3_IR == 1000) or (Port4_IR == 1000)):
        Port3_IR = Read_Port_IR_Touch(Port3, Option_IR)
        Port4_IR = Read_Port_IR_Touch(Port4, Option_IR)
    print("Port3_IR = ", Port3_IR)
    print("Port4_IR = ", Port4_IR)
```

Fig. 4.23 Functions **Read_IR_Sensor()** and **Read_Port_IR_Sensors()** in program "Follower_3.py".

4.2.6 Using Sound Counter

In this Sub-Section 4.2.6, we will get to use the Sound Counter facilities built-in the CM-150/151, i.e., Addresses 86 and 87 of their Control Table (https://emanual.robotis.com/docs/en/parts/controller/cm-151/). In R+BLOCK, Address 86 is referred to as "Controller Result of Sound Counter Value", and Address 87 is referred to as "Controller Real-Time Sound Counter Value" – see Fig. 2.6 of Section 2.3.

In the "CNT_Block.py" Module, there are two Functions created to interface with the Control Table Addresses 86 and 87:

- Utility Function **Reset_Read_Sound_Counter_Result(Option)** deals with Address/Parameter 86. This Function can reset Address/Parameter 86 to 0 if the argument **Option** is set to 0. If **Option** is set to 1, this Function returns the Current Value found at Address 86.
- Utility Function **Read_Sound_Counter_Real_Time()** returns the Current Value found at Address 87.

Fig. 4.24 lists the program "SCR_Throughput.py" illustrating a basic use of Addresses 86 (Sound Counter Result) and 87 (Sound Counter Real-Time):

- Line 6 shows that **Option_SCR** is set to 1, as we want to read the Current Value of Sound Counter Result (i.e., Address 86).
- Line 9 shows that the Current Value of the Sound Counter Real-Time (Address 87) is saved into Variable **SCR_RT**.
- Line 10 shows that the Current Value of the Sound Counter Result (Address 86) is saved into Variable **SCR**.
- Lines 11-12 print these values out on Thonny's Run Shell.
- Line 13 uses a Time Delay of only 5 ms, and surprisingly this Python code ran fine with the CM-151's BLE Receiver set at 57.6 Kbps.

```
5  from CNT_Block import *
6  Option_SCR = 1   # to read Sound Counter Result
7
8  while True:
9          SCR_RT = Read_Sound_Counter_Real_Time()
10         SCR = Reset_Read_Sound_Counter_Result(Option_SCR)
11         print("SCR_RT = ", SCR_RT)
12         print("SCR = ", SCR)
13         time.sleep(0.005)   # 5 ms
```

Fig. 4.24 Program "SCR_Throughput.py".

Below is a typical runtime output on Thonny's Run Shell:

SCR_RT = 0 ***// Previously the author had clapped 3 times***

SCR = 3

SCR_RT = 1 *// Here the author started to clap twice – SCR_RT jumps to 1*

SCR = 3 *// SCR is still showing the old value of 3*

....	*// 5 more sets of same numbers for SCR_RT and SCR*
SCR_RT = 1	*// Last time that SCR_RT = 1*
SCR = 3	*// SCR still shows 3*
SCR_RT = 2	*// First time that SCR_RT jumps to 2*
SCR = 3	*// SCR still shows 3*
....	*// 14 more sets of SCR_RT = 2*
SCR_RT = 2	*// Last time that SCR_RT = 2*
SCR = 3	*// SCR still shows 3*
SCR_RT = 0	*// SCR_RT cleared to 0 by CM-150/151 Firmware*
SCR = 2	***// Last Value of SCR_RT transferred to SCR***

Fig. 4.25 shows how Utility Function **Read_Sound_Counter_Result() is** used in program "Clapping_RC_1.py":

- The First Inner WHILE Loop (Lines 82-83) is used to reset Sound_Counter_Result (Address 86) to 0 at the beginning of each iteration of the Main Endless Loop (Lines 81-97):
    - First, please note that Utility Function **Read_Sound_Counter_Result()** is used with **Option=0** to clear this counter to ZERO.
    - Second, this Utility Function is used directly in the Conditional Expression controlling this WHILE Loop, meaning that Python will keep on repeating this resetting step until it receives a ZERO Value from the CM-150/151 – no matter how long this may take.
- Using the same technique, the Second Inner WHILE Loop (Lines 84-85) is designed for Python to read in the Current Value found at Address 86 (**Option=1**), and also to wait for the Operator to clap his/her hands, no matter how long it may have to wait (i.e., to wait as long as SCR reads in ZERO).
- Once the Operator performs the appropriate Clap Command, the Second Inner Loop exits to Line 86, and the final value of SCR is saved into Variable **Claps**.
- Lines 87-94 is an IF-ELSE-IF structure identical to the one used in the R+BLOCK equivalent program "Clapping_RC_1.rpb" in Fig. 2.36 of Sub-Section 2.5.7. This IF-ELSE-IF structure decides on the appropriate maneuver to set the robot in, depending on the Current Value of **Claps**.
- Line 95 waits 1 second for the robot's maneuver to be performed.
- Line 96 shuts down the motors and Line 97 waits for another 1 second for the motors to quiet down before working with the Sound Counter again at Line 82.

```
# Main Function
Set_Power()
Play_Musical_Melody(3)
time.sleep(3)
while True:
    while (Reset_Read_Sound_Counter_Result(0) != 0):
        time.sleep(0.01)
    while (Reset_Read_Sound_Counter_Result(1) == 0):
        time.sleep(0.01)
    Claps = Reset_Read_Sound_Counter_Result(1)
    if (Claps == 1):
        Forward()
    elif (Claps == 2):
        Backward()
    elif (Claps == 3):
        Rotate_Left()
    elif (Claps == 4):
        Rotate_Right()
    time.sleep(1)    # Motors activation time
    Motors_Off()
    time.sleep(1)  # to give time for motors to "quiet
```

Fig. 4.25 Main Function for Program “Clapping_RC_1.py”.

Fig. 4.26 shows the program “Clapping_RC_2.py” which is another solution using Parallel IFs.

```
# Main Function
Set_Power()
Play_Musical_Melody(3)
time.sleep(3)
while True:
    while (Reset_Read_Sound_Counter_Result(0) != 0):
        time.sleep(0.01)
    while (Reset_Read_Sound_Counter_Result(1) == 0):
        time.sleep(0.01)
    Claps = Reset_Read_Sound_Counter_Result(1)
    if (Claps == 1):
        Forward()
    if (Claps == 2):
        Backward()
    if (Claps == 3):
        Rotate_Left()
    if (Claps == 4):
        Rotate_Right()
    time.sleep(1)    # Motors activation time
    Motors_Off()
    time.sleep(1)  # to give time for motors to "quie
```

Fig. 4.26 Main Function for Program “Clapping_RC_2.py”.

4.2.7 Using Keyboard with Line Trackers

For the Python projects showcased in the rest of this Section 4.2, we will be using Keyboard Inputs using a simple OpenCV Function named "**waitKeyEx()**" (https://docs.opencv.org/3.4/d7/dfc/group__highgui.html#gafa15c0501e0ddd90918f17aa071d3dd0, https://www.geeksforgeeks.org/python-opencv-waitkeyex-function/?ref=rp).

Fig. 4.27 shows the complete listing of program "TestWaitKeyEx.py" which can be handy for the determination of the ASCII code attached to a specific Key on an Extended Keyboard:

- Line 1 shows that this program will use a Module named **cv2** which stands for OpenCV in Python practice.
- Line 3 defines Object **img_1** to contain a 2-D picture defined by a file named "Zeeb_On.jpg" which should be located on the same folder as the file "TestWaitKeyEx.py". It is included in the Example Codes package that comes with this book.
- Line 4 assigns "ZEEB" to a String Variable named **window1**.
- Lines 5 and 6 initialize Variables **Pos_X** and **Pos_Y** to zero.
- Lines 7 to 16 forms an Endless Loop that would move **img_1** diagonally down every time that the Operator presses any key on the keyboard. It would also display an integer which corresponds to the ASCII code for the pressed key:
    - Line 8 displays **img_1** within a window that is labeled with String **window1**.
    - Line 9 moves **window1** to Screen Position defined by **Pos_X** and **Pos_Y.**
    - Lines 10 and 11 increment **Pos_X** and **Pos_Y** by 10 pixels.
    - Line 12 invokes Function **cv2.waitKeyEx()** with a ZERO argument, meaning that Python will wait at Line 12 "forever", until the Operator presses on any key. At that time, Function **cv2.waitKeyEx()** returns the ASCII code of this pressed key into Variable **key**. If a non-zero value was used for this argument, 15 for example, then this program would wait 15 ms for a pressed key. If no key was pressed during this 15 ms, **waitKeyEx()** returns a "-1" into Variable **key** and then Python goes on to the next Line 13.
    - Lines 13-14 show that if the ESC key was pushed, this Endless Loop would break out into Line 18 and the program terminates.
    - Lines 15-16 check to see if **(key != -1)** which would evaluate TRUE if a valid key had been pressed. In this case, **key** is printed out by Line 16. And the WHILE Loop reiterates at Line 8 with new values for **Pos_X** and **Pos_Y.**

```python
import cv2

img_1 = cv2.imread("Zeeb_On.jpg") # display a test image
window_1 = "ZEEB"
Pos_X = 0
Pos_Y = 0
while True:
    cv2.imshow(window_1, img_1)
    cv2.moveWindow(window_1, Pos_X, Pos_Y)
    Pos_X += 10
    Pos_Y += 10
    key = cv2.waitKeyEx(0)    # 0 for indefinite wait on keyboard
    if key == 27:    # Esc key to stop
        break
    if key != -1:  # -1 is always returned from waitKeyEx()
        print (key)

cv2.destroyAllWindows()
```

Fig. 4.27 Program “TestWaitKeyEx.py”.

In Sub-Section 4.2.7, two “Line-Tracker” projects will be showcased: Simple Line Tracker and Yin-Yang Line Tracker.

The Python project ‘Line_Tracker_SC.py” is best compared with “Line_Tracker_SC.tsk3” of Sub-Section 3.2.6 which used the Buttons of the Virtual RC-100 to perform Stop/Start and Speed Up/Down for the Line Tracker robot. In Project “Line_Tracker_SC.py” (Figs. 4.28 through 4.32), we will combine the use of the **OpenCV** and **CNT_Block** modules.

Fig. 4.28 lists Part 1 of program “Line_Tracker_SC.py” which takes care of setting up various Constants used:

- Lines 155-158 show the ASCII codes for the Arrow keys. If the reader happens to use this program on a Windows PC, please note that the ASCII codes for the Arrow keys are different (see Lines 150-153).
- Lines 159-167 are the ASCII codes for the other keys used such as “Escape”, a/A, s/S, i/I and d/D.
- Variable **key** is defined for the user’s key-push (**only 1 key at a time**) – Line 168.
- Boolean Variable **stop_once** is set to FALSE initially, it is used with the Stopping and Restarting procedures for the robot (see Fig. 4.30).
- Lines 170-177 create a simple Menu for the user.

```
# Main Program starts here
# User keyboard interface section
# Windows PC Keyboard
#UP_ARROW = 2490368
#DOWN_ARROW = 2621440
#LEFT_ARROW = 2424832
#RIGHT_ARROW = 2555904
# Linux/Chromebook Keyboard
UP_ARROW = 65362
DOWN_ARROW = 65364
LEFT_ARROW = 65361
RIGHT_ARROW = 65363
ESC = 27
a = 97
A = 65
s = 115
S = 83
i = 105
I = 73
d = 100
D = 68
key = 0
stop_once = False
message1 = "Push ESC to quit program."
message2 = "!!! Push on s for Emergency Stop"
message3 = "Push a/A to restart"
message4 = "Push i/I to increase Power, d/D to decrease Power"
print(message1)
print(message2)
print(message3)
print(message4)
```

Fig. 4.28 Part 1 of program "Line_Tracker_SC.py".

Fig. 4.29 describes the Init sections for OpenCV and the robot's motion:

- Line 180 associates String Variable **im_Zeeb_ON** to the file named "Zeeb_On.jpg" which should be located on the same folder as the file "Line_Tracker_SC.py".
- Line 181 assigns "ZEEB" to another String Variable named **window_1**.
- Line 182 uses OpenCV's Function **cv2.imread()** to read in the File Image "Zeeb_On.jpg" via Variable **im_Zeeb_ON** and saves it into Image Object **img_1.**
- Lines 183-184 work in concert to display **img_1** at Location (100, 100) of the Chromebook Display Area.
- Next, the robot initialization steps are:
  - Stop the robot (Line 187).
  - Play Melody 2 (Line 193).
  - Execute a Time Delay of 2 s. (Line 194).

```
# OpenCV section init
im_Zeeb_ON = "Zeeb_On.jpg"   # 400x451 pixels
window_1 = "ZEEB"
img_1 = cv2.imread(im_Zeeb_ON)
cv2.imshow(window_1, img_1)
cv2.moveWindow(window_1, 100, 100)

# Robot Init Section
Stop()
#Set_Power()
#time.sleep(0.5)
#IR_Init()
#Ports_Init()
#time.sleep(1)
Play_Musical_Melody(2)
time.sleep(2)
```

Fig. 4.29 Part 2 of program "Line_Tracker_SC.py".

```
196 while True:
197 #     Read_IR_Sensors()
198 #     Read_Port_IR_Sensors()
199 #     print("Left_IR = ", Left_IR, "\tRight_IR = ", Right_IR)
200 #     print("Port3_IR = ", Port3_IR, "\tPort4_IR = ", Port4_IR)
201
202 # Keyboard processing section
203     key = cv2.waitKeyEx(11)
205     # Check on specific keys pushed by user
206     if (key == ESC):
207         print("User pressed Esc")
208         Stop()
209         break
210
211     if ((key == s) or (key == S)):
212         print("Emergency STOP")
213         Stop()
214         stop_once = True
215
216     if ((key == a) or (key == A)):
217         Power = 28
218         Forward()
219         stop_once = False
```

Fig. 4.30 Part 3 of program "Line_Tracker_SC.py".

Fig. 4.30 describes the beginning of the Main Endless Loop (Line 196):

- Line 203 picks up the Key pressed by the Operator and saves it in Variable **key**. If the Operator did not press any key within 11 ms., then **"-1"** is saved into **key**.

- Lines 206-209 check to see if the **key** pressed was the Escape key. If this is true, Line 207 prints out an informational message. Line 208 stops the robot and Line 209 breaks out of the Main Endless Loop, and this program terminates.
- Lines 211-214 are activated in case Key s/S is pushed, and an "Emergency Stop" is ensued with Lines 212 and 213. Line 214 sets Boolean Variable **stop_once** to TRUE.
- Lines 216-219 are activated when Key a/A is pressed, **Power** is set to 28% (Line 217) and the robot is commanded to go Forward (Line 218) and **stop_once** is set to FALSE (Line 219).

Fig. 4.31 describes the steps used for Key i/I or Key d/D which are set to be "mutually exclusive" via the use of an IF-ELSE-IF structure:

- For each Key i/I detected (Line 221), Variable **Power** is incremented by 2% (Line 222), but only up to **100%** (Lines 223-224).
- Else, for each Key d/D detected (Line 226), Variable **Power** is decremented by 2% (Line 227), but only down to **0%** (Lines 228-229).

```
221         if ((key == i) or (key == I)):
222             Power += 2
223             if (Power >= 100):
224                 Power = 100
225             print ("Power = ", Power)
226         elif ((key == d) or (key == D)):
227             Power -= 2
228             if (Power <= 0):
229                 Power = 0
230             print ("Power = ", Power)
```

Fig. 4.31 Part 4 of program "Line_Tracker_SC.py".

Fig. 4.32 describes the last part of the Main Endless Loop where the "Line Tracking" algorithm is implemented:

- It is only when Variable **stop_once** is found to be set to FALSE (Line 232), i.e., Key a/A had been pressed, that the "Line Tracking" algorithm is enabled:
    - Line 234 means that if Left IR Sensor sees Black and Right IR Sensor sees White, then the robot should Rotate Left (Line 235).
    - While Line 236 means that if Right IR Sensor sees Black and Left IR Sensor sees White, then the robot should Rotate Right (Line 237).
    - Otherwise, the robot should go Forward (Line 239).
- Lines 241-257 represent the "Line Tracking" algorithm when Ports 3 and 4 IR Sensors are used.

```
232         if (stop_once == False):
233             Read_IR_Sensors()
234             if ((Left_IR < Threshold) and (Right_IR > Threshold)):
235                 Rotate_Left()
236             elif ((Right_IR < Threshold) and (Left_IR > Threshold)):
237                 Rotate_Right()
238             else:
239                 Forward()
240
241 #           Read_Port_IR_Sensors()
242 #           if ((Port3_IR < Threshold) and (Port4_IR > Threshold)):
243 #               Rotate_Left()
244 #           elif ((Port4_IR < Threshold) and (Port3_IR > Threshold)):
245 #               Rotate_Right()
246 #           else:
247 #               Forward()
248
249 # End of Main Endless Loop
250 cv2.destroyAllWindows()
251 sys.exit()
```

Fig. 4.32 Part 5 of program "Line_Tracker_SC.py".

The Speed Control feature, described in Fig. 4.31, is used to see how high the Power to the Geared Motors to be set to, until the robot cannot stay "On-Track" anymore. The user will see that this "Python-Limit" is higher than the R+BLOCK equivalent solution described in Sub-Section 2.6.4, but it is lower than the "TASK-Limit" obtained in Sub-Section 3.2.6. As the physical robot remains the same in these 3 "solutions", it is **as expected** that the TASK runtime environment provides the Least I/O Systemic Time Delays for "any" robot.

The "Yin-Yang Line Tracker" project (see Fig. 2.56) has the same logical structure as the one used for the "Simple Line Tracker" project, thus only the "Yin-Yang Line Tracking" algorithm needs to be described in Fig. 4.33:

- Again, it is only when Variable **stop_once** is found to be set to FALSE (Line 232), i.e., Key a/A had been pressed, that the "Yin-Yang Line Tracking" algorithm is enabled, using 5 Event/Action Pairs:
    - This time, we need to read in four IR Sensor data: the Integral Left and Right IR Sensors (Line 233) and the Ports 3 and 4 IR Sensors (Line 234).
    - Line 238 means that if Left IR Sensor sees Black and Ports 3 and 4 IR Sensors see White, then the robot should Rotate Left (Line 239).
    - Line 240 illustrates that if Right IR Sensor sees Black and Ports 3 and 4 IR Sensors see White, then the robot should Rotate Right (Line 241).
    - Line 242 means that if Left IR Sensor sees White and Ports 3 and 4 IR Sensors see Black, then the robot should Rotate Left (Line 243).
    - Line 244 illustrates that if Right IR Sensor sees White and Ports 3 and 4 IR Sensors see Black, then the robot should Rotate Right (Line 245).
    - Otherwise, the robot should go Forward (Line 247).

```
if (stop_once == False):   # normally False
    Read_IR_Sensors()
    Read_Port_IR_Sensors()
    #print("Left_IR = ", Left_IR, "\tRight_IR = ", Right_IR)
    #print("Port3_IR = ", Port3_IR, "\tPort4_IR = ", Port4_IR)
    # Tracking the Yin-Yang Line
    if (Left_IR <= 100 and Port3_IR >= 600 and Port4_IR >= 400):   # White Background
        Rotate_Left()
    elif (Right_IR <= 150 and Port3_IR >= 600 and Port4_IR >= 400):   # White Background
        Rotate_Right()
    elif (Left_IR >= 400 and Port3_IR <= 300 and Port4_IR <= 150):   # Black Background
        Rotate_Left()
    elif (Right_IR >= 400 and Port3_IR <= 300 and Port4_IR <= 150):   # Black background
        Rotate_Right()
    else:
        Forward()
```

Fig. 4.33 Line Tracking algorithm for program "Yin_Yang_Line_Tracker.py".

The Python project 'Yin_Yang_Line_Tracker.py" is best compared with "Yin_Yang_Line_Tracker.tsk3" of Sub-Section 3.3.2, which of course yields the higher Power Limit that the robot can run at and still stays "On-Track".

4.2.8 Smart Avoider

As previously discussed in Sub-Sections 2.6.3 and Sub-Sections 3.2.7, the "Smart Avoider" project is a good example of mixing Remote-Control and Autonomous-Behavior.

This Python project "Smart_Avoider.py" is best compared with "Smart_Avoider.tsk3" of Sub-Section 3.2.7, in terms of programming structures and techniques (see Figs. 4.34 through Fig. 4.41).

Part 1 of Initializations needed in "Smart_Avoider.py" is shown in Fig. 4.34, mainly:

- Lines 20-22 list the 3 **Threshold** values used by the author, which should be adjusted for the user's situation.
- Line 23 initializes Boolean Variable **obstacle** to FALSE. This is an important Variable controlling the logic for the "Obstacle Avoidance" algorithm used.

Part 2 of Initializations needed in "Smart_Avoider.py" is shown in Fig. 4.35, mainly for:

- Boolean Variable **stop_once** is very important to keyboard processing techniques used later in this program. Line 46 initializes it to FALSE.

```python
from CNT_Block import *
# Dream II Section
Power = 28
Port1 = 0
Port2 = 1
CW = 1
CCW = 0
Port3 = 0
Port4 = 1
Port3_Init = 0
Port4_Init = 0
Port3_IR = 0
Port4_IR = 0
Option_IR = 0     # 0 for IR
threshold_1 = 400
threshold_2 = 350
threshold_3 = 300
obstacle = False
```

Fig. 4.34 Part 1 of Initializations in “Smart_Avoider.py”.

```python
key = 0
stop_once = False
message1 = "Use Arrow Keys UDLR to perform various RC operations."
message2 = "Toggle Key a/A ON/OFF for Autonomous Obstacle Avoidance feature."
message3 = "ESC to quit program."
message4 = "!!! Push on s for Emergency Stop"
message5 = "Push i/I to increase Power, d/D to decrease Power"
```

Fig. 4.35 Part 2 of Initializations in “Smart_Avoider.py”.

Fig. 4.36 lists the Init Actions for the robot and for the OpenCV procedures:

- Line 193 stops the robot.
- Line 194 sets the Robot’s Maneuvering Power as previously described in Sub-Section 4.2.5.
- Line 196 computes the Initial Values for Ports 3 and 4 IR Sensors as previously described in Sub-Section 4.2.5.
- OpenCV will be using two images, thus two String Variables are used: **im_Zeeb_ON** (associated with file “Zeeb_On.jpg”), and **im_Zeeb_OFF** (associated with file “Zeeb_Off.jpg”), via Lines 202 and 203 respectively.
- Display Window **window_1** is named “**ZEEB**”.

- Line 205 initializes **robot_1** to FALSE, this variable is used later for toggling ON/OFF the ZEEB images and the "Obstacle Avoidance" feature.
- Line 207 reads in the image file "**Zeeb_Off.jpg**" and saves it to Object **img_1.**
- Line 208 uses Function **cv2.imshow()** to display **img_1** inside the window named **window_1** (at Screen Location (0,0) - Top Left Corner).
- Line 209 uses Function **cv2.moveWindow()** to move **window_1** to its new Screen Location (100, 100) – in pixel units.
- Lines 210-213 display the simple menu described by **message1** through **message4** defined in Fig. 4.35.

```
191 # Main Program starts here
192 # Getting robot ready
193 Stop()
194 Set_Power()
195 time.sleep(0.5)
196 Ports_Init()
197 time.sleep(1)
198 Play_Musical_Melody(5)
199 time.sleep(2)
200
201 # OpenCV section init
202 im_Zeeb_ON = "Zeeb_On.jpg"    # 400x451 pixels
203 im_Zeeb_OFF = "Zeeb_Off.jpg"
204 window_1 = "ZEEB"
205 robot_1 = False
206 stop_once = False
207 img_1 = cv2.imread(im_Zeeb_OFF)
208 cv2.imshow(window_1, img_1)
209 cv2.moveWindow(window_1, 100, 100)
210 print(message1)
211 print(message2)
212 print(message3)
213 print(message4)
```

Fig. 4.36 Init Actions for Robot and OpenCV in "Smart_Avoider.py".

Fig. 4.37 describes Part 1 of the Main Endless Loop in "Smart_Avoider.py" (starting at Line 215):

- Line 217 uses OpenCV Function **cv2.waitKeyEx()** to pick up the ASCII code of the Key pressed by the Operator if he/she so chooses to and saves it in Variable **key**. If the Operator did not press any Key, then "-1" is saved into Variable **key**. Line 217 also shows that the Operator is allowed only ONE Key to press on at any one time (which is the case for Linux with the author's experiences so far).

```
215 while True:
216 # Keyboard processing section
217     key = cv2.waitKeyEx(11)   # wait 11 ms for a key stroke
218 
219     # Check on specific keys pushed by user
220     if (key == ESC):
221         print("User pressed Esc")
222         break
223 
224     if ((key == s) or (key == S)):
225         print("Emergency STOP")
226         Stop()
227 
228     if ((key == a) or (key == A)): # Key a/A for Autonomous Obstacle Avoidance (ON/OFF)
229         if (robot_1 == False):
230             img_1 = cv2.imread(im_Zeeb_ON) <-
231             cv2.imshow(window_1, img_1)
232             robot_1 = True <-
233         else:
234             img_1 = cv2.imread(im_Zeeb_OFF) <-
235             cv2.imshow(window_1, img_1)
236             robot_1 = False <-
237         cv2.waitKey(1)  # to refresh display
```

Fig. 4.37 Part 1 of Main Endless Loop in "Smart_Avoider.py".

- Lines 220-222 shows that if the **Escape** Key is pressed, then this program breaks out of the Main Endless Loop at Line 222, and subsequently this program terminates.
- Lines 224-226 shows that Key s/S is used for Emergency Stop, thus the robot is stopped by Line 226.
- Lines 228-237 is activated when Key a/A is pressed:
    - An IF-ELSE structure is used to toggle between two images (ON or OFF) of robot Zeeb.
    - Note that **robot_1** was initialized to FALSE back in Line 205 of Fig. 4.36. Thus, when the Operator pushes on **Key a/A** for the first time after this program starts, the IF structure at Line 229 evaluates TRUE, and the image **im_Zeeb_ON** is displayed via Lines 230-231. Then **robot_1** is set to TRUE (Line 232).
    - Thus, the next time that the Operator pushes on **Key a/A**, the ELSE branch at Line 233 will be activated instead, and the image **im_Zeeb_OFF** is displayed via Lines 234-235. Then **robot_1** is set to FALSE (Line 236).
    - In short, this IF-ELSE structure allows **window_1** to flip-flop between the ON and OFF images of robot Zeeb, each time that the Operator pushes on Key a/A. This flip-flop action is used to provide visual feedback to the Operator to let him/her know whether the "Obstacle Avoidance" feature is ON or OFF. Later Variable **robot_1** is used in the "Obstacle Avoidance" algorithm.

- Lastly, in Line 237, Function **cv2.waitKey(1)** is used to refresh the Display Screen so that the Operator can see the image flip-flop process "right-away", instead of waiting until the next iteration of the Main Endless Loop for the Line 217 to do its job.

```
250     if ((key == -1) and stop_once == False):    # when all keys are released first time
251         Stop()
252         stop_once = True
253
254 # Check on Obstacle Presence and switch on RC Mode or Escape Mode
255     Read_Port_IR_Sensors()  # Smart Sensors Read Procedure
256     if ( ((Port3_IR - Port3_Init) < threshold_1) and ((Port4_IR - Port4_Init) < threshold_1) ):
257         Keyboard_RC(key)
258     else:
259         if (robot_1 == True):  # when Zeeb's Image is ON
260             Escape()
261         else:
262             print("Watch out for OBSTACLE !!!")
263             Stop()
```

Fig. 4.38 Part 2 of Main Endless Loop in "Smart_Avoider.py".

Fig. 4.38 shows Part 2 of the Main Endless Loop:

- Lines 250-252 are designed to Stop the robot ONCE, when the Operator releases all keys of the keyboard, using the fact that Variable **key** is assigned a "**-1**" value when no key is pushed. The "ONCE" feature is accomplished by using Variable **stop_once** as a toggling device: the IF structure is activated only when **stop_once** is set to FALSE (Line 250), but after stopping the robot (Line 251), **stop_once** is set to TRUE (Line 252). Thus, preventing the Lines 250-252 from being re-activated in the next iteration of the Main Endless Loop, unless **stop_once** is set to FALSE somewhere else in the program (see Fig. 4.39).
- Recalling that Line 196 in Fig. 4.36 activated Function **Ports_Init()** to provide Variables **Port3_Init** and **Port4_Init**, Line 255 now calls Function **Read_Port_IR_Sensors()** to get the current **Port3_IR** and **Port4_IR** values.
- The IF structure in Line 256 can then be tested to see if both Port IR Sensor's Net Value is below **threshold_1 (= 400).** If this condition evaluates TRUE, i.e., the obstacle is still "far enough" from the robot, then the Operator can use the keyboard to move the robot around via Function **Keyboard_RC(key)** (Line 257 and see Fig. 4.39). If this condition evaluates FALSE, then the ELSE branch is taken (Line 258) and an Inner IF-ELSE structure (Line 259-263) implements two possible "Obstacle Avoidance" actions:
    - If **robot_1** is set to TRUE, meaning that the Operator did toggle ON the "Obstacle Avoidance" feature by pressing on key a/A previously (see Fig. 4.37), then Function **Escape()** is invoked. (Line 260 and see Fig. 4.40).
    - Otherwise, the program prints out a Warning Message (Line 262) and stops the robot with Line 263.

- After letting the Functions **Keyboard_RC()** and **Escape()** perform their tasks, the Main Endless Loop would then reiterates once more.

Fig. 4.39 displays the programming steps for Function **Keyboard_RC()** which uses an IF-ELSE-IF structure to implement the 4 mutually-exclusive Event/Action Pairs depending on which Arrow Key was pushed: Up, Down, Left or Right (Lines 145-156):

- Please note that Function **Keyboard_RC()** will perform its Actions (i.e. to move the robot around) **if and only if** any of the UDLR Arrow keys was pushed (while the other keys are still operating as "normal" – see Fig. 4.37).
- Furthermore, it is only within Function **Keyboard_RC()** that Variable **stop_once** is reset to FALSE (please review Lines 250-252 in Fig. 4.38 also).

```
140 def Keyboard_RC(key):
141     global stop_once
142     global UP_ARROW, DOWN_ARROW, LEFT_ARROW, RIGHT_ARROW
143 
144     # only one of the UDLR keys is allowed
145     if (key == UP_ARROW):
146         Forward()
147         stop_once = False
148     elif (key == DOWN_ARROW):
149         Backward()
150         stop_once = False
151     elif (key == LEFT_ARROW):
152         Rotate_Left()
153         stop_once = False
154     elif (key == RIGHT_ARROW):
155         Rotate_Right()
156         stop_once = False
```

Fig. 4.39 Function **Keyboard_RC()** in "Smart_Avoider.py".

Function **Escape()** is described in Figs. 4.40 and 4.41:

- First, Variable **obstacle** is set to TRUE (Line 162), and the robot is stopped by Line 163.
- The Escaping Maneuvers are contained within the OUTER WHILE Loop (Lines 165-179) which has a Conditional Expression that is "complementary" to the one used on Line 256 of Fig. 4.38. This important feature had been explained twice in Sub-Sections 2.6.3 and 3.2.7, so it won't be repeated here.
- Once inside the OUTER WHILE Loop, the issue is to figure out whether the Obstacle is Frontal, on the robot's Left or on the robot's Right.
  - The Obstacle is considered "Frontal" when the Ports 3 and 4 IR Sensor readings are within ± 10 units of each other (Lines 166 and 167). In this case, the robot should drive Backward for 0.5 seconds (Lines 168-170 in Fig. 4.40).

```python
def Escape():
    global obstacle, threshold_1, threshold_2, threshold_3
    global Port3_Init, Port4_Init, Port3_IR, Port4_IR
    # start of obstacle evading procedure
    obstacle = True
    Stop()  # Stop the robot
    #Read_Port_IR_Sensors()   # optional
    while ( ((Port3_IR - Port3_Init) > threshold_2) or ((Port4_IR - Port4_Init) > threshold_2) ):
        NIR_Diff = (Port3_IR - Port3_Init) - (Port4_IR - Port4_Init)
        if ((NIR_Diff >= -10) and (NIR_Diff <= 10)):
            print("Evading Backward")  # once only
            Backward()
            time.sleep(0.5)
```

Fig. 4.40 Part 1 of Function **Escape()** in “Smart_Avoider.py”.

- Next, **new** Ports 3 and 4 IR Sensor readings are taken by Line 171 of Fig. 4.41.
- Then, if the Obstacle happens to be on the robot’s Left, then the first INNER WHILE Loop applies (Lines 172-175):
  - An informational message is printed for the Operator’s to monitor (Line 173).
  - Line 174 uses Function **Rotate_Right()** to turn the robot to the Right (for a little bit).
  - Line 175 then reads in the **latest** Ports 3 and 4 IR Sensor readings so that the Conditional Expression in Line 172 can be re-evaluated afresh.
  - The First Inner Loop would stay “active” until the robot clears the Obstacle, i.e., when both Ports 3 and 4 IR Sensors report values that are below **threshold_3**.
- Or, if the Obstacle happens to be on the robot’s Right, then the second INNER WHILE Loop applies (Lines 176-179):
  - An informational message is printed for the Operator’s to monitor (Line 177).
  - Line 178 uses Function **Rotate_Left()** to turn the robot to the Left (for a little bit).
  - Line 179 then reads in the **latest** Ports 3 and 4 IR Sensor readings so that the Conditional Expression in Line 176 can be re-evaluated afresh.
  - The Second Inner Loop would stay “active” until the robot clears the Obstacle, i.e., when both Ports 3 and 4 IR Sensors report values that are below **threshold_3**.

```
171         Read_Port_IR_Sensors()
172         while ( ((Port3_IR - Port3_Init) > threshold_3) and ((Port3_IR - Port3_Init) > (Port4_IR - Port4_Init)) ):
173             print("Evading Right")
174             Rotate_Right()
175             Read_Port_IR_Sensors()
176         while ( ((Port4_IR - Port4_Init) > threshold_3) and ((Port4_IR - Port4_Init) > (Port3_IR - Port3_Init)) ):
177             print("Evading Left")
178             Rotate_Left()
179             Read_Port_IR_Sensors()
180     # end of obstacle evading procedure
181     Stop()
182     obstacle = False
183     print("Obstacle Cleared!")
184     Play_Musical_Melody(10) # upon hearing this tune, operator must release keyboard if not done so
185     time.sleep(2)
186     # waiting on Operator to release ALL Keys (i.e. when key input = -1)
187     while (cv2.waitKeyEx(10) != -1): # wait 10 ms for a key stroke
188         time.sleep(0.010)
```

Fig. 4.41 Part 2 of Function **Escape()** in "Smart_Avoider.py".

- Once cleared of the Obstacle, the robot is stopped (Line 181), next using the screen display (Line 183) and the robot's buzzer (Line 184), this "success" is announced to the Operator. At this point, the Operator should release all keys.
- But for "just in case", Lines 187-188 are used to keep the program "stuck" at this location until Function **cv2.waitKeyEx()** reads in "**-1**", i.e., when no key is pushed by the Operator. The reader can compare this solution to the one used in TASK – see Lines 77-81 in Fig. 3.43.

4.2.9 Smart TriCycle

In Sub-Sections 4.2.9 and 4.2.10, we will be using the Servo Motor SM-10 (https://emanual.robotis.com/docs/en/parts/motor/servo_motor/). The SM-10 can be used in Position Control mode (used in this book) or in Continuous Turn mode (not used in this book). The program "Basic_Servo.py" was written to show first-time users how to set up this Position Control mode, and to flip-flop between two Servo Positions (-50 and +50 degrees).

The two relevant Utility Functions from the "CNT_Block.py" Module are:

- **Turn_Servo_Motor (PortNo, Direction, Power),** whereas:
  - **PortNo** = **0** for Port 3 and **PortNo** = **1** for Port 4.
  - **Direction** = **0** for CCW and **Direction** = **1** for CW (if using SM-10 in Position Control mode use CCW only).
  - **Power** is in the range **[0-100]** (see Line 14 in Fig. 4.42).
- **Set_Servo_Position (PortNo, GoalPosition)**, whereas:
  - **PortNo** = **0** for Port 3 and **PortNo** = **1** for Port 4.
  - **Goal Position** in the range **[-150, 150]** degrees (see Line 15 in Fig. 4.42).

- **Read_Servo_Position (PortNo)**, whereas:
    - **PortNo = 0** for Port 3 and **PortNo = 1** for Port 4.
    - This function returns the Current Servo Position as a Value between -150 and +150 degrees.
    - *Please note that this Function does not exist in the current R+BLOCK version of 1.1.3.*

Please also note that these three Utility Functions were designed to stay close to the R+BLOCK syntax as much as possible (see Fig. 2.5). Fig. 4.42 lists the Variable Initializations used in program "Basic_Servo.py".

```
from CNT_Block import *
# Dream II Section
Power = 28
Port1 = 0
Port2 = 1
CW = 1
CCW = 0    # used for Position Control
Port3 = 0
Port4 = 1
Port3_Speed = 0    # [0-100]
Port3_Angle = 0    # [-150, 150] to ma
Port4_Init = 0
Port3_IR = 0
Port4_IR = 0
Option_IR = 0    # 0 for IR Sensor, 1
```

Fig. 4.42 Variables Initialization in program "Basic_Servo.py".

Fig. 4.43 describes the setup to put an SM-10 connected to Port 3 into Position Control mode:

- Line 22 is used for safety, essentially put Port 3 Servo Motor into Position Control mode with 0 Power (also see Line 11 in Fig. 3.55 for the equivalent in TASK code).
- Line 23 sets the Goal Position of Port 3 Servo Motor to 0 degree (see Fig. 2.58 also). At this point in time, the embedded controller on the SM-10 would try to send the servo to the 0 location, but because Power=0, this Servo Motor actually stays at its Initial Position (wherever this may be) when this program starts at runtime.

```
21 # Init Port 3 Servo Motor SM-10 to Goal Position 512 and @ 50% of Max Speed
22 Turn_Servo_Motor(Port3, CCW, 0)  # First Step  - always use CCW and 0 power/speed for safety
23 Set_Servo_Position(Port3, 0)   # 2nd Step - set Goal Angle = 0 degrees [-150, 150] (raw GP = 512)
24 Port3_Speed = 50 # Percentage of Max Speed  # 3rd Step - always positive value
25 Turn_Servo_Motor(Port3, CCW, Port3_Speed)  # 4th Step - always use CCW
26 # this is when the Port 3 servo actually moves to GP = 512 (or 0 degrees)
27 Play_Musical_Scale(3,1)
28 time.sleep(1)
```

Fig. 4.43 Setting Up Servo 3 in Position Control mode in program "Basic_Servo.py".

- Line 24 sets Variable **Port3_Speed** (actually Rotational Power) to 50%.
- Line 25 sets Servo 3 Speed to **Port3_Speed** which is now non-zero, this is when the Servo Motor moves to Goal Position 0. Please note to **always use the CCW option**, if not the Servo Motor may switch to its Continuous Turn mode. Please see the equivalent setup for TASK code in Fig. 3.55 (Lines 11-13).
- After this initial setup, the user can just use these two Utility Functions **Turn_Servo_Motor()** and **Set_Servo_Position()** as needed (just remember to use the CCW Option for **Turn_Servo_Motor()**). More details will be explained in Sub-Section 4.3.3.

```
30 while True:
31     Set_Servo_Position(Port3, -50)
32     time.sleep(1)
33     print("Present Goal Angle [-50] = ", Read_Servo_Position(Port3))
34     Set_Servo_Position(Port3, 50)
35     time.sleep(1)
36     print("Present Goal Angle [50] = ", Read_Servo_Position(Port3))
37     Set_Servo_Position(Port3, 0)
38     time.sleep(1)
39     print("Present Goal Angle [0] = ", Read_Servo_Position(Port3))
```

Fig. 4.44 Main Endless Loop used in program "Basic_Servo.py".

Fig. 4.44 lists the Main Endless Loop used to flip-flop Servo 3 between Goal Positions "-50" and "+50" degrees:

- Line 31 sets Servo 3's Goal Position to -50 degrees, and Line 32 waits 1 second for it to get there.
- Line 33 prints out the Present Position of Servo 3 using Utility Function **Read_Servo_Position(Port3)**. The reader would see that this value would be off the "-50" value by several degrees (the SM-10 is just not an accurate servo motor to use).
- Line 34 sets Servo 3's Goal Position to +50 degrees, and Line 35 waits 1 second for it to get there.

- Line 36 prints out the Present Position of Servo 3 using Utility Function **Read_Servo_Position(Port3)**. The reader would also see that this value would be off the "+50" value by several degrees as the SM-10 is not an accurate servo motor to use.
- Line 37 now sets Servo 3's Goal Position to 0 degree, and Line 38 waits 1 second for it to get there.
- Line 39 prints out the Present Position of Servo 3 using Utility Function **Read_Servo_Position(Port3)**. The reader would see that this value would be off the "0" value by several degrees also.

We can now get into the details for the Smart TriCycle Project ("Smart_Tricycle.py") through Figs. 4.45 to 4.53.

```
6  from CNT_Block import *
7  # Dream II Section
8  Power = 36
9  SpeedL = 0
10 SpeedR = 0
11 SpeedDif = 0
12 Port1 = 0
13 Port2 = 1
14 CW = 1
15 CCW = 0
16 Port3 = 0
17 Port4 = 1
18 Port3_Init = 0
19 Port4_Init = 0
20 Port3_IR = 0
21 Port4_IR = 0
22 Option_IR = 0     # 0 for IR Sen
23 Port3_Speed = 0    # [0-100]
24 Port3_Angle = 0    # [-150, 150]
25 threshold_1 = 335
26 #threshold_2 = 350
27 threshold_3 = 10
28 obstacle = False
```

Fig. 4.45 Constants and Variables used in program "Smart_Tricycle.py".

Fig. 4.45 shows the Constants and Variables used in program "Smart_Tricycle.py":

- Only **threshold_1** and **threshold_2** are used for the TriCycle as compared to the Smart Avoider (Lines 25 and 27).
- **Obstacle** is used in the same way as for the Smart Avoider.

Fig. 4.46 lists the Initial Actions for the robot:

- Line 266 stops the robot, and Lines 271-272 play Melody 5 to announce that it is ready for action.

- Lines 274-277 set Servo 3 into Position Control mode (already explained in Fig. 4.43), announced with a Low DO for 1 second (Lines 279-280).

```
265 # Getting robot ready
266 Stop()
267 #Set_Power()
268 #time.sleep(0.5)
269 #Ports_Init()
270 #time.sleep(1)
271 Play_Musical_Melody(5)
272 time.sleep(2)
273 # Init Port 3 Servo Motor SM-10 to Goal Pos
274 Turn_Servo_Motor(Port3, CCW, 0)    # First S
275 Set_Servo_Position(Port3, 0)     # 2nd Step
276 Port3_Speed = 70 # Percentage of Max Speed
277 Turn_Servo_Motor(Port3, CCW, Port3_Speed)
278 # this is when the Port 3 servo actually mc
279 Play_Musical_Scale(3,1)
280 time.sleep(1)
```

Fig. 4.46 Initial Actions for Robot in program "Smart_Tricycle.py".

The "Obstacle Avoidance" algorithm (inside the Main Endless Loop) is described in three parts via Figs. 4.47 to 4.49. Part 1 is shown in Fig. 4.47:

- After reading in the latest IR Sensor data **Port4_IR** via Function **Read_Port_IR_Sensors()** (already described in Sub-Section 4.2.5), Line 337 is used to check if an Obstacle is still far away, by testing **Port4_IR** against **threshold_1**:
    - If this is TRUE, Function **Keyboard_RC(key)** is activated, and the Operator's UDLR Arrow commands for the robot are processed via this Function.
    - If this is FALSE, i.e., an obstacle is close by, then a big ELSE structure is in-play (Lines 339-378), implementing a "nested-conditions" situation/solution:
        - If the Obstacle Avoidance feature is set (Line 340), i.e., (**robot_1==TRUE**), Variable **obstacle** is set to TRUE, and the robot is Stopped (Lines 341-342).
        - The Present Position of Servo 3 is read in next and saved into **Servo3_Pos** (Line 343).
        - If (**Servo3_Pos <= -12**) (see Line 344) this means that the Steering Wheel is already turned to the Right of the robot, thus the robot should "Escape Right" (Line 345).
        - Else If (**Servo3_Pos >= 12**) (see Line 346) this means that the Steering Wheel is already turned to the Left of the robot, thus the robot should "Escape Left" (Line 347).

```python
# Check on Obstacle Presence and switch on RC Mode or Escape Mode
    Read_Port_IR_Sensors()  # Smart Sensors Read Procedure
    if ((Port4_IR - Port4_Init) < threshold_1):
        Keyboard_RC(key)
    else:   # Escape Procedures
        if (robot_1 == True):  # when Zeeb's Image is ON
            obstacle = True
            Stop()
            Servo3_Pos = Read_Servo_Position(Port3)
            if (Servo3_Pos <= -12):  # already steered to the Right
                Escape_Right()
            elif (Servo3_Pos >= 12):  # already steered to the Left
                Escape_Left()
```

Fig. 4.47 Part 1 of "Obstacle Avoidance" Algorithm used in program "Smart_Tricycle.py".

```python
348         else:  # steering wheel straight forward but robot's body @ an angle w
349             Servo3_Pos_L = Servo3_Pos + 9
350             Servo3_Pos_R = Servo3_Pos - 9
351             Set_Servo_Position(Port3, Servo3_Pos_L)
352             time.sleep(0.25)
353             Read_Port_IR_Sensors()
354             ObstacleDistance_L = Port4_IR
355             Set_Servo_Position(Port3, Servo3_Pos_R)
356             time.sleep(0.25)
357             Read_Port_IR_Sensors()
358             ObstacleDistance_R = Port4_IR
359             if (ObstacleDistance_L > ObstacleDistance_R):
360                 Escape_Right()
361             elif (ObstacleDistance_L < ObstacleDistance_R):
362                 Escape_Left()
363             else:
364                 Backward()
365                 time.sleep(0.1)
366                 Escape_Right()  # just a p          for turning right
367         # end of obstacle evading procedures
```

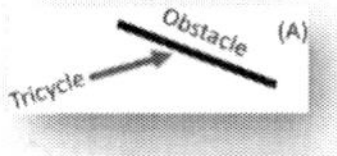

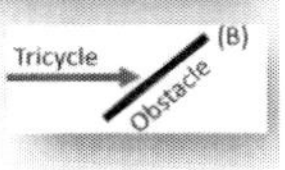

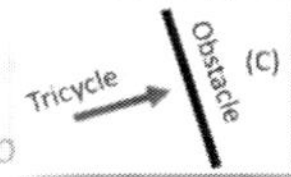

Fig. 4.48 Part 2 of "Obstacle Avoidance" Algorithm used in program "Smart_Tricycle.py".

- Otherwise, the Steering Wheel is mostly straight forward (Lines 348-366 in Fig. 4.48), but it is still unknown about the relative position of the robot's Longitudinal Axis with respect to the Obstacle Plane. Thus, Servo 3 swings to the Left of Center Line to record **ObstacleDistance_L** (Lines 349, 351-354), then it swings to the Right of Center Line to record **ObstacleDistance_L** (Lines 350, 355-358).
- These two Variables are then used to determine the 3 cases A, B and C:
  - ✓ Case A – the robot's Left is closer to the Obstacle Plane (Line 359), then the robot should "Escape Right" (Line 360).
  - ✓ Case B - the robot's Right is closer to the Obstacle Plane (Line 361), then the robot should "Escape Left" (Line 362).

- ✓ Case C – the robot is mostly perpendicular to the Obstacle Plane (Line 363), then it should go Backward for 0.1 s (Lines 364-365), and then "Escape Right" (Line 366).

- At this point in the algorithm, the robot should have "cleared" the Obstacle (see Fig. 4.49), so:
  - ✓ The robot stops (Line 368).
  - ✓ Variable **obstacle** is reset to FALSE (Line 369).
  - ✓ Text and audio announcements of this result are then activated (Lines 370-372).
  - ✓ Subsequently, the Main Endless Loop reiterates.
- However, if Variable **robot_1** was set to FALSE when checking at Line 340 earlier, then the robot is just stopped (Lines 377-378). **In this case, the Operator can next press Key a/A to activate the "Obstacle Avoidance" algorithm and see the previously described process unfold.**

```
368         Stop()
369         obstacle = False
370         print("Obstacle Cleared!")
371         Play_Musical_Melody(10)  # upon hear
372         time.sleep(2)
373         # waiting on Operator to release ALL
374         while (cv2.waitKeyEx(10) != -1):  #
375             time.sleep(0.010)
376     else:   # when Zeeb's Image is OFF
377         print("Watch out for OBSTACLE !!!")
378         Stop()
```

Fig. 4.49 Part 3 of "Obstacle Avoidance" Algorithm used in program "Smart_Tricycle.py".

Figs. 4.50 and 4.51 describe Functions **Keyboard_RC()** and **CheckSpeeds()** which follow similar logic to matching Functions used in the Smart Avoider Project (Sub-Section 4.2.8) so those explanations won't be repeated here.

```python
def Keyboard_RC(key):
    global stop_once, SpeedL, SpeedR, SpeedDif, Power, Port3
    global UP_ARROW, DOWN_ARROW, LEFT_ARROW, RIGHT_ARROW

    # only one of the UDLR keys is allowed
    if (key == UP_ARROW):
        Set_Servo_Position(Port3, 0)
        SpeedDif = 0
        SpeedL = Power - SpeedDif
        SpeedR = Power + SpeedDif
        Forward()
        stop_once = False
    elif (key == DOWN_ARROW):
        Set_Servo_Position(Port3, 0)
        SpeedDif = 0
        SpeedL = Power - SpeedDif
        SpeedR = Power + SpeedDif
        Backward()
        stop_once = False
```

Fig. 4.50 Part 1 of Function **Keyboard_RC()** used in program "Smart_Tricycle.py".

```python
    elif (key == LEFT_ARROW):
        Set_Servo_Position(Port3, Read_Servo_Position(Port3) + 20)  # 20 degrees
        SpeedDif += 2  # 2% Power step each time
        SpeedL = Power - SpeedDif
        SpeedR = Power + SpeedDif
        CheckSpeeds()
        Rotate_Left()
        stop_once = False
    elif (key == RIGHT_ARROW):
        Set_Servo_Position(Port3, Read_Servo_Position(Port3) - 20)  # 20 degrees
        SpeedDif += 2  # 2% Power step each time
        SpeedL = Power - SpeedDif
        SpeedR = Power + SpeedDif
        CheckSpeeds()
        Rotate_Right()
        stop_once = False
```

```python
def CheckSpeeds():
    global SpeedL, SpeedR
    if (SpeedL < 0):
        SpeedL = 0
    elif (SpeedL > 100):  # in Percentage Power
        SpeedL = 100
    if (SpeedR < 0):
        SpeedR = 0
    elif (SpeedR > 100):
        SpeedR = 100
```

Fig. 4.51 Part 2 of Function **Keyboard_RC()** and Function **CheckSpeeds()** used in program "Smart_Tricycle.py".

Fig. 4.52 describes Functions **Escape_Left()** and **Escape_Right()**:

```
206 def Escape_Left():
207     global obstacle, threshold_1, threshold_2, threshold_3, Port3
208     global Port3_Init, Port4_Init, Port3_IR, Port4_IR
209     # start of obstacle LEFT evading procedure
210     while (obstacle == True):
211         Set_Servo_Position(Port3, 60)  # turn steering wheel left
212         time.sleep(0.25)
213         Forward()
214         time.sleep(0.30)
215         Stop()
216         Check_NIR()
217
218 def Escape_Right():
219     global obstacle, threshold_1, threshold_2, threshold_3
220     global Port3_Init, Port4_Init, Port3_IR, Port4_IR
221     # start of obstacle RIGHT evading procedure
222     while (obstacle == True):
223         Set_Servo_Position(Port3, -60)  # turn steering wheel right
224         time.sleep(0.25)
225         Forward()
226         time.sleep(0.30)
227         Stop()
228         Check_NIR()
```

Fig. 4.52 Functions **Escape_Left()** and Function **Escape_Right()** in "Smart_Tricycle.py".

- Both follow a similar logic and structure which uses a WHILE Loop depending on Variable **obstacle** which was set to TRUE back in Line 341 of Fig. 4.47. Thus, the WHILE Loops (Lines 210-216 or Lines 222-228) are for sure activated when either Function is called.
- As an example, let us go through Function **Escape_Left()**:
    - Line 211 turns the Steering Wheel (Servo 3) 60 degrees Left of Center Line and Line 212 waits for 0.25 s for Servo 3 to get there.
    - The robot is next commanded to go Forward for 0.3 s (Lines 213-214).
    - The robot is then stopped (Line 215), and Function **Check_NIR()** is called (Line 216) to see if the Obstacle is cleared (**obstacle** set to FALSE) or not yet (**obstacle** still set to TRUE). Function **Check_NIR()** is listed in Fig. 4.53.
    - If **obstacle** is reported as TRUE from **Check_NIR()**, then the WHILE Loop reiterates from Line 211 once again.
    - Eventually, **obstacle** should be set to FALSE by **Check_NIR()**, then this WHILE Loop exits, and Function **Escape_Left()** returns back to where it was called from (see Figs. 4.47 and 4.48).

```
193 def Check_NIR():
194     global obstacle, threshold_1, threshold_2, threshold_3, Port3
195     global Port3_Init, Port4_Init, Port3_IR, Port4_IR
196 
197     Set_Servo_Position(Port3, 0)  # straighten up steering wheel
198     time.sleep(0.25)
199     Read_Port_IR_Sensors()
200     ObstacleDistance = Port4_IR
201     if (ObstacleDistance > threshold_3):
202         obstacle = True
203     else:
204         obstacle = False
```

Fig. 4.53 Function **Check_NIR()** in “Smart_Tricycle.py”.

Fig. 4.53 illustrates the steps implemented in Function **Check_NIR()**:

- Line 197 straightens up the Steering Wheel, and Line 198 waits 0.25 s for it to get there.
- Line 199 reads in the latest Port 4 IR Sensor data and Line 200 saves it in Variable **obstacleDistance**.
- The IF-ELSE structure (Lines 201-204) is then used to decide whether the Obstacle is cleared out of the robot’s LOS or not.

The reader has probably observed that this Python solution is very close to the TASK solution described in Sub-Section 3.4.2, except for the way that the Operator’s runtime inputs are handled: for TASK, it was through the Virtual RC-100 Buttons, and for Python, it was through the Keyboard directly.

4.2.10 Dowel Scanner

First, let’s use the program “Servo_Scan.py” to get some basic Dowel scanning data so that we can compare between the TASK and Python performances for this type of operation.

Fig. 4.54 shows the Variables definitions used in “Servo_Scan.py”:

- The only “new” Variables are the NumPy numerical arrays used on Lines 7 and 8 (https://pythonnumericalmethods.berkeley.edu/notebooks/chapter02.07-Introducing_numpy_arrays.html).
- Line 7 defines an Array named “NIR_data” with 25 elements initially set to ZEROs.
- Line 8 defines an Array named “GP_data” with 25 elements initially set to ZEROs.

Fig. 4.55 shows the Main Endless Loop (Lines 36-52) used in program “Servo_Scan.py”, which basically repeatedly swings Servo 3 from Goal Position -50 to Goal Position +50, while recording pairs of **NIR_data** and **GP_data** into their respective arrays:

- Line 37 sets Servo 3’s Speed to 75% Power.
- Line 38 sends Servo 3 to Goal Position -50 and Line 39 waits 1 s for it to get there.

```
from CNT_Block import *
# NumPy Numerical arrays section
NIR_data = np.zeros(25)   # 25 ele
GP_data = np.zeros(25)
#for i in range(25):
#     print("i = ", i, "\tNIR_data

# Dream II Section
Power = 28
Port1 = 0
Port2 = 1
CW = 1
CCW = 0    # used for Position Cor
Port3 = 0
Port4 = 1
Port3_Speed = 0    # [0-100]
Port3_Angle = 0    # [-150, 150] t
Port4_Init = 0
Port3_IR = 0
Port4_IR = 0
Option_IR = 0     # 0 for IR Sensc
```

Fig. 4.54 Variable Definitions in "Servo_Scan.py".

```
36 while True:
37     Turn_Servo_Motor(Port3, CCW, 75)   # speed Servo 3 up to get to Init Scan Position
38     Set_Servo_Position(Port3, -50)
39     time.sleep(1)
40     print("Present Goal Angle [-50] = ", Read_Servo_Position(Port3))
41     Background = Read_Port_IR_Touch(Port4, Option_IR)
42     print("Background NIR = ", Background)
43     Turn_Servo_Motor(Port3, CCW, 40)   # slow Servo 3 down for NIR scanning
44     Set_Servo_Position(Port3, 50)
45     for i in range(25):
46         NIR_data[i] = Read_Port_IR_Touch(Port4, Option_IR)
47         GP_data [i] = Read_Servo_Position(Port3)
48 #        time.sleep(0.01)
49     for i in range(25):
50         print("i = ", i, "    NIR_data = ", (NIR_data[i]-Background), "     GP_data = ", GP_data[i])
51     print("Present Goal Angle [50] = ", Read_Servo_Position(Port3))
52     time.sleep(5)
```

Fig. 4.55 Main Endless Loop in "Servo_Scan.py".

- Line 40 uses Utility Function **Read_Servo_Position()** to read in and print out the Present Position of Servo 3 (which should be in the neighborhood of "-50").

- Line 41 reads in the current Port 4 IR Sensor data and saves it into Variable **Background**, while Line 42 prints out this Value.
- Line 43 slows Servo 3 down to 40% Power before starting the Dowel scanning process.
- Please remember that at this time, Servo 3 is at Goal Position **-50**, so Line 44 then sends Servo 3 to Goal Position **+50**. Thus, starting here, Servo 3 is on its trip to Position **+50**.
- The first FOR Loop (Lines 45-47) recorded, as fast as possible, 25 data pairs (Port 4 IR Sensor and Port 3 Present Position) into Arrays **NIR_data[]** and **GP_data[]**.
- The second FOR Loop (Lines 49-50) prints out the values contained in these two Arrays. Please note that **Background** is subtracted from **NIR_data[i]** before printing out this adjusted value.
- Line 51 prints the Present Position of Servo 3 to check it against **+50**.
- Line 52 waits for 5 seconds before reiterating the FOR Loop.

A typical runtime output is shown below for a CM-150 version of the Dowel Scanner robot (with inserted comments from the author):

%Run Servo_Scan.py

*Present Goal Angle [-50] = -50 // **at start of scan***

Background NIR = 17

Empty String Received from Port No 3 // Communication Errors during recording

Empty String Received from Port No 4 // these messages come from Utility Functions

Empty String Received from Port No 3 // created inside "CNT_Block.py" Module

*i = 0 NIR_data = **-2.0** GP_data = -50.0*

i = 1 NIR_data = 0.0 GP_data = -48.0

i = 2 NIR_data = 0.0 GP_data = -44.0

*i = 3 NIR_data = **-1.0** GP_data = **1000.0** // First Comm. Error for Servo 3*

i = 4 NIR_data = 52.0 GP_data = -34.0 // Picking up First Dowel

i = 5 NIR_data = 168.0 GP_data = -29.0

i = 6 NIR_data = 309.0 GP_data = -24.0 // Center Line of First Dowel

i = 7 NIR_data = 169.0 GP_data = -20.0

i = 8 NIR_data = 72.0 GP_data = -16.0

i = 9 NIR_data = 1.0 GP_data = -11.0

*i = 10 NIR_data = **-1.0** GP_data = -7.0*

i = 11 NIR_data = 0.0 GP_data = -3.0

*i = 12 NIR_data = **-1.0** GP_data = 3.0*

i = 13 NIR_data = 1.0 GP_data = 7.0

*i = 14 NIR_data = **983.0** GP_data = 14.0 // Comm. Error for Port 4 IR Sensor*

i = 15 NIR_data = 245.0 GP_data = 18.0 // Center Line of Second Dowel

i = 16 NIR_data = 179.0 GP_data = 23.0

i = 17 NIR_data = 74.0 GP_data = 26.0

i = 18   NIR_data = ***-1.0***   *GP_data = 31.0*

i = 19   NIR_data = ***-1.0***   *GP_data = 34.0*

i = 20   NIR_data = ***-2.0***   *GP_data = 38.0*

i = 21   NIR_data = ***-2.0***   *GP_data = 40.0*

i = 22   NIR_data = ***-1.0***   *GP_data = 43.0*

i = 23   NIR_data = ***-1.0***   *GP_data = 44.0*

i = 24   NIR_data = 1.0   GP_data = ***1000.0*** *// Second Comm. Error for Servo 3*

Present Goal Angle [50] = ***45***

As shown above, the CM-150 using the BT-410 UART Receiver had occasional communications issue with the USB BT-410 Dongle (this situation did not happen with the CM-151 for the author). Also please review Fig. 3.72 for comparison with TASK code performance.

With Servo 3's Power set at 40, this program was getting about 4-5 readings for each Dowel used. Other runs showed that the author could get up to 6 readings per Dowel. Eventually, this Scanning Power is reduced to 22% in Project "Dowel_Scanner.py" to get more "Dowel" data points.

Project "Dowel_Scanner.py" uses the same programming approach and techniques used with its TASK-equivalent "Dowel_Scanner.tskx/tsk3", **thus the reader is referred back to Sub-Section 3.4.3 for most Project descriptions and explanations.** Figs. 4.56 to 4.62 show the Python implementation of key Variables and Functions used: the key difference is that there is no equivalent CALLBACK facility in Python.

Fig. 4.56 list Variables definitions used in program "Smart_Avoider.py".

```
Option_IR = 0        # 0 for IR Sensor, 1 for Touch Sensor
Port3_Speed = 0       # [0-100]
Port3_Angle = 0       # [-150, 150] to match R+BLOCK usage
SpeedL = 25
SpeedR = 25
LeftLimit = 60
RightLimit = -60
ProximityLimit = 0
PreviousNIR = 0
CurrentNIR = 0
DowelDistance1 = 0
DowelDistance2 = 0
Position1 = 0
Position2 = 0
FoundDowel = 0
BackgroundNIR = 0
DowelThreshold = 0
message1 = "Clap when ready for Dowel Scanning!"
```

Fig. 4.56 Variables definitions used in "Dowel_Scanner.py".

Fig. 4.57 shows how User Function **Read_Port_PP()** uses Utility Function **Read_Servo_Position()** to read and do error checking on the Present Position of Servo 3:

- Line 85 reads Servo 3 Present Position and saves it into **Port3_PP**.
- Lines 88-89 form an error checking loop that keeps on reading in the latest Servo Position from Servo 3/Port 3 if this value happens to be equal to "1000", which is a Special Value that Utility Function **Read_Servo_Position()** returns, when there is some communications error with data received from the CM-150/151 (such as those shown in the earlier runtime printout of program "Servo_Scan.py").
- Thus, Line 91 would always print out a valid value for **Port3_PP,** and consequently whenever Variable **Port3_PP** is used in the rest of program "Dowel_Scanner.py". *This feature is not available in R+BLOCK, and TASK does not seem to need it!*

```
83 def Read_Port_PP():  # Read Ports 3 & 4 Present Position with Error Checking
84     global Port3, Port4, Port3_PP, Port4_PP
85     Port3_PP = Read_Servo_Position(Port3)  # Port 3 is used for Servo SM-10
86 #    Port4_PP = Read_Servo_Position(Port4)
87 #    while ((Port3_PP == 1000) or (Port4_PP == 1000)):
88     while (Port3_PP == 1000):
89         Port3_PP = Read_Servo_Position(Port3)
90 #        Port4_PP = Read_Servo_Position(Port4)
91     print("Port3_PP = ", Port3_PP)
92 #    print("Port4_PP = ", Port4_PP)
```

Fig. 4.57 Function **Read_Port_PP()** used in "Dowel_Scanner.py".

The details of Function **FindDowel()** are shown in Fig. 4.58:

- Please remember from Sub-Section 3.4.3, that by this point in the Dowel Search algorithm, Servo 3 is scanning for a Dowel by starting out from "empty space".
- Lines 146-148 form a WHILE Loop that keeps on iterating as long as it reads in a **Port4_IR** value that is less or equal to **DowelThreshold** (Line 147), meaning that Port 4 IR Sensor is still "seeing empty space". As soon as a Dowel edge gets within the viewport of Port 4 IR Sensor, **Port4**_IR gets larger than **DowelThreshold** making this WHILE Loop exit to the next procedure which is looking for the Center Line of this Dowel (Lines 150-155).
- As Servo 3 now scans more towards the Center Line of this Dowel, **Port4_IR** would keep on increasing its Consecutive Values until it passes over the Center Line. This event will make the WHILE Loop in Lines 153-156 exit to Line 157. Thus, strictly speaking, Function **Read_Port_PP()** on Line 157 reads in a Servo Position that is **a bit off** the Center Line, but fortunately this Servo Position is not relevant to the rest of the Dowel Scan algorithm.

```
143 def FindDowel():
144     global FoundDowel, DowelThreshold, BackgroundNIR, Port4_IR, Port3_PP
145 
146     Read_Port_IR_Sensors()
147     while (Port4_IR <= DowelThreshold):  # Waiting Loop until Port4_IR sees Left Edge of Dowel
148         Read_Port_IR_Sensors()
149 
150     PreviousNIR = BackgroundNIR   # Procedure to find "Center Line" of Dowel
151     Read_Port_IR_Sensors()
152     NIRDifference = Port4_IR - PreviousNIR
153     while (NIRDifference >= 0):
154         PreviousNIR = Port4_IR
155         Read_Port_IR_Sensors()  # updating Port4_IR
156         NIRDifference = Port4_IR - PreviousNIR
157     Read_Port_PP()  # here "Center Line" already passed
158     FoundDowel += 1   # update the number of Dowels found
159     print("Port3_PP = ", Port3_PP, "     Dowel Distance = ", PreviousNIR, "\tFoundDowel = ", FoundDowel)
160     return PreviousNIR
```

Fig. 4.58 Function **FindDowel()** used in "Dowel_Scanner.py".

- Lines 158-159 update and print out some informational data, and Line 160 returns the value of **PreviousNIR** which will be needed to figure out which Dowel (Left or Right) is closer to the robot, in later parts of the Dowel Scan algorithm.

Fig. 4.59 lists the steps of Function **ServoScan()** which manages the overall operations for scanning **Dowel 1** (on robot's Left) and **Dowel 2** (on robot's Right):

- Lines 166-171 send Servo 3 to the **LeftLimit** Goal Position and use a Conditional WHILE Loop to detect when Servo 3 is within **6 degrees** of **LeftLimit**. When this location is reached, the program continues to Line 173 which sets **FoundDowel** to 0.
- Line 174 slows Servo 3 down to 22% Power for the upcoming Dowel Scanning work.
- Line 175 then sends Servo 3 to Goal Position **RightLimit**, which would take Servo 3 a few seconds to get there. It is during this time that Servo 3 and Port 4 IR Sensor work together to determine the Location and Distance of each Dowel to the robot.
- Line 176 uses Function **FindDowel** to locate Dowel 1 and its distance to the robot (IR Sensor reading) is recorded into Variable **DowelDistance1**.
- Next, Lines 178-180 implement a procedure to scan the "remaining half" of Dowel 1 (i.e., **Port4_IR > BackgroundNIR**).
- Line 181 uses Function **FindDowel()** once more to determine the Location and Distance of Dowel 2 which is saved into Variable **DowelDistance2**.
- Line 183 immediately stops Servo 3 at this Servo Position by setting its Power to 0. However, in Servo 3's firmware it is still commanded to go further to **RightLimit** (Line 175), thus the next two Lines 184 and 185 are very important and needed to "change its mind" and "make it stay" at its Current Servo Position, when Power is set to a non-zero value later in this Dowel Scan algorithm.

- Line 186 prints out the results of the search for the two Dowels.

```
162 def ServoScan():
163     global LeftLimit, RightLimit, FoundDowel, DowelDistance1, DowelDistance2
164     global Port3, CCW, Port4, Port3_PP, Port4_IR
165 
166     print("Start scanning for dowels")
167     Set_Servo_Position(Port3, LeftLimit) # Send Servo3 to LeftLimit at "regular" spee
168     LeftLimit -= 6
169     Read_Port_PP() # to get current Port3_PP
170     while (Port3_PP < LeftLimit): # Waiting Loop until Servo3 is about 6 deg. below L
171         Read_Port_PP()
172 
173     FoundDowel = 0   # getting ready to find First Dowel
174     Turn_Servo_Motor(Port3, CCW, 22)  # Slow Servo3 to 22% for dowel scanning
175     Set_Servo_Position(Port3, RightLimit) # Send Servo3 to RightLimit
176     DowelDistance1 = FindDowel()  # Use Function FindDowel() to locate First Dowel
177 
178     Read_Port_IR_Sensors() # to get current Port4_IR
179     while (Port4_IR > BackgroundNIR): # Waiting Loop until Servo3 goes into empty spa
180         Read_Port_IR_Sensors()  # to get current Port4_IR
181     DowelDistance2 = FindDowel()  # Use Function FindDowel() to locate Second Dowel
182 
183     Turn_Servo_Motor(Port3, CCW, 0)  # stop Servo 3 because 2nd Dowel located
184     Read_Port_PP() # to get current Port3_PP
185     Set_Servo_Position(Port3, Port3_PP) # Because Servo 3 was previoulsy set to Right
186     print("DowelDistance1 = ", DowelDistance1, "\tDowelDistance1 = ", DowelDistance2)
```

Fig. 4.59 Function **ServoScan()** used in "Dowel_Scanner.py".

After Function **ServoScan()** has done its job of locating the two Dowels, Function **VerifyDowels()** is invoked next. Function **VerifyDowels()** decides on which Dowel is closer to the robot and sends it closer to this Dowel as a verification move, then it makes the robot perform the same task for the other Dowel:

- Line 227 shows that if **DowelDistance2** is **greater** than **DowelDistance1**, this means that the Right Dowel is closer to the robot, then the appropriate maneuvers to be performed by the robot are implemented by Lines 228-232:
    - Function **CenterSensor()** uses Servo 3 to center Port 4 IR Sensor (Line 228).
    - Function **Rotate_Right()** makes the robot turn right (Line 229).
    - As the robot turns right, Line 230 activates Function **FindApproachDowel()** which starts scanning for the Right Dowel. When it finds this Dowel, it will make the robot approach this Dowel up to a given distance, stops and plays a Melody while waiting for the Operator to promptly remove this Right Dowel.

- Next, Line 231 makes the robot rotate back to the Left and Line 232 activates Function **FindApproachDowel()** again, but this time to locate and approach the Left Dowel (in the same way as for the Right Dowel previously).

```
224 def VerifyDowels():
225     global DowelDistance1, DowelDistance2
226 
227     if (DowelDistance2 > DowelDistance1):
228         CenterSensor()
229         Rotate_Right()
230         FindApproachDowel()
231         Rotate_Left()
232         FindApproachDowel()
233     else:  # Left Dowel is closest to Robc
234         CenterSensor()
235         Rotate_Left()
236         FindApproachDowel()
237         Rotate_Right()
238         FindApproachDowel()
```

Fig. 4.60 Function **VerifyDowels()** used in "Dowel_Scanner.py".

- Otherwise, meaning that the Left Dowel is closer to the robot, the ELSE branch starting at Line 233 is activated, and the appropriate maneuvers to be performed by the robot are implemented by Lines 234-238:
    - Function **CenterSensor()** uses Servo 3 to center Port 4 IR Sensor (Line 234).
    - Function **Rotate_Left()** makes the robot turn left (Line 235).
    - As the robot turns left, Line 236 activates Function **FindApproachDowel()** which starts scanning for the Left Dowel. When it finds this Dowel, it will make the robot approach this Dowel up to a given distance, stops and plays a Melody while waiting for the Operator to remove the Left Dowel.
    - Next, Line 237 makes the robot rotate back to the Right and Line 238 activates Function **FindApproachDowel()** again, but this time to locate and approach the Right Dowel (in the same way as for the Left Dowel previously).

Fig. 4.61 shows steps used in Function **CenterSensor()**:

- Line 191 puts Servo 3 to 50% Power and Line 192 sends Servo 3 to the 0-degree Goal Position.
- Line 193 waits for 0.5 s for Servo 3 to get there.

```
def CenterSensor():
    global Port3, CCW

    Turn_Servo_Motor(Port3, CCW, 50)
    Set_Servo_Position(Port3, 0)  #
    time.sleep(0.5)
```

Fig. 4.61 Function **CenterSensor()** used in “Dowel_Scanner.py”.

Fig. 4.62 describes the procedures used in Function **FindApproachDowel()**:

```
def FindApproachDowel():
    global DowelThreshold, BackgroundNIR
    global Port4, Port4_IR

    Read_Port_IR_Sensors()
    while (Port4_IR <= DowelThreshold):  # Wait
        Read_Port_IR_Sensors()

    PreviousNIR = BackgroundNIR   # Procedure t
    Read_Port_IR_Sensors()
    NIRDifference = Port4_IR - PreviousNIR
    while (NIRDifference >= 0):
        PreviousNIR = Port4_IR
        Read_Port_IR_Sensors()  # updating Port
        NIRDifference = Port4_IR - PreviousNIR
    time.sleep(0.01)  # waiting for Geared Moto
    Stop()

    Read_Port_IR_Sensors()  # To determine how
    ProximityLimit = Port4_IR
    ProximityLimit += 150
    Forward()  # approaching target dowel
    Read_Port_IR_Sensors()
    while (Port4_IR < ProximityLimit):  # Waiti
        Read_Port_IR_Sensors()
    Stop()
    Play_Musical_Melody(3)
    time.sleep(5)   # waiting for Operator to r
```

Fig. 4.62 Function **FindApproachDowel()** used in “Dowel_Scanner.py”.

- Please remember that the robot is rotating either Left or Right when this Function is called, meaning that Port 4 IR Sensor is "seeing empty space" when Line 199 is activated. The WHILE Loop in Lines 200-201 lets the robot continue its rotational maneuver as long as Port 4 IR Sensor reports a value less than **DowelThreshold**. When Port 4 IR Sensor picks up the Dowel's edge, this WHILE Loop would exit to Line 203.
- Lines 203-209 implement a search for the Center Line of this Dowel (same approach as for Function **FindDowel()** described earlier). When this Center Line is found, this WHILE Loop exits to Line 210 which is a short Time Delay, designed to let the robot "overshoot" the Dowel a little bit. This Time Delay can be adjusted by the reader for his/her robot. Line 211 stops the robot.
- Lines 213-214 are used to determine the "Current Distance" between the dowel and the robot and assign this value to Variable **ProximityLimit**. Line 215 adds 150 IR Sensor units to it as we want the robot to approach the dowel within 150 units at a later time.
- Line 216 then moves the robot Forward.
- Lines 217-219 make sure that the robot won't get any closer than 150 units away from the dowel.
- Line 220 stops the robot, and Line 221 plays Melody 3.
- Line 222 waits for 5 seconds so the Operator can pick up the Dowel that just had been "verified".

It will be interesting for the reader to compare the runtime performances of this Python version to the TASK version of Sub-Section 3.4.3.

At this point, the reader should be very comfortable with using the Utility Functions provided in the Python Module "CNT_Block.py", so let's start looking "under the hood" for this Python Module and its companion TASK code, so that the reader can add his/her own Utility Functions or create his/her own Modules to work with the DREAM II system, and actually **with any other ROBOTIS system that uses the Remocon Protocol**, such as BIOLOID or ENGINEER.

4.3 "Looking Under The Hood"

The reader should be aware that the ROBOTIS Remocon Protocol is implemented in all "Edutainment" CM controllers (CM-5 to CM-550, including OpenCM-904, and probably in future ROBOTIS CM controllers also). Thus, the information described in Section 4.3 has potential and wide applications to other ROBOTIS robotics kits, even though this book was written for the CM-150/151 Controllers.

The format of the Remocon Packet is described at this link https://emanual.robotis.com/docs/en/parts/communication/rc-100/#communication-packet. Since it was introduced in Sub-Section 3.1.6, we have used only the first 10 bits of the 16 bits that are available to the user. The "design" of the "CNT_Block.py" and "CNT_Block.tskx/tsk3" modules relied on the use of all 16 bits.

In Python, the **Remocon Packet** is implemented as a **bytearray** of 6 elements, for a total of 6 bytes, although the actual **Remocon Data** is only 16 bits or 2 bytes long (see Fig. 4.63):

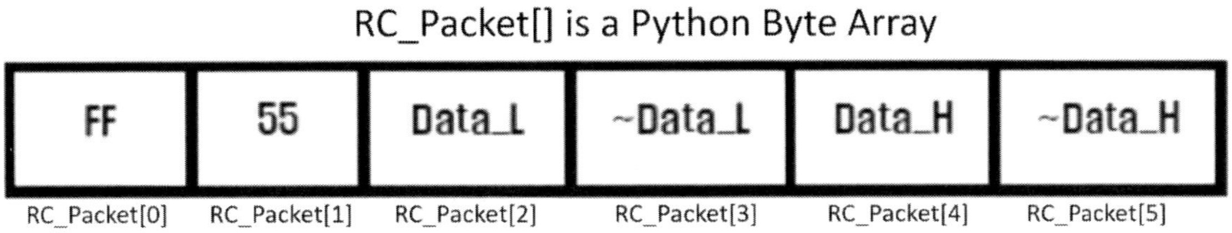

Fig. 4.63 “Anatomy” of the ROBOTIS Remocon Packet.

- **RC_Packet[0]** and **RC_Packet[1]** form the **Header** of the **Remocon Packet**, so they stay “the same” for all Remocon Packets.
- The actual 16-bit **Remocon Data** of course changes per programmer needs:
  - This **Data** is actually chopped into a Low Byte **Data_L** and a High Byte **Data_H**.
  - **Data_L** contains only information of the lower 8-bits of **Data**. **Data_L** is stored in **RC_Packet[2]**, while its 1’s complement **~Data_L** is stored in **RC_Packet[3]** (https://www.tutorialspoint.com/1-s-complement-vs-2-s-complement).
  - **Data_H** contains only information of the higher 8-bits of **Data**. **Data_H** is stored in **RC_Packet[4]**, while its 1’s complement **~Data_H** is stored in **RC_Packet[5]**.

The reader can see that the ROBOTIS Remocon Packet has built-in error-checking features as wireless data transmission is not 100% reliable at all times. Later, the Utility Functions in the “CNT_Block” Module will demonstrate how to use these features to sort out “good” from “bad” packets.

4.3.1 Using 16-bit Remocon Packet

In this work, the highest bits (**Bits 12-15**) of the **Remocon Data** are reserved for Parameter **comm_type** (i.e., Command Type). As 4 bits are used, **comm_type** can have up to 16 values (0 to 15), the “**CNT_Block.py**” Module uses the values **0 to 10** (i.e., 11 values), thus there is room for the reader to add 5 more Command Types if he/she wishes to!

Each **comm_type** corresponds to a specific Python Utility Function which also matches with a particular R+BLOCK Block command:

1. comm_type=0 – Turn_Gear_Motor (PortNo, Direction, Power).
2. comm_type=1 – Turn_Servo_Motor (PortNo, Direction, Power).
3. comm_type=2 – Set_Servo_Position (PortNo, Goal Position).
4. comm_type=3 – Set_LED_Module (PortNo, Orange_LED, Blue_LED).
5. comm_type=4 – Play_Musical_Scale (ScaleIndex, TimeDuration)
6. comm_type=5 – Play_Musical_Melody (MelodyIndex).
7. comm_type=6 – Read_Integrated_IR (Location).
8. comm_type=7 – Read_Port_IR_Touch (PortNo, Sensor Option).
9. comm_type=8 – Reset_Read_Sound_Counter_Result (Option).
10. comm_type=9 – Read_Sound_Counter_Real_Time ().
11. comm_type=10 – Read_Servo_Position (PortNo) – this one has no R+BLOCK equivalent.

As the reader can see, each Utility Function has a different number of Parameters as Inputs, and each Parameter also has a different range of numerical values: for example, “Power” is from 0 to 100, but “Goal Position” is from -150 to 150. Thus, the remaining bits of the **Remocon Data** (**Bits 0-14**) are designed “adaptively” depending on the particular **comm_type**.

Appendix C and the provided source file "CNT_Block.py" contains the complete details for each **comm_type** Utility Function and the format of its matching **Remocon Data**. In Sub-Sections 4.3.2 and 4.3.3, only selected **comm_type** Utility Functions and their corresponding **Remocon Data** will be explained in detail. The reader can figure out the rest on his/her own, as an exercise towards the reader's progress of creating his/her own **comm_type** facilities in the "near future"!

4.3.2 "CNT_Block.py"

Fig. 4.64 shows the IMPORT Section of the Module "CNT_Block.py":

- Lines 8 and 9 show that this Module uses the Standard Time and Sys modules.
- Line 10 shows that OpenCV is used.
- Line 11 shows that PySerial is used.
- Line 12 shows that NumPy is also used.

```
 8 import time
 9 import sys
10 import cv2
11 import serial
12 import numpy as np
```

Fig. 4.64 IMPORT Section of "CNT_Block.py".

Fig. 4.65 shows the Variables to be used in Communications Programming between the Chromebook and the CM-150/151:

- Lines 18 and 19 show that Variable **highbyte** and **lowbyte** are initialized to 0, in binary numerical format **0bXXXXXXXX**.
- Lines 20 and 21 define **TxD_packet** and **RxD_packet** as **bytearray** with 6 elements (https://www.w3schools.com/python/ref_func_bytearray.asp). **TxD_packet** is used to send Commands from Python (i.e., Chromebook) to the CM-150/151. While **RxD_packet** is used to receive Actuator/Sensor data sent back from the CM-150/151 to the Chromebook.

```
14 #
15 #  Global Initialization Sections start here
16 #
17 # PySerial section
18 highbyte = 0b00000000
19 lowbyte = 0b00000000   # lowbyte + highbyte to make complete Data
20 TxD_packet = bytearray(6)
21 RxD_packet = bytearray(6)
22 TxD_packet[0] = 0xff
23 TxD_packet[1] = 0x55
24 TxD_packet[2] = lowbyte
25 TxD_packet[3] = ~lowbyte & 0xff  # to make sure that ~lowbyte is contained within 1 byte
26 TxD_packet[4] = highbyte
27 TxD_packet[5] = ~highbyte & 0xff  # to make sure that -highbyte is contained within 1 byte
28 RxD_packet[0] = 0
29 RxD_packet[1] = 0
30 RxD_packet[2] = 0
31 RxD_packet[3] = 0
32 RxD_packet[4] = 0
33 RxD_packet[5] = 0
```

```
41 # RC-100 section
42 TxData = 0
43 Comm_Type = -1  # Command Type
```

Fig. 4.65 “Communications” Variables used in “CNT_Block.py”.

- Lines 22 and 23 set the (constant) Header component of the Remocon Packet.
- Lines 24-25 show how **lowbyte** and **~lowbyte** are saved into elements [2] and [3] of the **TxD_packet bytearray**. The “~” operator does the **1’s complement** conversion, while the “&” operator performs the **bit-wise AND** operation.
- Lines 26-27 show how **highbyte** and **~highbyte** are saved into **TxD_packet[4]** and **TxD_packet[5]**.
- Lines 28-33 shows that the whole **RxD_packet** is initialized to zeros.
- Line 42 sets Variable **TxData** to zero, later this Variable will contain the **comm_type** (initialized to -1 by Line 43) and other parameters needed for a given Utility Function.

Fig. 4.66 shows the proper Initialization of the PySerial Communications Port on the Chromebook to work with the USB BT-410 Dongle:

- The “CNT_Block.py” file will be provided to the reader with Line 35 “enabled” and Line 37 “commented out”, so that the reader can use this Python Module “right away” on the Chromebook (CB). Of course, this is assuming that the reader is using the setup described in Fig. 4.2, whereas the only USB device used with the CB is the USB BT-410 Dongle. In this case, this Dongle is mapped by Linux as “**ttyUSB0**”. If the reader is using other USB devices, he/she will have to use the Linux Utility named “**dmesg**” to figure out which “**ttyUSBx**” Port is actually connected to the BT-410 Dongle which will be listed as “FTDI USB Serial Device”.

- **btser** is the PySerial Device/Handle for the programmer to use for sending and receiving various Remocon Packets corresponding to each Utility Function of Module **CNT_Block**.
- Lines 38 and 39 are used to "flush" the associated serial buffers of any older data that may still be in there when the user's Python program starts.
- As the "CNT_Block.py" Module uses Standard Python, the reader can use it on a Windows OS device as well, in that case Line 35 must be "commented out" and Line 37 "enabled", and the user will need to check with Windows Settings for the appropriate COM Port to use with his/her USB BT-410 Dongle. The author's setup just happened to be using COM24.

```
34 # USB-410 Dongle on Chromebook - BT-410 receiver on CM-150 or BLE on CM-151
35 btser = serial.Serial("/dev/ttyUSB0", 57600, timeout=0.1, write_timeout=0.1)
36 # USB-410 Dongle on Windows PC - BT-410 receiver on CM-150 or BLE on CM-151
37 btser = serial.Serial("COM24", 57600, timeout=0.01, write_timeout=0.01)
38 btser.reset_input_buffer()   # required
39 btser.reset_output_buffer()  # required
```

```
48 def send_data():
49     global TxData, TxD_packet, lowbyte, highbyte
50     lowbyte = TxData & 0xff
51     TxD_packet[2] = lowbyte & 0xff
52     TxD_packet[3] = ~lowbyte & 0xff
53     highbyte = (TxData >> 8) & 0xff
54     TxD_packet[4] = highbyte & 0xff
55     TxD_packet[5] = ~highbyte & 0xff
56     if (btser.write (TxD_packet) != 6):
57         print("Transmit Error!!!")
58     time.sleep(0.010)
```

Fig. 4.66 Init of PySerial Com Port and Function **send_data()** used in "CNT_Block.py".

- Fig. 4.66 also contains the details for Function **send_data()** which does the job of "packaging" the actual Remocon Data represented by Variable **TxData** into the appropriate elements of the **bytearray TxD_packet[]**:
    - In Line 50, **TxData** is **bit-wise ANDed** with the hexadecimal constant **0xff** to extract the lower 8 bits of **TxData** and saves this result into Variable **lowbyte**.
    - In Lines 51 and 52, the **&** operator is again used to make sure that only the lower 8 bits of the 2 versions of **lowbyte** are saved into the respective array elements **TxD_packet[2]** and **TxD_packet[3]**.
    - Line 53 shifts **TxData** right by 8 bits with **">> 8"** to retain only the higher 8 bits of **TxData** and saves this result in Variable **highbyte**.
    - Lines 54-55 are used to save the two versions of **highbyte** into the appropriate array elements **TxD_packet[4]** and **TxD_packet[5]**.
    - Line 56 shows that PySerial Function **write()** is used to send the **bytearray TxD_packet** from the CB to the CM-150/151. If an error occurs, then Line 57

prints out an informational message, but the previous Remocon Packet is not re-sent to the CM-150/151.
- Line 58 waits for 10 ms for the BT Transmission Process to finish up before exiting the **send_data()** Function.

Fig. 4.67 shows the details of Utility Function **Turn_Gear_Motor()** and the equivalent R+BLOCK command:

- This function corresponds to a **comm_type** = **0** (Line 62).
- Input Parameter **port_no** can be set to **0** to designate Port 1 on the CM-150/151 or it can be set to **1** for Port 2. **port_no** is recorded into **Bit 0** of Variable **TxData** (Line 63) using the bitwise AND Operator **&**.
- Input Parameter **cw_ccw** can be set to **0** to mean CCW, or set to **1** to mean CW. **cw_ccw** is recorded into **Bit 1** of Variable **TxData** (Line 64), using a combination of bitwise AND (**&**) and OR (|) operators, and also the Shift-Left operator (<<).
- Input Parameter **power** is expected to be in the range of [0-100] and its value is recorded in **Bits 2 to 8** of Variable **TxData** (Line 65), also using a combination of bitwise AND (**&**) and OR (|) operators, and also the Shift-Left operator (<<).
- The last component **comm_type** is shifted 12 bits to the left and "inserted" into **Bits 12-15** of **TxData** using the OR Operator (Line 66).
- **TxData** now contains 4 pieces of information and Function **send_data()** is used to transmit it from the CB to the CM-150/151 (Line 67).
- Function **Turn_Gear_Motor()** does not expect a packet to be returned from the CM-150/151.

```
60 def Turn_Gear_Motor(port_no, cw_ccw, power):
61     global TxData
62     comm_type = 0 # local variable
63     TxData = port_no & 0x01 # PUT port_no's value into Bit 0 of TxData
64     TxData = TxData | ((cw_ccw & 0x01) << 1)  # ADD cw_ccw's value into Bit 1 of previous TxData
65     TxData = TxData | ((power & 0x07f) << 2)  # ADD power's value into Bits 2-8 of previous TxData
66     TxData = TxData | (comm_type << 12) # ADD comm_type = 0 into Bits 12-15 of previous TxData
67     send_data()
```

Fig. 4.67 Function **Turn_Gear_Motor()** used in "CNT_Block.py".

Fig. 4.68 shows the details of Utility Function **Turn_Servo_Motor()** and the equivalent R+BLOCK command:

- This function corresponds to a **comm_type** = **1** (Line 71).

- Input Parameter **port_no** can be set to **0** to designate Port 3 on the CM-150/151 or it can be set to **1** for Port 4. **port_no** is recorded into **Bit 0** of Variable **TxData** (Line 72) using the bitwise AND Operator **&**.
- Input Parameter **cw_ccw** can be set to **0** to mean CCW, or set to **1** to mean CW. **cw_ccw** is recorded into **Bit 1** of Variable **TxData** (Line 73).
- Input Parameter **power** is expected to be in the range of [0-100] and its value is recorded in **Bits 2 to 8** of Variable **TxData** (Line 74).
- The last component **comm_type** is shifted 12 bits to the left and "inserted" into **Bits 12-15** of **TxData** using the OR Operator (Line 75).
- **TxData** is now "complete" and Function **send_data()** is used to transmit it from the CB to the CM-150/151 (Line 76).
- Function **Turn_Servo_Motor()** does not expect a packet to be returned from the CM-150/151.

```
69 def Turn_Servo_Motor(port_no, cw_ccw, power):
70     global TxData
71     comm_type = 1 # local variable
72     TxData = port_no & 0x01 # PUT port_no's value into Bit 0 of TxData
73     TxData = TxData | ((cw_ccw & 0x01) << 1)  # ADD cw_ccw's value into Bit 1 of previous TxData
74     TxData = TxData | ((power & 0x07f) << 2)  # ADD cw_ccw's value into Bits 2-8 of previous TxData
75     TxData = TxData | (comm_type << 12) # ADD comm_type = 1 into Bits 12-15 of previous TxData
76     send_data()
```

Fig. 4.68 Function **Turn_Servo_Motor()** used in "CNT_Block.py".

Fig. 4.69 describes the details of Utility Function **Set_LED_Module()** and the equivalent R+BLOCK command:

- This function corresponds to a **comm_type** = **3** (Line 138).
- Input Parameter **port_no** can be set to **0** to designate Port 3 or it can be set to **1** for Port 4. **port_no** is recorded into **Bit 0** of Variable **TxData** (Line 139).
- Input Parameter **o_led** (Orange LED) can be set to **0** to mean OFF or to **1** to mean ON. **o_led** is recorded into **Bit 1** of Variable **TxData** (Line 140).
- Input Parameter **b_led** (Blue LED) can be set to **0** to mean OFF or to **1** to mean ON. **b_led** is recorded into **Bit 2** of Variable **TxData** (Line 141).
- The last component **comm_type** is shifted 12 bits to the left and "inserted" into **Bits 12-15** of **TxData** using the OR Operator (Line 142).
- Function **send_data()** is used to transmit the completed **TxData** to the CM-150/151 (Line 143).

- Function **Set_LED_Module()** does not expect a packet to be returned from the CM-150/151.

```
def Set_LED_Module(port_no, o_led, b_led):
    global TxData
    comm_type = 3 # local variable
    TxData = port_no & 0x01 # PUT port_no's val
    TxData = TxData | ((o_led & 0x01) << 1)  #
    TxData = TxData | ((b_led & 0x01) << 2)  #
    TxData = TxData | (comm_type << 12) # ADD c
    send_data()
```

Fig. 4.69 Function **Set_LED_Module()** used in "CNT_Block.py".

Fig. 4.70 describes the details of Utility Function **Play_Musical_Scale()** and the equivalent R+BLOCK command:

- This function corresponds to a **comm_type** = **4** (Line 147).
- Input Parameter **index** can be set as an integer from **0** to **51.** Fig. 4.1 can be used to map some of these integers to musical notes. **index** is recorded into **Bits 0-5** of Variable **TxData** (Line 148).
- Input Parameter **duration** is an integer from **0** to **5** (i.e., seconds) is recorded into **Bits 6-8** of Variable **TxData** (Line 149).
- The last component **comm_type** is shifted 12 bits to the left and "inserted" into **Bits 12-15** of **TxData** using the OR Operator (Line 150).
- Function **send_data()** is used to transmit the completed **TxData** to the CM-150/151 (Line 151).

```
145 def Play_Musical_Scale(index, duration):
146     global TxData
147     comm_type = 4 # local variable
148     TxData = index & 0x3f # PUT index's value i
149     TxData = TxData | ((duration & 0x07) << 6)
150     TxData = TxData | (comm_type << 12) # ADD c
151     send_data()
```

Fig. 4.70 Function **Play_Musical_Scale()** used in "CNT_Block.py".

Fig. 4.71 describes the details of Utility Function **Play_Musical_Melody()** and the equivalent R+BLOCK command:

- This function corresponds to a **comm_type** = **5** (Line 155).
- Input Parameter **index** can be set as an integer from **0** to **24. index** is recorded into **Bits 0-4** of Variable **TxData** (Line 156).
- The last component **comm_type** is shifted 12 bits to the left and "inserted" into **Bits 12-15** of **TxData** using the OR Operator (Line 157).
- Function **send_data()** is used to transmit the completed **TxData** to the CM-150/151 (Line 158).

```
153 def Play_Musical_Melody(index):
154     global TxData
155     comm_type = 5  # local variable
156     TxData = index & 0x1f # PUT index's va
157     TxData = TxData | (comm_type << 12) #
158     send_data()
```

Fig. 4.71 Function **Play_Musical_Melody()** used in "CNT_Block.py".

Fig. 4.72 describes the details of Utility Function **Set_Servo_Position()** and the equivalent R+BLOCK command:

- This function corresponds to a **comm_type** = **2** (Line 80).
- Input Parameter **port_no** can be set to **0** to designate Port 3 or it can be set to **1** for Port 4. **port_no** is recorded into **Bit 0** of Variable **TxData** (Line 82).
- Input Parameter **goal_angle** can be set as an integer from **-150** to **150** degrees. But it is transformed into a number between 0 and 300 by Line 81 and saved into Variable **shifted_goal_angle,** which is then recorded into **Bits 1-9** of Variable **TxData** (Line 83).
- The last component **comm_type** is shifted 12 bits to the left and "inserted" into **Bits 12-15** of **TxData** using the OR Operator (Line 84).
- Function **send_data()** is used to transmit the completed **TxData** to the CM-150/151 (Line 85).

```
def Set_Servo_Position(port_no, goal_angle):  # goal_angle [-150, 150]
    global TxData
    comm_type = 2  # local variable
    shifted_goal_angle = goal_angle + 150   # shifted_goal_angle [0, 300]
    TxData = port_no & 0x01 # PUT port_no's value into Bit 0 of TxData
    TxData = TxData | ((shifted_goal_angle & 0x01ff) << 1)  # ADD shifted
    TxData = TxData | (comm_type << 12) # ADD comm_type = 2 into Bits 12-
    send_data()
```

Fig. 4.72 Function **Set_Servo_Position()** used in "CNT_Block.py".

Figs. 4.73 and 4.74 show the details of Utility Function **Read_Servo_Position(port_no)** which **currently does not have an R+BLOCK equivalent command**:

- This Function corresponds to a **comm_type = 10** (Line 89).
- Input parameter **port_no** can be set to **0** to designate Port 3 or to **1** to designate Port 4, and it is recorded into **Bit 0** of Variable **TxData** (Line 90).
- The last component **comm_type** is shifted 12 bits to the left and "inserted" into **Bits 12-15** of **TxData** using the OR Operator (Line 91) and Line 92 sends it away.

```
87  def Read_Servo_Position(port_no):
88      global TxData, RxD_packet, lowbyte, highbyte
89      comm_type = 10  # local variable
90      TxData = port_no & 0x01 # PUT port_no's value into Bit 0 of TxData
91      TxData = TxData | (comm_type << 12) # ADD comm_type = 10 into Bits 12-15
92      send_data()

94      # receive & process packet sent back from CM-150
95      received_packet = btser.read(size=6)
96      #print("received_packet = ", received_packet)
97      if (received_packet):  # received string not empty
98          RxD_packet = bytearray(received_packet)
99          #for i in range(6):
100         #    print("i = ", i, "\tbyte = ", RxD_packet[i])
101         if ((RxD_packet[0] == 0xff) and (RxD_packet[1] == 0x55)):
102             #print("Valid Remocon Packet received from Robot")
103             LB = False
104             if (RxD_packet[3] == (~(RxD_packet[2]) & 0xff)):
105                 LB = True
106                 lowbyte = RxD_packet[2] & 0xff
107             HB = False
108             if (RxD_packet[5] == (~(RxD_packet[4]) & 0xff)):
109                 HB = True
110                 highbyte = RxD_packet[4] & 0xff
```

Fig. 4.73 Part 1 of Function **Read_Servo_Position()** used in "CNT_Block.py".

- This function does expect a return Remocon Packet from the CM-150/151 containing the Servo's Present Position. This task is implemented with Lines 95 through 134.
- Line 95 uses PySerial's Function **read()** to pick up 6 bytes from the CB Communications Buffer and saves it into String Variable **received_packet**.
- If **received_packet** is **empty** via the IF check in Line 97, then this Function jumps to Line 131 of Fig. 4.74, prints out the error found (Line 132), clears the CB input buffer (Line 133), and returns the special value of **1000** at Line 134.

```
111             if ((LB == True) and (HB == True)):
112                 RxData = ((highbyte << 8) + lowbyte) & 0xffff  # reassemble RxData sent
113                 rec_comm_type = ((RxData & 0xf000) >> 12)  # extract received comm_type
114                 rec_port_no = RxData & 0x01  # extract received Port No - Bit 0
115                 rec_goal_angle = ((RxData & 0x03fe) >> 1)  # extract received Goal Angle
116                 if ((rec_comm_type == comm_type) and (rec_port_no == port_no)):  # matc
117                     #print("received Goal Angle [0-300] = ", rec_goal_angle, "\t for Por
118                     return (rec_goal_angle - 150)  # returning angle [-150, 150]
119                 else:
120                     print("Invalid Goal Angle Data Received for Port No ", port_no + 3)
121                     btser.reset_input_buffer()
122                     return 1000
123             else:
124                 print("LB and/or HB received corrupted for Port No ", port_no + 3)
125                 btser.reset_input_buffer()
126                 return 1000
127         else:
128             print("Invalid Remocon Packet received for Port No ", port_no + 3)
129             btser.reset_input_buffer()
130             return 1000
131     else:
132         print("Empty String Received from Port No ", port_no + 3)
133         btser.reset_input_buffer()
134         return 1000
```

Fig. 4.74 Part 2 of Function **Read_Servo_Position()** used in "CNT_Block.py".

- If **received_packet** is **not empty** via Line 97, then a complex procedure (Lines 98 to 130) is started to look into the contents of **received_packet**:
  - First, **received_packet** is converted into a **bytearray** and saved into **RxD_packet[]** (Line 98). Now the 6 bytes of **RxD_packet[]** can be checked individually, first for proper packet format, then for appropriateness of data received (i.e., a Servo Present Position).
  - Line 101 checks to see if the Header component is valid. **If it is not**, this Function jumps to Line 127, prints the error message (Line 128), resets the CB input buffer (Line 129) and returns the **1000** value.
  - If the **Header component is valid**, then **RxD_packet[2]** and **RxD_packet[3]** are compared to each other to see if one is the 1's complement of the other by Line 104. If this test is successful, then **LB** is set to TRUE (Line 105),

and **lowbyte** receives its "correct" value **RxD_packet[2]** (Line 106). If this test turns out unsuccessful, then **LB** stays FALSE as set by Line 103 previously.
  - Similarly, **RxD_packet[4]** and **RxD_packet[5]** are compared to each other to see if one is the 1's complement of the other by Line 108. If this test is successful, then **HB** is set to TRUE (Line 109), and **highbyte** receives its "correct" value **RxD_packet[4]** (Line 110). If this test turns out unsuccessful, then **HB** stays FALSE as set by Line 107 previously.
  - At Line 111 if **LB** and **HB** are both TRUE, then **RxData** is "reconstructed" from **lowbyte** and **highbyte** (Line 112) and further data analysis is pursued (Lines 113-122). If either **LB** or **HB** is FALSE, then this Function jumps to Line 123, prints out the error message (Line 124), resets input buffer (Line 125) and return the **1000** value (Line 126).
  - If further data analysis on **RxData** is warranted:
    - Line 113 extracts the **comm_type** information and saves it in **rec_comm_type**.
    - Line 114 extracts the **port_no** information and saves it in **rec_port_no**.
    - Line 115 extracts the **goal_angle** information (in **Bits 1-9**) and saves it in **rec_goal_angle** which should be in the range [0-300].
    - Then Line 116 checks to see if **rec_comm_type** matches with **comm_type** and if **rec_port_no** matches with **port_no**. If this is TRUE, then **rec_goal_angle** is transformed into a value between -150 and 150, before returning this corrected value to the rest of the Python code (Line 118). If this test turns out FALSE, then the ELSE branch at Line 119 takes over to print the error message (Line 120), resets input buffer (Line 121), and return the "special" value of **1000** (Line 122).
- In real practice, the "Empty String" type of error happened occasionally when the CM-150 is used with the UART BT-410 Receiver, but not with the CM-151. "Invalid Remocon Packet" errors were rare.

Fig. 4.75 lists the steps used in Function **Read_Integrated_IR()** and its R+BLOCK equivalent command:

- This function corresponds to a **comm_type** = **6** (Line 164).
- Input Parameter **location** can be set to **0** to designate Left IR Sensor, or to **1** for Middle IR Sensor, or to **2** for Right IR Sensor. **location** is recorded into **Bits 0-1** of Variable **TxData** (Line 165).
- The last component **comm_type** is shifted 12 bits to the left and "inserted" into **Bits 12-15** of **TxData** using the OR Operator (Line 166).
- Function **send_data()** is used to transmit the completed **TxData** to the CM-150/151 (Line 167).
- This Function does receive a return packet from the CM-150/151 for the current reading of the specified IR Sensor (**Bits 2-11**), using a similar procedure to the one described in detail previously for Function **Read_Servo_Position()**. Thus, the reader is referred to the actual code for the details which won't be repeated here for lack of space and also as an exercise for the reader.

```
def Read_Integrated_IR(location):    # location = 0 for Left IR,
    global TxData, RxD_packet, lowbyte, highbyte
    #btser.reset_input_buffer()
    #btser.reset_output_buffer()
    comm_type = 6 # local variable
    TxData = location & 0x03 # PUT location's value into Bits 0-1
    TxData = TxData | (comm_type << 12) # ADD comm_type = 6 into
    send_data()

    # receive & process packet sent back from CM-150
    received_packet = btser.read(size=6)
    #print("received_packet = ", received_packet)
    if (received_packet): # received string not empty
        RxD_packet = bytearray(received_packet)
```

Fig. 4.75 Function **Read_Integrated_IR()** used in "CNT_Block.py".

Fig. 4.76 describes the steps used in Function **Read_Port_IR_Touch()** and its R+BLOCK equivalent command:

- This function corresponds to a **comm_type** = 7 (Line 215).
- Input Parameter **port_no** can be set to **0** to designate Port 3 or to **1** for Port 4. **port_no** is recorded into **Bit 0** of Variable **TxData** (Line 216).
- Input Parameter **option** can be set to **0** to designate an IR Sensor or to **1** for a Touch Sensor. **option** is recorded into **Bit 1** of Variable **TxData** (Line 217).
- The last component **comm_type** is shifted 12 bits to the left and "inserted" into **Bits 12-15** of **TxData** using the OR Operator (Line 218).
- Function **send_data()** is used to transmit the completed **TxData** to the CM-150/151 (Line 219).
- This Function does receive a return packet from the CM-150/151 for the current reading of the specified Sensor (**Bits 2-11**), using a similar procedure to the one described in detail previously for Function **Read_Servo_Position()**. Thus, the reader is referred to the actual code for the details which won't be repeated here for lack of space and also as an exercise for the reader.

```
211 def Read_Port_IR_Touch(port_no, option):
212     global TxData, RxD_packet, lowbyte, highbyte
213     #btser.reset_input_buffer()
214     #btser.reset_output_buffer()
215     comm_type = 7 # local variable
216     TxData = port_no & 0x01 # PUT Port No's value into Bits
217     TxData = TxData | ((option & 0x01) << 1)  # ADD option's
218     TxData = TxData | (comm_type << 12) # ADD comm_type = 7
219     send_data()
220
221     # receive & process packet sent back from CM-150
222     received_packet = btser.read(size=6)
223     #print("received_packet = ", received_packet)
224     if (received_packet):  # received string not empty
225         RxD_packet = bytearray(received_packet)
```

Fig. 4.76 Function **Read_Port_IR_Touch()** used in "CNT_Block.py".

Fig. 4.77 shows the steps used in Function **Reset_Read_Sound_Counter_Result()** and it combines the features of two R+BLOCK equivalent commands (Reset and Read SCR):

- This function corresponds to a **comm_type** = **8** (Line 268).
- Input Parameter **option** can be set to **0** to **reset** Sound_Counter_Result (SCR) to **0**, or to **1** to read the Current Value of SCR. **option** is recorded into **Bit 0** of Variable **TxData** (Line 217).
- The last component **comm_type** is shifted 12 bits to the left and "inserted" into **Bits 12-15** of **TxData** using the OR Operator (Line 218).
- Function **send_data()** is used to transmit the completed **TxData** to the CM-150/151 (Line 219).
- This Function does receive a return packet from the CM-150/151 for the current reading of the SCR (**Bits 1-3**), using a similar procedure to the one described in detail previously for Function **Read_Servo_Position()**. Thus, the reader is referred to the actual code for the details which won't be repeated here for lack of space and also as an exercise for the reader.

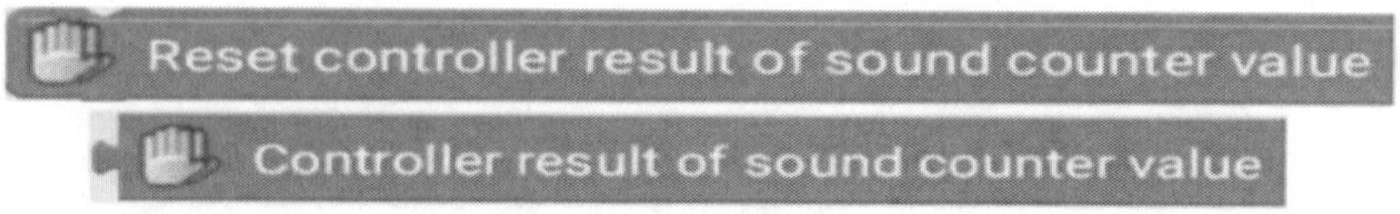


```
def Reset_Read_Sound_Counter_Result(option):
    global TxData, RxD_packet, lowbyte, highbyte
    #btser.reset_input_buffer()
    #btser.reset_output_buffer()
    comm_type = 8 # local variable
    TxData = option & 0x01 # PUT Option's value into Bit 0 (
    TxData = TxData | (comm_type << 12) # ADD comm_type = 8
    send_data()

    # receive & process packet sent back from CM-150
    received_packet = btser.read(size=6)
    #print("received_packet = ", received_packet)
    if (received_packet):  # received string not empty
        RxD_packet = bytearray(received_packet)
```

Fig. 4.77 Function **Reset_Read_Sound_Counter_Result()** used in "CNT_Block.py".

Fig. 4.78 shows the steps used in Function **Read_Sound_Counter_Real_Time()** and its R+BLOCK equivalent command:

- This function corresponds to a **comm_type** = **9** (Line 319).
- There is no Input Parameter needed for this Function, so **TxData** is set to 0 (Line 320).
- Only **comm_type** is needed so it is shifted 12 bits to the left and "inserted" into **Bits 12-15** of **TxData** using the OR Operator (Line 321).
- Function **send_data()** is used to transmit the completed **TxData** to the CM-150/151 (Line 322).
- This Function does receive a return packet from the CM-150/151 for the current reading of the SCRT (**Bits 0-2**), using a similar procedure to the one described in detail previously for Function **Read_Servo_Position()**. Thus, the reader is referred to the actual code for the details which won't be repeated here for lack of space and also as an exercise for the reader.

```
315 def Read_Sound_Counter_Real_Time():
316     global TxData, RxD_packet, lowbyte, highbyte
317     #btser.reset_input_buffer()
318     #btser.reset_output_buffer()
319     comm_type = 9 # local variable
320     TxData = 0
321     TxData = TxData | (comm_type << 12) # ADD comm_type = 9
322     send_data()
323 
324     # receive & process packet sent back from CM-150
325     received_packet = btser.read(size=6)
```

Fig. 4.78 Function **Read_Sound_Counter_Real_Time()** used in "CNT_Block.py".

Let us switch now to the CM-150/151 side to see how its TASK code handles these 11 different Remocon Packets.

4.3.3 "CNT_Block.tskx/tsk3"

The overall structure of the TASK program "CNT_Block.tskx/tsk3" is rather straightforward, just using a "lengthy" IF-ELSE-IF structure to handle each of the 11 types of Commands sent over from the Chromebook, some needing no return packet and some requiring some more work to acquire Actuator or Sensor readings, and then sending them back to the CB via the same Remocon Protocol.

From Sub-Section 4.3.2, the reader is aware of the need for bit wise operators AND **(&)**, OR **(|)**, Left-Shift **(<<)** and Right-Shift **(>>)** in Python. The TASK Tool supports the bit wise operators **&** and | explicitly, but we will have to replace the Left-Shift operator with the Integer Multiplication operator, and the Right-Shift operator with the Integer Division operator.

Fig. 4.79 shows the beginning of the program "CNT_Block.tsk3":

```
1   //  CNT_Block.tsk3 - Block-like Implementation
2   //  CM-150/151 RC Client Program designed to receive Remocon commands from Chromebooks
3   //  Copyright (C) CNT Robotics LLC. All rights reserved. 2022
4   START PROGRAM
5   {
6       //  Init Action
7       Buzzer Timer = Melody Time
8       Buzzer Index = Melody No.2 (2)
9       WAIT WHILE ( Buzzer Timer > 0.0sec )
10      Speed = 1023
11      Data_In = 0
12      DelayTime = 250
13      Comm_Type = -1
14      FirstComm_1 = TRUE (1)
15      FirstComm_2 = TRUE (1)
```

Fig. 4.79 Beginning of "CNT_Block.tsk3".

- Lines 7-9 play Melody 2.
- Lines 10-15 initializes some of the Variables used in TASK.

Fig. 4.80 displays the beginning of the Main Endless Loop for "CNT_Block.tsk3":

- Line 19 waits for a NEW Remocon Packet to arrive which would set the Flag "Remocon Data Arrived" to TRUE and therefore terminates this WAIT WHILE Loop.
- Line 20 retrieves the Received **Remocon Data** and saves it in **Data_In**.

```
17  ENDLESS LOOP
18  {
19      WAIT WHILE ( Remocon Data Arrived == FALSE (0) )
20      Data_In = Remocon RXD
21      Comm_Type = Data_In & 0000 0000 0000 0000 1111 0000 0000 0000
22      Comm_Type = Comm_Type / 0000 0000 0000 0000 0001 0000 0000 0000
23      // Print Screen with Line = Comm_Type
24      IF ( Data_In == -- )
25      {
26          CALL Stop
27      }
28      ELSE
29      {
30          IF ( Comm_Type == 0 )
31          {
32              PortNo = Data_In & 0000 0000 0000 0000 0000 0000 0000 0001
33              Direction = Data_In & 0000 0000 0000 0000 0000 0000 0000 0010
34              Direction = Direction / 0000 0000 0000 0000 0000 0000 0000 0010
35              Power = Data_In & 0000 0000 0000 0000 0000 0001 1111 1100
36              Power = Power / 0000 0000 0000 0000 0000 0000 0000 0100
```

Fig. 4.80 Beginning of Main Endless Loop in "CNT_Block.tsk3".

- Line 21 extracts **Comm_Type** from **Bits 12-15** of **Data_In.**
- **Comm_Type** is then "right-shifted" by 12 bits to its "proper place" by Line 22.
- Lines 24-27 take care of the situation when a "0" is received for **Data_In**, whereas the robot is stopped.
- Otherwise, the ELSE logical branch is taken, starting at Line 28. This ELSE branch contains a lengthy IF-ELSE-IF structure that handles each case of the 11 possible values for **Comm_Type** [0-10].
- Line 30 marks the beginning of the case **(Comm_Type == 0)** which corresponds to Function **Turn_Gear_Motor()** in the Python Module "CNT_Block.py" (see Fig. 4.67):
    - Line 21 extracts **PortNo** from Bit 0 of **Data_In.**
    - Line 33 extracts **Direction** from Bit 1 of **Data_In**, and Line 34 shifts it right to its proper bit location.
    - Line 35 extracts **Power** from **Bits 2-8** of **Data_In**, and Line 36 shifts it right to its proper bit location.
    - Fig. 4.81 shows the remaining steps for the case **(Comm_Type == 0).**

```
37        // If Direction is CCW then Speed Range [0-1023]
38        Speed = (Power * 1023) / 100
39        // If Direction is CW then Speed is "negative" by adding 1024 to it
40        IF ( Direction == 1 )
41        {
42            Speed = 1024 + Speed
43        }
44        IF ( PortNo == 0 )
45        {
46            PORT[1]:Geared Motor (Left_Geared_Motor) = Speed
47        }
48        ELSE IF ( PortNo == 1 )
49        {
50            PORT[2]:Geared Motor (Right_Geared_Motor) = Speed
51        }
52    }
```

Fig. 4.81 Part 2 of case **(Comm_Type == 0)** in "CNT_Block.tsk3".

- As **Power** is received as a value between 0 and 100, Line 38 converts it into a number between 0 and 1023 (as required internally for the CM-150/151) and saves this result into Variable **Speed**.
- Next, if **(Direction == 1)** (Line 40), this Geared Motor needs to turn CW, then **Speed** is incremented by 1024 (Line 42), to make it a "negative" value as internally the CM-150/151 uses the 2's complement scheme for negative numbers.
- Lines 44-51 uses an IF-ELSE-IF structure to turn on either Port 1 or Port 2 Geared Motor to the proper **Speed** and **Direction** as commanded by Function **Turn_Gear_Motor()** transmitted from the CB via BT communications.

Figs. 4.82 and 4.83 describe the steps used for the case **(Comm_Type == 1)** which corresponds to Function **Turn_Servo_Motor()** in the Python Module "CNT_Block.py" (see Fig. 4.68):

- Line 56 of Fig. 4.82 extracts **PortNo** from **Bit 0** of **Data_In**.
- Line 57 extracts **Direction** from **Bit 1** of **Data_In**, and Line 58 shifts it right to its proper bit location.
- Line 59 extracts **Power** from **Bits 2-8** of **Data_In**, and Line 60 shifts it right to its proper bit location.
- If this is the **very first Comm_Type 1** packet received by the CM-150/151 since "CNT_Block.tsk3" **started to run**, then the IF structure (Lines 62-73) applies and **once only** (see Line 14 of Fig. 4.79 and Line 62 of Fig. 4.82):
    - It is used to set the Servo Motor, connected to either Port 3 or Port 4, into **Continuous Turn** mode (Line 66 or Line 70).
    - Variable **FirstComm_1** is then set to FALSE for the remaining runtime of this TASK program (Line 72).

```
54 ELSE IF ( Comm_Type == 1 )
55 {
56     PortNo = Data_In & 0000 0000 0000 0000 0000 0000 0000 0001
57     Direction = Data_In & 0000 0000 0000 0000 0000 0000 0000 0010
58     Direction = Direction / 0000 0000 0000 0000 0000 0000 0000 0010
59     Power = Data_In & 0000 0000 0000 0000 0000 0001 1111 1100
60     Power = Power / 0000 0000 0000 0000 0000 0000 0000 0100
61     // Setting Drive Mode to False only ONCE
62     IF ( FirstComm_1 == TRUE (1) )
63     {
64         IF ( PortNo == 0 )
65         {
66             PORT [3] :Servo Motor Drive Mode (PortNickName) = FALSE (0)
67         }
68         ELSE IF ( PortNo == 1 )
69         {
70             PORT [4] :Servo Motor Drive Mode (PortNickName) = FALSE (0)
71         }
72         FirstComm_1 = FALSE (0)
73     }
```

Fig. 4.82 Part 1 of case **(Comm_Type == 1)** in "CNT_Block.tsk3".

```
74     // If Direction is CCW (=0) then Speed Range [0-1023]
75     Speed = (Power * 1023) / 100
76     // If Direction is CW (=1) then Speed is "negative" by adding 1024 to it
77     IF ( Direction == 1 )
78     {
79         Speed = 1024 + Speed
80     }
81     IF ( PortNo == 0 )
82     {
83         PORT [3] :Servo Joint Speed (PortNickName) = Speed
84     }
85     ELSE IF ( PortNo == 1 )
86     {
87         PORT [4] :Servo Joint Speed (PortNickName) = Speed
88     }
89 }
```

Fig. 4.83 Part 2 of case **(Comm_Type == 1)** in "CNT_Block.tsk3".

- Line 75 converts **Power** [0-100] to **Speed** [0-1023].

- Next, if **(Direction == 1)** (Line 77), this Servo Motor (in Continuous Turn mode) needs to turn CW, then **Speed** is incremented by 1024 (Line 79), to make it a "negative" value as internally the CM-150/151 uses the 2's complement scheme for negative numbers. Please note that **Turn_Servo_Motor()** (i.e., **Comm_Type 1**) can also be used to change the speed of Servo Motors when they are in **Position Control**, just remember to use the **CCW Direction** option. *This feature is not available in R+BLOCK.*
- Lines 81-88 uses an IF-ELSE-IF structure to turn on either Port 3 or Port 4 Servo Motor to the proper **Speed** and **Direction** as commanded by Function **Turn_Servo_Motor()** sent over from the CB via BT communications.

Figs. 4.84 and 4.85 describe the steps used for the case **(Comm_Type == 2)** which corresponds to Function **Set_Servo_Position()** in the Python Module "CNT_Block.py" (see Fig. 4.72):

- Line 93 of Fig. 4.84 extracts **PortNo** from **Bit 0** of **Data_In**.
- If this is the **very first Comm_Type 2** packet received by the CM-150/151 since "CNT_Block.tsk3" **started to run**, then the IF structure (Lines 95-108) applies and **once only** (see Line 15 of Fig. 4.79 and Line 95 of Fig. 4.84):
  - It is used to set the Servo Motor, connected to either Port 3 or Port 4, into **Position Control** mode (Line 99 or Line 104). By default, R+BLOCK sets the speed of all Servo Motors to 1023 (fixed) when they are in Position Control mode (Lines 100 and 105 are commented out as a reminder).

```
91  ELSE IF ( Comm_Type == 2 )
92  {
93    PortNo = Data_In & 0000 0000 0000 0000 0000 0000 0000 0001
94    // Setting Drive Mode to True only ONCE, Joint Speed defaulting to 1023
95    IF ( FirstComm_2 == TRUE (1) )
96    {
97      IF ( PortNo == 0 )
98      {
99        PORT[3]:Servo Motor Drive Mode (PortNickName) = TRUE (1)
100       // PORT[3]:Servo Joint Speed (PortNickName) = 1023
101     }
102     ELSE IF ( PortNo == 1 )
103     {
104       PORT[4]:Servo Motor Drive Mode (PortNickName) = TRUE (1)
105       // PORT[3]:Servo Joint Speed (PortNickName) = 1023
106     }
107     FirstComm_2 = FALSE (0)
108   }
109   GoalAngle = Data_In & 0000 0000 0000 0000 0000 0011 1111 1110
110   GoalAngle = GoalAngle / 0000 0000 0000 0000 0000 0000 0000 0010
```

Fig. 4.84 Part 1 of case **(Comm_Type == 2)** in "CNT_Block.tsk3".

  - Variable **FirstComm_2** is then set to FALSE for the remaining runtime of this TASK program (Line 107).

- Line 109 extracts **GoalAngle** from **Bits 1-9** of **Data_In**, and Line 110 shifts it right to its proper bit location. At this point **GoalAngle** has a value between **0** and **300** (degrees).
- Line 112 of Fig. 4.85 converts this **GoalAngle** value **[0-300]** to an equivalent value **[0-1023]** to satisfy requirements of the CM-150/151 firmware and saves it into Variable **GoalPosition**.
- The IF-ELSE-IF structure in Lines 113-120 (Fig. 4.85) is used to set either Port 3 or Port 4 Servo Motor to its proper Goal Position.

```
111            // Convert [0-300] degrees to [0-1023] raw goal position
112            GoalPosition = (GoalAngle * 1023) / 300
113            IF ( PortNo == 0 )
114            {
115                PORT[3]:Servo Joint Position (PortNickName) = GoalPosition
116            }
117            ELSE IF ( PortNo == 1 )
118            {
119                PORT[4]:Servo Joint Position (PortNickName) = GoalPosition
120            }
121        }
```

Fig. 4.85 Part 2 of case **(Comm_Type == 2)** in "CNT_Block.tsk3".

Fig. 4.86 lists the steps used for the case **(Comm_Type == 10)** which corresponds to Function **Read_Servo_Position()** in the Python Module "CNT_Block.py" (see Figs. 4.73 and 4.74). **This Function has no R+BLOCK equivalent command**:

- For this case, the CM-150/151 needs to send a Remocon Packet back to the CB at the end of this procedure, thus **Data_In** is duplicated into **Data_Out** by Line 322.
- Line 323 extracts **PortNo** from **Bit 0** of **Data_In**.
- The IF-ELSE-IF structure in Lines 324-331 is used to read the **Present Position** of either Port 3 or Port 4 Servo Motor and saves it into Variable **GoalPosition** (Line 326 or 330).
- Line 333 converts **GoalPosition [0-1023]** into a value **[0-300]** and saves it into Variable **GoalAngle**.
- Line 334 makes sure to "fit" **GoalAngle** to **Bits 0-8**, and Line 335 shifts it left by one bit (i.e., as **Bits 1-9**) as per protocol for this type of Packet, defined in "CNT_Block.py" Module.
- Line 336 combines **GoalAngle** into the rest of **Data_Out**, and Line 337 sends it back to the CB.

```
ELSE IF ( Comm_Type == 10 )
{
    // Data_Out Keeps the same Comm_Type and Port No as for Data_In
    Data_Out = Data_In
    PortNo = Data_In & 0000 0000 0000 0000 0000 0000 0000 0001
    IF ( PortNo == 0 )
    {
        GoalPosition = PORT [3]:Servo Joint Position (PortNickName)
    }
    ELSE IF ( PortNo == 1 )
    {
        GoalPosition = PORT [4]:Servo Joint Position (PortNickName)
    }
    // Convert [0-1023] raw goal position to [0-300] goal angle
    GoalAngle = (GoalPosition * 300) / 1023
    GoalAngle = GoalAngle & 0000 0000 0000 0000 0000 0001 1111 1111
    GoalAngle = GoalAngle * 0000 0000 0000 0000 0000 0000 0000 0010
    Data_Out = GoalAngle | Data_Out
    Remocon TXD = Data_Out
}
```

Fig. 4.86 Procedure for case **(Comm_Type == 10)** in "CNT_Block.tsk3".

```
ELSE IF ( Comm_Type == 3 )
{
    PortNo = Data_In & 0000 0000 0000 0000 0000 0000 0000 0001
    O_LED = Data_In & 0000 0000 0000 0000 0000 0000 0000 0010
    O_LED = O_LED / 0000 0000 0000 0000 0000 0000 0000 0010
    B_LED = Data_In & 0000 0000 0000 0000 0000 0000 0000 0100
    B_LED = B_LED / 0000 0000 0000 0000 0000 0000 0000 0100
    IF ( PortNo == 1 )
    {
        IF ( B_LED == FALSE (0) && O_LED == FALSE (0) )
        {
            PORT [4]:LED Module (PortNickName) = [Left OFF, Right OFF] (0)
        }
        ELSE IF ( B_LED == FALSE (0) && O_LED == TRUE (1) )
        {
            PORT [4]:LED Module (PortNickName) = [Left OFF, Right ON] (1)
        }
        ELSE IF ( B_LED == TRUE (1) && O_LED == FALSE (0) )
        {
            PORT [4]:LED Module (PortNickName) = [Left ON, Right OFF] (2)
```

Fig. 4.87 Procedure for case **(Comm_Type == 3)** in "CNT_Block.tsk3".

Fig. 4.87 displays the steps used for the case **(Comm_Type == 3)** which corresponds to Function **Set_LED_Module()** in the Python Module "CNT_Block.py" (see Fig. 4.69):

- Line 125 extracts **PortNo** from **Bit 0** of **Data_In**.
- Line 126 extracts **O_LED** from **Bit 1** of **Data_In**, and Line 127 shifts it right to its proper bit location.
- Line 128 extracts **B_LED** from **Bit 2** of **Data_In**, and Line 129 shifts it right to its proper bit location.
- Lines 130-142 shows only the beginning of a nested IF-ELSE-IF structure designed to turn ON/OFF the Orange and Blue LEDs on the LED Module SM-10 (https://emanual.robotis.com/docs/en/parts/display/lm-10/#introduction), based on the values contained in Variables **O_LED** and **B_LED**. For example, Line 134 turns both LEDs OFF, while Line 138 turns ON the Orange (Right) LED, while keeping the Blue (Left) LED OFF.

Fig. 4.88 displays the steps used for the cases **(Comm_Type == 4)** and **(Comm_Type == 5)** which correspond respectively to Function **Play_Musical_Scale()** and Function **Play_Musical_Melody()** in the Python Module "CNT_Block.py" (see Figs. 4.70 and 4.71):

```
170  ELSE IF ( Comm_Type == 4 )
171  {
172      WAIT WHILE ( Buzzer Timer > 0.0sec )
173      Time_Duration = Data_In & 0000 0000 0000 0000 0000 0001 1100 0000
174      Time_Duration = Time_Duration / 0000 0000 0000 0000 0000 0000 0100 0000
175      Buzzer Timer = Time_Duration * 10
176      Scale_Index = Data_In & 0000 0000 0000 0000 0000 0000 0011 1111
177      Buzzer Index = Scale_Index
178      Time_Duration = 0
179      Scale_Index = 0
180  }
181  // End Comm_Type == 4
182  ELSE IF ( Comm_Type == 5 )
183  {
184      WAIT WHILE ( Buzzer Timer > 0.0sec )
185      Buzzer Timer = Melody Time
186      Melody_Index = Data_In & 0000 0000 0000 0000 0000 0000 0001 1111
187      Buzzer Index = Melody_Index
188      Melody_Index = 0
189  }
```

Fig. 4.88 Procedures for cases **(Comm_Type == 4)** and **(Comm_Type == 4)** in "CNT_Block.tsk3".

- In the case for **(Comm_Type == 4)**:
  - Line 172 is used in case the Buzzer is still "busy" with its last operation, basically just to wait for the Buzzer Timer to get to 0.

    - Line 173 and 174 extract **Time_Duration** out of **Data_In** and shift it down to its proper bit location.
    - Line 175 multiplies **Time_Duration** by 10, because the Buzzer Timer uses 0.1 seconds time intervals.
    - Line 176 extracts **Scale_Index** from **Data_In** and Line 177 assigns this value to **Buzzer Index** to enable the requested Music Scale for the Time Duration specified earlier.
    - Lines 178 and 179 are activated to clear the two Music-Scale Variables used.
- In the case for **(Comm_Type == 5)**:
    - Line 184 is used in case the Buzzer is still "busy" with its previous operation, basically just to wait for the Buzzer Timer to wind down to 0.
    - Line 185 sets **Buzzer_Timer** to **Melody_Time** (i.e., 5 seconds).
    - Line 186 extracts **Melody_Index** from **Data_In** and Line 187 assigns this value to **Buzzer Index** to enable the requested Music Melody for 5 seconds.
    - Line 188 clears **Melody_Index** to zero.

Fig. 4.89 describes the steps used for the case **(Comm_Type == 6)** which corresponds to Function **Read_Integrated_IR()** in the Python Module "CNT_Block.py" (see Fig. 4.75):

- Only the steps used for IR Left Sensor are shown herein, the reader is referred to the source code for the other two Integrated IR Sensors.
- Line 193 extracts **Location** out of **Data_In**.
- Line 195 duplicates **Data_In** into **Data_Out** in preparation for the Return Packet to the CB.
- If **(Location == 0)**, i.e., IR Left Sensor, Line 198 saved the current reading of Left IR Sensor into Variable **IR_Left**.
- Lines 200 and 202 prepare and record **IR_Left** into its proper bit locations.
- Line 204 inserts **IR_Left** into **Data_Out**, and Line 205 sends it to the CB via Bluetooth.

Fig. 4.90 describes the steps used for the case **(Comm_Type == 7)** which corresponds to Function **Read_Port_IR_Touch()** in the Python Module "CNT_Block.py" (see Fig. 4.76):

- Only the steps used for Port 3 IR Sensor are shown herein, the reader is referred to the source code for the other cases.
- Line 233 extracts **PortNo** out of **Data_In**.
- Line 234 extracts **Option** out of **Data_In**.
- Line 236 duplicates **Data_In** into **Data_Out** in preparation for the Return Packet to the CB.
- If **(Option == 0) and (PortNo == 0)**, i.e., an IR Sensor being used on Port 3, Lines 242-250 apply:
    - Line 242 saves the current reading of Port 3 IR Sensor into Variable **IR_Port3**.
    - Lines 244 and 246 prepare and record **IR_Port3** into its proper bit locations.
    - Line 248 inserts **IR_Port3** into **Data_Out**, and Line 249 sends it to the CB via Bluetooth.

```
ELSE IF  (  Comm_Type  ==  6  )
{
    Location  =  Data_In  &  0000 0000 0000 0000 0000 0000 0000 0011
    //  Data_Out Keeps the same Comm_Type and Location as for Data_In
    Data_Out  =  Data_In  &  0000 0000 0000 0000 1111 0000 0000 0011
    IF  (  Location  ==  0  )
    {
        IR_Left  =  IR Left
        //  Make sure that only the LOWER 10 bits of IR_Left are used from now on
        IR_Left  =  IR_Left  &  0000 0000 0000 0000 0000 0011 1111 1111
        //  Next, IR_Left is Shifted Left by 2 bits
        IR_Left  =  IR_Left  *  0000 0000 0000 0000 0000 0000 0000 0100
        //  Insert shifted IR_Left into Bits 2-11 of Data_Out
        Data_Out  =  Data_Out  |  IR_Left
        Remocon TXD  =  Data_Out
    }
```

Fig. 4.89 Procedure for case **(Comm_Type == 6)** in "CNT_Block.tsk3".

```
ELSE IF  (  Comm_Type  ==  7  )
{
    PortNo  =  Data_In  &  0000 0000 0000 0000 0000 0000 0000 0001
    Option  =  Data_In  &  0000 0000 0000 0000 0000 0000 0000 0010
    //  Data_Out Keeps the same Comm_Type, Option and Port No as for Data_In
    Data_Out  =  Data_In  &  0000 0000 0000 0000 1111 0000 0000 0011
    IF  (  Option  ==  0  )
        //  IR Sensors being used on Ports 3 & 4
    {
        IF  (  PortNo  ==  0  )
        {
            IR_Port3  =  PORT [3] :IR Sensor (PortNickName)
            //  Make sure that only the LOWER 10 bits of IR_Port3 are used from now on
            IR_Port3  =  IR_Port3  &  0000 0000 0000 0000 0000 0011 1111 1111
            //  Next, IR_Port3 is Shifted Left by 2 bits
            IR_Port3  =  IR_Port3  *  0000 0000 0000 0000 0000 0000 0000 0100
            //  Insert shifted IR_Port3 into Bits 2-11 of Data_Out
            Data_Out  =  Data_Out  |  IR_Port3
            Remocon TXD  =  Data_Out
        }
```

Fig. 4.90 Procedure for case **(Comm_Type == 7)** in "CNT_Block.tsk3".

Fig. 4.91 describes the steps used for the case **(Comm_Type == 8)** which corresponds to Function **Reset_Read_Sound_Counter_Result()** in the Python Module "CNT_Block.py" (see Fig. 4.77):

- Line 291 extracts **Option** out of **Data_In**.
- Line 293 duplicates **Data_In** into **Data_Out** in preparation for the Return Packet to the CB.
- If **(Option == 0)**, i.e., the **Sound Counter Result** (or **Detected Clap Count** in TASK) needs to be reset to 0 and this result to be sent to the CB for cross-checking:
    - Line 296 sets **Detected Clap Count** to **0**, and Line 297 waits for 5 ms for this process to happen.
    - Line 298 reads in **Detected Clap Count** (which should be 0), shifts this value left by 1 bit, and saves the result into Variable **SCR**.
    - Line 299 inserts **SCR** into **Data_Out**, and Line 300 sends it to the CB via Bluetooth.
- Else If **(Option == 1)**, i.e., the **Sound Counter Result** (i.e., **Detected Clap Count**) needs to be read and this result to be sent back to the CB for use in the user's Python code:
    - Line 304 reads in the current value of **Detected Clap Count**, shifts this value left by 1 bit, and saves the result into Variable **SCR**.
    - Line 305 inserts **SCR** into **Data_Out**, and Line 306 sends it to the CB via Bluetooth.

```
289  ELSE IF ( Comm_Type == 8 )
290  {
291    Option = Data_In & 0000 0000 0000 0000 0000 0000 0000 0001
292    // Data_Out Keeps the same Comm_Type and Option as for Data_In
293    Data_Out = Data_In & 0000 0000 0000 0000 1111 0000 0000 0001
294    IF ( Option == 0 )
295    {
296      Detected Clap Count = 0
297      Delay = 0.005sec
298      SCR = Detected Clap Count * 0000 0000 0000 0000 0000 0000 0000 0010
299      Data_Out = Data_Out | SCR
300      Remocon TXD = Data_Out
301    }
302    ELSE IF ( Option == 1 )
303    {
304      SCR = Detected Clap Count * 0000 0000 0000 0000 0000 0000 0000 0010
305      Data_Out = Data_Out | SCR
306      Remocon TXD = Data_Out
307    }
308  }
```

Fig. 4.91 Procedure for case **(Comm_Type == 8)** in "CNT_Block.tsk3".

Fig. 4.92 describes the steps used for the case **(Comm_Type == 9)** which corresponds to Function **Read_Sound_Counter_Real_Time()** in the Python Module "CNT_Block.py" (see Fig. 4.78):

- Line 313 duplicates **Data_In** into **Data_Out** in preparation for the Return Packet to the CB.
- Line 314 reads in the current value of **Sound Counter Real Time** (i.e., **Clap Counter**), records this value into **Bits 0-2**, and saves the final result into Variable **SCR**.

- Line 315 inserts **SCR** into **Data_Out**, and Line 316 sends it to the CB via Bluetooth.

```
310        ELSE IF ( Comm_Type == 9 )
311        {
312            // Data_Out Keeps the same Comm_Type as for Data_In
313            Data_Out = Data_In & 0000 0000 0000 0000 1111 0000 0000 0000
314            SCR = Clap Counter & 0000 0000 0000 0000 0000 0000 0000 0111
315            Data_Out = Data_Out | SCR
316            Remocon TXD = Data_Out
317        }
```

Fig. 4.92 Procedure for case **(Comm_Type == 9)** in "CNT_Block.tsk3".

4.4 Beyond Chromebook and Dream

At present, Spring 2022, there is only a limited number of USB devices that are supported via Linux on Chromebooks, for example a USB webcam is supported at the Chrome OS level, but it is not recognized by the Linux Virtual Environment. Thus, we cannot **yet** use all the Image Processing features of OpenCV with the current "CNT_Block.py" Module. But with 5 more command types still available to be defined, the author will be ready to update the **CNT_Block** Module whenever USB cameras (and other USB devices) are supported at the Linux level for Chromebooks.

However, because the **CNT_Block** Python Module is written in Standard Python, the reader can **right now** use it on a Windows OS PC or on a Linux Single-Board-Computer like the RPi Zero 2W which can support a multitude of USB devices (including video camera) and other GPIO devices also.

Furthermore, on the CM-150/151 side, the TASK tool is available for all CM Controllers from ROBOTIS, thus the DREAM II user can transfer all the robotics programming concepts, techniques and skills learned in this book to other ROBOTIS systems such as BIOLOID and ENGINEER very efficiently. When the new USB BT Dongle becomes available in late 2022 from ROBOTIS, we should have a boost in runtime performance of the CM-151 system.

In summary, this book followed a unique approach in developing the same 8 DREAM II projects in three ways (R+BLOCK, R+TASK and Python+TASK). This approach should have exposed the reader to the principle/fact that "Robotics Programming" needs to go beyond the robot per-se and seeks to integrate/adapt/optimize the "Supervisory Platform" through its Hardware and Software capabilities and limitations.

Appendix A: Using Chromebooks

A.1 Choosing a Chromebook Model

Of course, the Web should be searched for the most recent survey of the "best" Chromebooks available that suit the reader's budget. The author used 3 Chromebook models to develop/check materials used in this book, listed in descending order for overall runtime performances:

1. Acer Spin 713 (64-bit) with 4-core Intel i5-10210U, 8 GB RAM, 128 GB SSD.
2. Lenovo Chromebook 3 (64-bit) with 8-core MediaTek 8183, 4 GB RAM, 64 GB eMMC.
3. ASUS C101P (32-bit) with Rockchip RK3399, 4 GB RAM, 16 GB eMMC.

Surprisingly, the ASUS device with 16 GB of overall storage was able to handle R+BLOCK, R+TASK V.2 and V.3, including the Linux Virtual Environment along with THONNY IDE, on 4.6-4.7 GB for Linux storage at 50% usage. Of course, it offered only 2.6-2.7 GB of "free" memory to run Linux applications. Thus, if the reader is looking for a Chromebook to buy, the author would recommend a "median" model with 4 GB RAM and 32 GB of storage, with the fastest CPU that the reader's budget allows. Bluetooth 4.2 should be sufficient to connect to the BT-210/410 modules. The reader should also invest in a USB A or C hub to use the USB BT-410 dongle(s) which are needed for Linux applications for single or multiple CM-150/151 robots. Micro SD cards are of course handy to transfer/backup user files from/to the Chromebook's internal eMMC or SSD.

Official web support for Chromebooks is available at https://support.google.com/chromebook#topic=3399709, and it has good materials for Chromebook's first-time users. Chrome OS is based on Debian and the Debian Wiki can be viewed at https://wiki.debian.org/. The author also uses the Chrome Remote Desktop (https://remotedesktop.google.com/) to access his Chromebooks from a Windows PC to create video tutorials used for this book.

ROBOTIS Apps, such as R+BLOCK and R+TASK, are available for installation via the Google Play Store (for example - https://play.google.com/store/apps/details?id=com.robotis.rblock).

For Python Programming, the user first needs to set up Linux on the Chromebook: see instructions at this web link https://support.google.com/chromebook/answer/9145439?hl=en. For more information about this Linux container, please visit https://chromium.googlesource.com/chromiumos/docs/+/HEAD/containers_and_vms.md#Runtime-Features. More information about the installation and usage of the THONNY IDE is provided in the upcoming Section A.2.2.

The author encountered a "strange" issue with the Acer Spin 713 when using TASK V.2 and V.3, as these two Apps would not read in example TASK codes created by the author for this book (meaning that the reader would have to "retype" these example codes when using the Spin 713). However, the Spin 713 has no problems in loading and editing the author-provided R+BLOCK and Python example codes. On the other hand, the other Chromebooks did not exhibit these TASK related issues.

Another "usage quirk" to watch for if the reader uses a mouse with ROBOTIS Apps (especially for R+BLOCK): the reader has to "slow" his/her button clicking rate down when choosing the wanted "programming block" and to move it around on the computer screen. This is because these Apps are originally tailored to work on touchscreens and touch interactions are much slower than mouse clicks for most users.

A.2 User Files Storage and Management

The Chromebook's "Files" tool is like "Windows Explorer" (see Fig. A.1), but only the user-created files are accessible under the Main Folder named "My files" at first, but the user can enable the display of "hidden" Play folders and files under the "More" icon ("3 vertical dots" icon on the right of the top menu row). This step is necessary to enable the transfer of this book's example programs into their proper folders on the Chromebook's internal storage device.

A.2.1 For ROBOTIS Apps

When a user installs any ROBOTIS App via Google Play Store, this App creates a Folder named "RoboPlus" (normally hidden) inside the "Play files" Folder which itself is located inside the Main Folder named "My files" (see Fig. A.1).

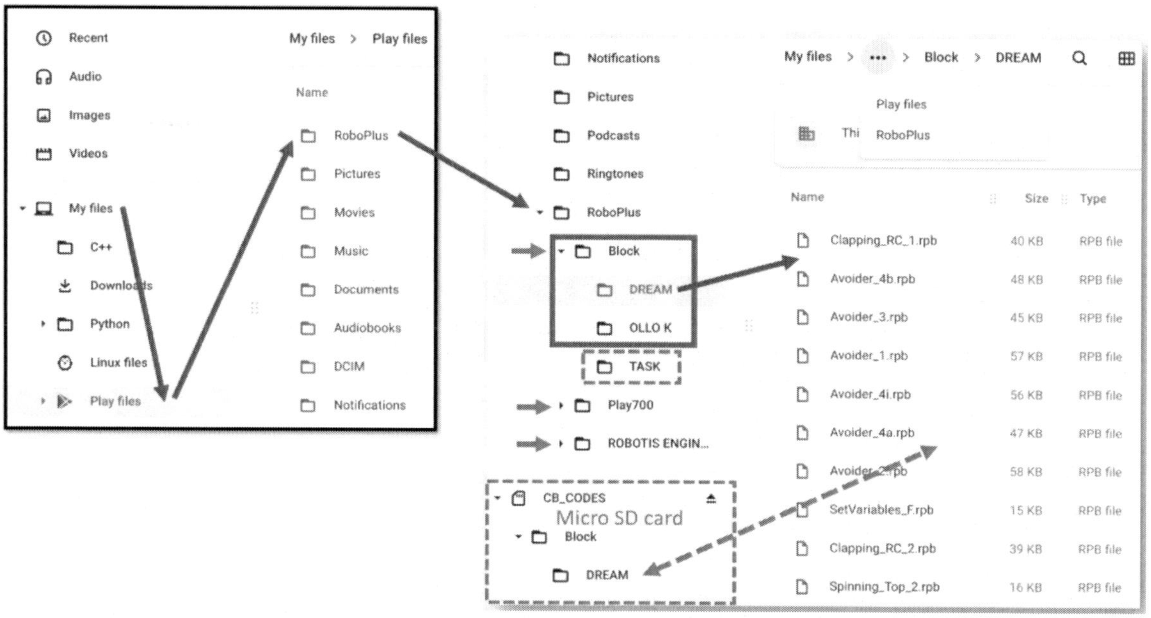

Fig. A.1 Author's Folders Structure for ROBOTIS example program files.

For example, if R+BLOCK was installed, then the Folder "Block" would be created inside "RoboPlus", along with 2 sub-folders "DREAM" (for CM-150/151) and "OLLOK" (for CM-50). When the user uses R+BLOCK to create new BLOCK projects (*.rpb), they are saved inside the DREAM sub-folder. The author also uses an external Micro SD card to back up his BLOCK projects. For the readers, a reverse procedure would be needed to transfer this book's example programs from an external Micro SD card to the DREAM sub-folder on the Chromebook to enable their further use, as R+BLOCK and other ROBOTIS Apps cannot reach the Micro SD card directly.

Please contact the author at roboteer@comcast.net for any possible issue in using these sample programs.

A.2.2 For THONNY IDE

For Python Programming, the user first needs to set up Linux on his/her Chromebook if not done so: see instructions at this web link https://support.google.com/chromebook/answer/9145439?hl=en. Once this installation is done, the Folder named "Linux files" would appear under "My files", and the "Terminal" App would appear in the "Launcher" tool (see Fig. A.2). The "Linux files" folder stores all the files needed for the user's Desktop and installed Apps. We will need to use the "Terminal" tool to install the THONNY IDE as it is a Linux App.

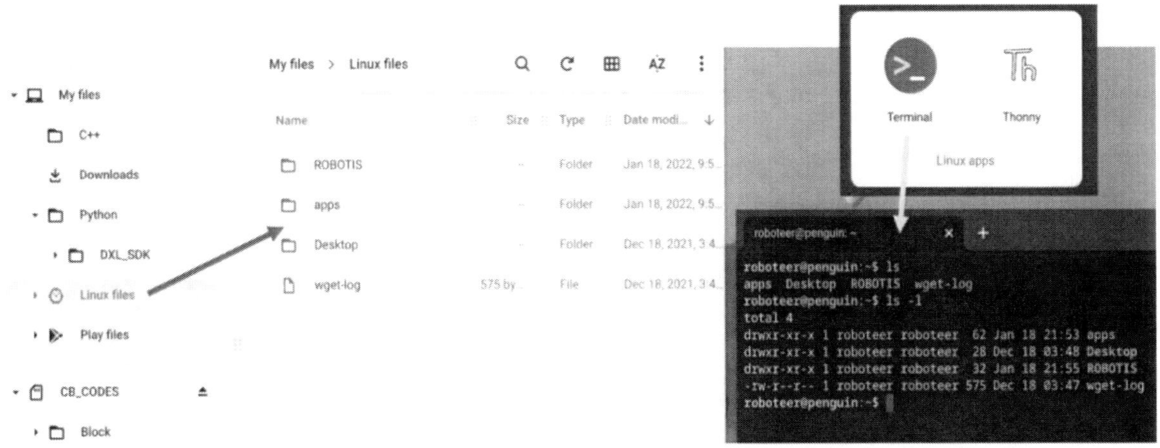

Fig. A.2 "Linux files" Folder and Linux "Terminal" App.

The Thonny IDE is described at https://thonny.org/ and the reader can scroll down to the bottom of that web page to see links for installation instruction, FAQ and Wiki. For Chromebooks, users have two options to install THONNY using the Linux "Terminal" tool:

1. If the Chromebook's CPU is of an Intel type, then issue the following command line in the "Terminal" window (to install the latest stable versions for both Thonny and Python):

   bash <(wget -O - https://thonny.org/installer-for-linux)

2. If the Chromebook's CPU is of another type (most likely an ARM chip), then issue the following command line in the "Terminal" window (to install the "Distro" packages for both Python and Thonny – used by the author):

   ```
   sudo apt install python3-tk thonny
   ```

   Or issue the following command line in the "Terminal" window to install the latest bundle for Python and Thonny:

   ```
   bash <(curl -s https://thonny.org/installer-for-linux)
   ```

More installation information is available at https://github.com/thonny/thonny/wiki/Linux. If the installation was successful, the Launcher would add the "Thonny" icon to the "Linux apps" button (see Fig. A.2).

Next, the user should run the Thonny App and click on the "Manage Packages" item in the "Tools" option on the Top Menu Bar. Then, the user should check if the "pyserial" module is already installed, if not just enter "pyserial" in the blank input field for "Search on PyPi" and follow the on-screen instructions to install it.

The user will also need to install ninja, numpy, opencv-python and pygame - in this order for best results – to use the example Python programs provided with this book.

On his Chromebook, the author created a folder named "Python" and shared it with "Linux" so that THONNY can also access it as it is a Linux App (see Fig. A.3). In Linux, the actual path to this folder is "/mnt/chromeos/MyFiles/Python". The author used this folder to save Python projects listed in this book and put their backup copies onto the Micro SD card named "CB_CODES". Again, the reader would need to perform a reverse procedure to transfer the author-provided example Python codes from his/her Micro SD card onto this Linux-shared "Python" folder to try them out.

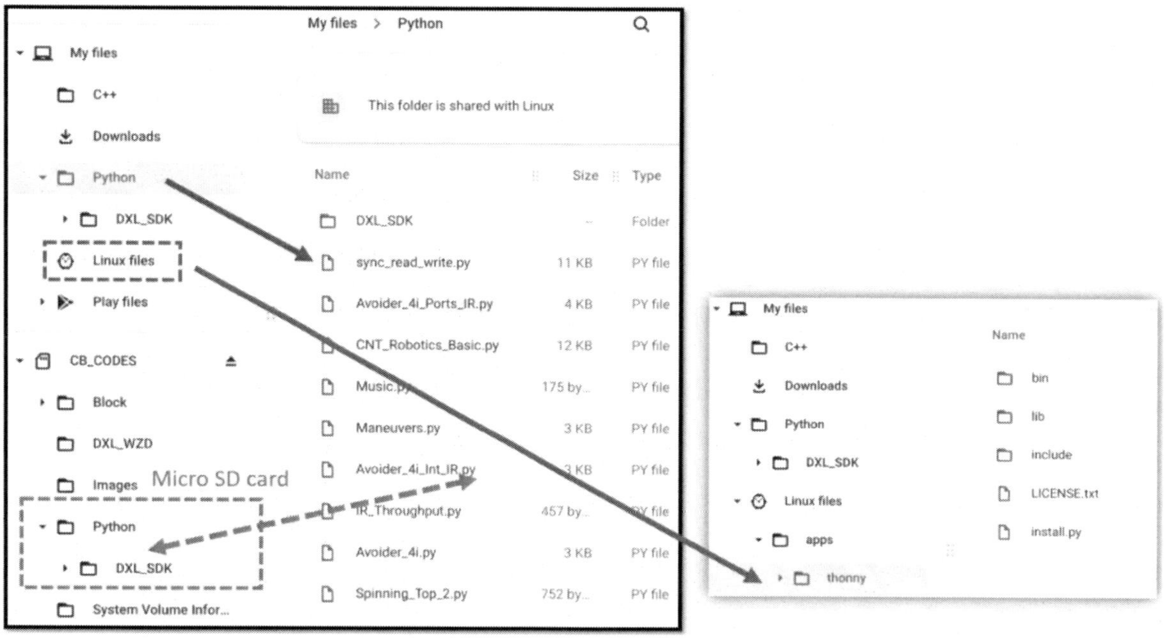

Fig. A.3 Author's Folders Structure for Python example program files.

Please contact the author at roboteer@comcast.net for any possible issue in using these sample programs.

Appendix B: Robot Construction Notes

B.1 Avoider/Follower/Line Tracker

Please watch Video at https://www.youtube.com/watch?v=TORJnsplPYI about how to use the R+DESIGN App to construct this robot, but only up to **Step 201 of the DREAM L2 Avoider design**. The final steps are to add two IRSS-10 sensors to the Back Side of this robot and to Port 3 and Port 4 of the CM-150/151 (see Fig. B.1)

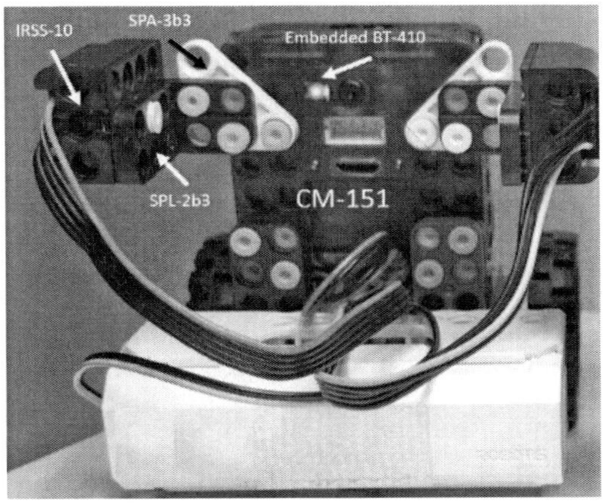

Fig. B.1 Connecting IRSS-10s to Avoider/Follower.

Fig. B.2 Front Wheel for Line Tracker.

B.2 TriCycle

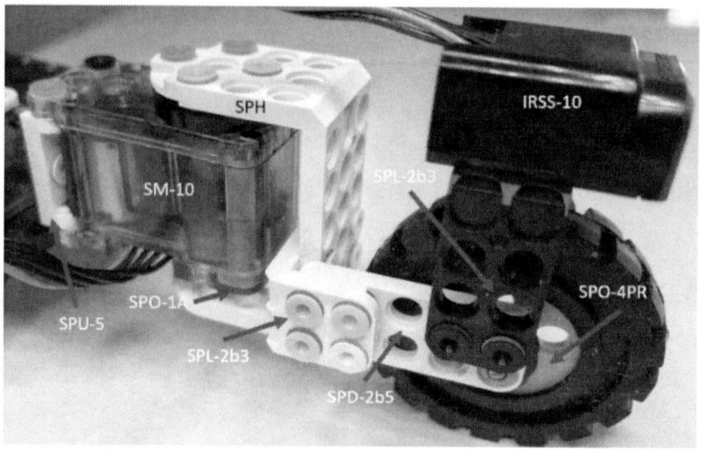

Fig. B.3 Front Wheel Assembly of TriCycle.

B.3 Dowel Scanner

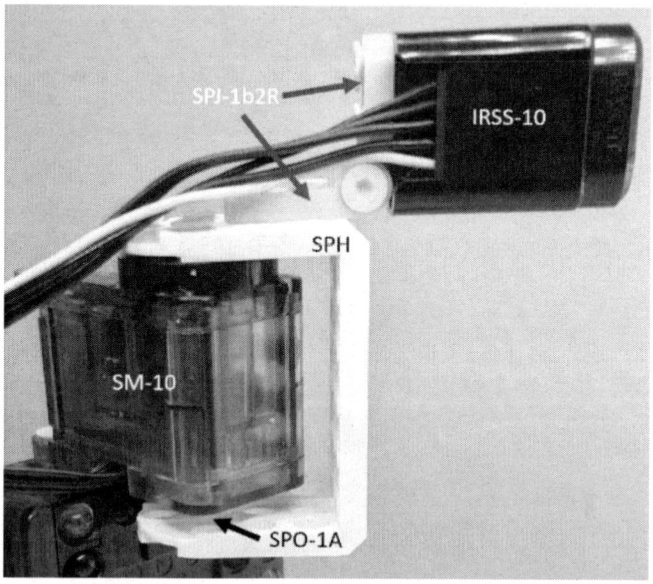

Fig. B.3 Scanning Head for Dowel Scanner.

Appendix C: CNT_Block Modules

C.1 CNT_Block.py

The requirements/features of the Module "CNT_Robotics_Basic.py" are listed below:

1. The Programmer can use BLOCK-like single Python statements to create basic control algorithms for controlling geared and servo motors and for reading sensors. Each Python statement corresponds to a Specific Utility Function that sends an Appropriate 16-bit Remocon packet to the CM-150/151. This Module is designed for interacting with a single CM-150/151 based robot.
2. The format for the 16-bit Remocon packet (Bits 0 to 15) to be sent from the Chromebook is defined as follows:
    a. **Bits 12-15** contain the information for Command Type (**Comm_Type**). Thus, **Comm_Type** can have up to 16 values [0-15]:
        i. **Comm_Type=0** – Turn_Gear_Motor (PortNo > Bit 0, CW_CCW > Bit 1, Power > Bits 2-8).
        ii. **Comm_Type=1** – Turn_Servo_Motor (PortNo > Bit 0, CW_CCW > Bit 1, Power > Bits 2-8 [0-100]).
        iii. **Comm_Type=2** – Set_Servo_Position (PortNo > Bit 0, GoalPosition input as [-150, 150] degrees converted to [0, 300] and recorded in Bits 1-9).
        iv. **Comm_Type=3** – Set_LED_Module (PortNo > Bit 0, Orange_LED > Bit 1, Blue_LED > Bit 2).
        v. **Comm_Type=4** – Play_Musical_Scale (ScaleIndex > Bits 0-5 [0-51], TimeDuration > Bits 6-8 [0-5 seconds]).
        vi. **Comm_Type=5** – Play_Musical_Melody (MelodyIndex > Bits 0-4 [0-24]).
        vii. **Comm_Type=6** – Read_Integrated_IR (Location > Bits 0-1 [0-2]).
        viii. **Comm_Type=7** – Read_Port_IR_Touch (PortNo > Bit 0, Option > Bit 1 [0 for IR, 1 for Touch]).
        ix. **Comm_Type=8** – Reset_Read_Sound_Counter_Result (Option > Bit 0 [0 to reset SCR to 0, 1 to read Current SCR).
        x. **Comm_Type=9** – Read_Sound_Counter_Real_Time ().
        xi. **Comm_Type=10** – Read_Servo_Position(PortNo > Bit 0). GoalPosition coming back from CM-150 > Bits 1-9 – in degrees [0, 300] then converted to [-150, +150] as returned value).

C.2 CNT_Block.tskx/tsk3

The requirements/features of the TASK Program "CNT_Robotics_Basic.tskx/tsk3" are listed below:

1. The format of the **16-bit** Remocon Message sent from the CM-150/151 to the Chromebook as Sensor Packet is defined as follows:
    a. **Bits 12-15** contain the information for Command Type (**Comm_Type**). Thus, **Comm_Type** can have up to 16 values [0-15]:
        i. **Comm_Type=0 through 5** > No return packet/value.
        ii. **Comm_Type=6** – Read_Integrated_IR() returning Packet with (Location > Bits 0-1 [0-2]) and (Current IR data in Bits 2-11 [0-1023]).
        iii. **Comm_Type=7** – Read_Port_IR_Touch() returning Packet with (PortNo > Bit 0, Option > Bit 1 [0 for IR, 1 for Touch]) and (Current IR or Touch data in Bits 2-11 [0-1023]).
        iv. **Comm_Type=8** – Reset_Read_Sound_Counter_Result() returning Packet with Option in Bit 0 and Current Sound Counter Result data in Bits 1-3 [0-7].
        v. **Comm_Type=9** – Read_Sound_Counter_Real_Time() returning Current Sound Counter Real Time data in Bits 0-2 [0-7].
        vi. **Comm_Type=10** – Read_Servo_Position() returning Present Goal Position as an angle [0-300] in Bits 1-9. Bit 0 contains the original PortNo.

Made in United States
North Haven, CT
05 November 2026

21760268R00152